Fiber-Optic Sensors for Biomedical Applications

Daniele Tosi
Guido Perrone

ARTECH
HOUSE

BOSTON | LONDON
artechhouse.com

Library of Congress Cataloging-in-Publication Data
A catalog record for this book is available from the U.S. Library of Congress.

British Library Cataloguing in Publication Data
A catalog record for this book is available from the British Library.

ISBN-13: 978-1-63081-152-5

Cover design by John Gomes

10 9 8 7 6 5 4 3 2 1

Contents

Introduction *xiii*

CHAPTER 1

Fundamentals of Wave Optics and Optical Fibers 1

1.1 Introduction 1
1.2 Electromagnetic Waves 1
1.3 Reflection and Refraction of Plane Waves 5
1.4 Dielectric Waveguides 7
1.5 Optical Fibers 10
1.6 Practical Aspects in Using Optical Fibers 14
Selected Bibliography 17

CHAPTER 2

Devices for Fiber-Optic Sensing Applications 19

2.1 Introduction 19
2.2 Light Sources 19
2.2.1 Light-Emitting Diodes 20
2.2.2 Laser Diodes 22
2.3 Photodiodes 27
2.4 Isolators and Circulators 29
2.5 Couplers 30
2.5.1 Wavelength-Insensitive Couplers 31
2.5.2 Wavelength-Sensitive Couplers 33
2.6 Polarization Optics 33
2.7 Optical Spectrum Analyzers and Spectrometers 35
References 37

CHAPTER 3

Principles of Fiber-Optic Sensing 39

3.1 Definitions 39
3.2 Classification 40
3.3 Working Principles 41
3.4 Sensor Performance Analysis 44

3.5 Application-Integrated Design 46
References 47

CHAPTER 4
Intensity-Based Sensors 49
4.1 Introduction 49
4.2 Sensors Exploiting the Modulation of Losses 51
4.3 Sensors Based on Coupling Loss 53
References 57

CHAPTER 5
Fiber Bragg Gratings 59
5.1 Description 59
5.2 Uniform FBGs 59
5.3 FBG Thermal and Mechanical Response 63
5.4 FBG Arrays 66
5.5 Chirped and Apodized FBG 68
5.6 Tilted FBG 73
5.7 Fabrication of FBGs 76
5.7.1 Photosensitivity 76
5.7.2 Phase Mask Inscription 77
5.7.3 Interfering Beams 79
5.7.4 Draw-Tower Method 80
5.7.5 Point-by-Point 81
5.8 FBG Package 82
5.9 Recent Advances 83
References 84

CHAPTER 6
Distributed Sensors 87
6.1 Introduction 87
6.2 OFDR Theory 88
6.2.1 Demonstration 88
6.2.2 Detection 90
6.3 Microwave Photonics 91
6.4 Sensing Elements 92
6.5 Practical Considerations 94
6.6 Perspectives 96
References 97

CHAPTER 7
Fabry-Perot Interferometers 99
7.1 Fabry-Perot Interferometer Principle 99
7.2 FPI-Based Sensors 102
7.3 FPI/FBG Dual Sensors 105

7.4 Fabrication of FPI Sensors 106
7.5 Self-Mixing Interferometry 110
7.6 Other Interferometers 112
References 112

CHAPTER 8
Fiber-Optic Biosensor Principles 115
8.1 Introduction 115
8.2 Sensors Exploiting the Evanescent Field Absorption 115
8.3 Sensors Exploiting Surface Plasmon Resonance 118
8.3.1 SPR Sensors in Bulk Optics 118
8.3.2 SPR Sensors in Optical Fibers 121
8.3.3 SPR Sensors in Photonic Crystal Fibers 123
References 124

CHAPTER 9
Optical Fiber Spectroscopy 129
References 131

CHAPTER 10
Fiber-Optic Sensor Networks 133
10.1 Introduction 133
10.2 Amplitude Detection Methods 134
10.2.1 FBG and Fabry-Perot Interferometers 134
10.2.2 Plastic Fiber-Based Systems 136
10.3 White Light-Based Setup 137
10.3.1 Setup and Instrumentation 137
10.3.2 Time/Wavelength Division Multiplexing 139
10.3.3 Cepstrum Division Multiplexing 140
10.4 Scanning-Source Interrogators 140
10.5 Multiparametric Sensors 142
10.6 Distributed Sensing Units 143
10.7 Commercial Devices 145
10.8 Conclusions 148
References 148

CHAPTER 11
Interrogation Software 151
11.1 FBG Tracking Direct Methods 151
11.1.1 Centroid 151
11.1.2 Bandwidth Tracking 152
11.1.3 Polynomial Fitting 153
11.2 EFPI Direct Tracking 154
11.2.1 Short-Cavity EFPI 154
11.2.2 Long-Cavity EFPI 155

11.3 Direct Karhunen-Loeve Transform 156
11.4 Multi-EFPI Systems 158
11.5 Tilted and Chirped FBG 159
11.5.1 TFBG 159
11.5.2 CFBG 160
11.6 LabVIEW Implementation 161
11.7 Conclusions 161
References 161

CHAPTER 12
Standards for Medical Sensors 163
12.1 Main Standards 163
12.2 ISO 10993 163
12.3 ISO 13485 165
12.4 IEC 60601 165
12.5 Other Standards 167
12.6 CE Mark 168
References 169

CHAPTER 13
Protocols and Tools for Validation 171
13.1 Moral Norms 171
13.2 Relevant Definitions 171
13.2.1 In Vivo 172
13.2.2 Ex Vivo 172
13.2.3 In Vitro 172
13.2.4 Phantom 173
13.2.5 Principle of 3Rs 173
13.2.6 Ethical Issues 174
13.3 Research and Test Methodologies 175
13.4 Investigational Device Exemption 176
13.5 Useful Tools 177
References 180

CHAPTER 14
Sensor Catheterization 181
14.1 Characteristics of Medical Catheters 181
14.1.1 Invasiveness and Size 181
14.1.2 Form Factor and Rigidity 182
14.1.3 Insertion 183
14.1.4 Positioning 183
14.1.5 FOS Protection and Functionality 183
14.1.6 Metrologic Issues 184
14.1.7 Disposable Format and Sterilization 184
14.2 FOS Catheterizations 185

14.2.1 Urologic Pressure and Multipressure Catheter 185
14.2.2 Fiber-Optic Manometer 186
14.2.3 Force-Sensing Device for Epidural Anesthesia 186
14.2.4 Cranial Intraventricular Catheter 187
14.2.5 Needle Thermotherapy Probe 187
14.2.6 Fiber-Optic Guidewire 188
14.2.7 Biosensor Catheterization 188
14.3 Perspectives 189
References 189

CHAPTER 15
Cardiovascular Sensors 191
15.1 Pressure Measurement 191
15.2 Blood Pressure Measurement 192
15.3 Fractional Flow Reserve 194
15.4 Heart-Assistive Devices 196
15.5 Blood Temperature Sensor 198
15.6 Conclusions 199
References 199

CHAPTER 16
Diagnostics in Gastroscopy, Urology, and Neurology 201
16.1 Gastroscopy 201
16.1.1 Sensing Element 202
16.1.2 Esophageal Detection 203
16.1.3 Colonic Analysis 204
16.1.4 Dual Sensing 204
16.1.5 Present and Future Perspectives 204
16.2 Urology 206
16.2.1 Urodynamics 206
16.2.2 Traditional Diagnostic 207
16.2.3 Differential Diagnostic 208
16.2.4 Current Outlook and Future Prospects 210
16.3 Neurology 210
16.4 Epidural 212
16.5 Conclusions 213
References 213

CHAPTER 17
Sensing in Thermal Ablation 215
17.1 Thermal Ablation: Procedures and Rationale 215
17.1.1 Radio Frequency Ablation 216
17.1.2 Microwave Ablation 217
17.1.3 Laser Ablation 217
17.1.4 High-Intensity Focused Ultrasound 217

17.1.5 Other Techniques 217
17.2 Sensing in Thermal Ablation 217
17.3 FBG Arrays 219
17.4 High-Density Temperature Sensing 222
17.4.1 Chirped FBG 222
17.4.2 Distributed Sensing 224
17.5 Pressure Sensing 226
17.6 Concluding Remarks and Future Perspectives 227
References 229

CHAPTER 18

Application and Detection of SPR Sensors 231

18.1 Genesis and Present Status 231
18.2 Lung Cancer Biomarker Detection 232
18.3 Thrombin Detection 233
18.4 DNA Detection 235
18.5 Celiac Disease Diagnostic 235
18.6 Perspectives 236
References 237

CHAPTER 19

Minimally Invasive Robotic Surgery 239

19.1 Robotic Microsurgery and Its Feedback 239
19.2 Vitreoretinal Surgery: Case Study 241
19.2.1 Tool-Tip FBG Force Detection 241
19.2.2 Multi-FPI Force Detection 242
19.2.3 Tool-Shaft and Transverse Force Detection 243
19.3 Intensity-Based Sensors 244
19.4 Distributed Sensing: An Industrial Prospect 245
19.5 Future Evolution 246
References 247

CHAPTER 20

Smart Textiles and Wearable Sensors 249

20.1 Introduction 249
20.2 Smart Medical Textiles 250
20.3 Wearable Sensors 253
20.4 Future Prospects 254
References 255

CHAPTER 21

Fiber-Optic Spectrometric Sensors 257

21.1 Detection of Liver Tumors 257
21.2 Detection of Lung Tumors 258

21.3 Portable Tissue Scanner 258
21.4 Application Perspective 259
References 259

CHAPTER 22
Conclusions and Future Perspectives 261

Appendix A
FBG and EFPI Generators 267
FBG Generator: FBGgeneration.m 267
EFPI Generator (2-Mirror Extrinsic Structure): EFPIgeneration.m 268
FBG Array Analyzer: FBGarray.m 269
LCFBG Analyzer: LCFBG.m 270
Dual FBG/EFPI Sensor Analyzer: FBGandEFPI.m 271

Appendix B
FBG and EFPI Interrogation 275
FBG Wavelength Shift Tracking: FBGtracking.m 275
Short-Cavity EFPI Tracking: EFPIshort.m 276
Long-Cavity EFPI Tracking: EFPIlong.m 276
KLT for FBG and EFPI Tracking: KLT.m 276
Capon Estimator: CaponPSD.m 277

About the Authors 279
Index 281

Introduction

Galileo Galilei, the Italian polymath of the sixteenth century who founded experimental science and pioneered engineering disciplines, reported in one of his famous quotes: "Measure what is measurable, and make measurable what is not so." Technologists have long worked on sensors, as the key element of measurement science: from microelectromechanical systems (MEMS) and nanoelectromechanical systems (NEMS) to piezoelectric sensors, from lab-on-chip to biosensors, sensors appear in everyday life to measure what is measurable. We can think of the impact of sensors in measurement science, and the impact of measurement science in our activities, looking at the recent examples. Smart, self-driving cars are able to combine sensing and imaging with actuators, record the physical conditions of the road and the traffic, and automate driving through actuators. Smart heating systems record the temperature in different parts of the environment and are able to adjust the heat distribution based on sensor's reading. Sensors have paved a long road in measurements in harsh environment such as smart structures, oil and gas pipelines, and environmental detection.

Nevertheless, there is one environment that poses strong challenges to measurement science: the human body. Medical measurement is a science with decades of history; sensors, usually embedded in medical-grade catheters or medical devices, support clinicians in making key decisions during interventional surgery and minimally invasive surgery; diagnostics of diseases and conditions are also carried out on the basis of sensors readout. Within human body, the requirement for minimal invasiveness and to reduce discomfort for patients, the need to not alter the measurement environment with the external devices, biocompatibility issues, and the device-tissue interactions limits require ad hoc design of each medical probe. A set of standards for measurement and commercialization and ethical principles regulate the process of development and test of biomedical sensors.

Fiber-optic sensors (FOSs), also known as optical fiber sensors (OFSs), are emerging as a key enabling technology for several applications in biomedical engineering and biomedicine. FOSs have several advantages, compared to other traditional methods such as MEMS or piezoelectric sensors.

- Miniature size and invasiveness: Optical fibers are small and lightweight, with diameter as low as 80 to 200 μm (about the size of 1 to 3 human hairs). Sensing devices based on optical fibers are often fabricated within the fiber itself or with a thin film that maintains a small size. Having an extremely miniature size for both the fiber cable and sensing devices is a key asset, as it allows minimizing the invasiveness and the alteration of tissues during the medical procedures. The simple cylindrical structure of an optical fiber makes it easily embeddable in microcatheters and is a significant reduction of size over traditional MEMS technologies.

- Electromagnetic immunity: Medical procedures are often carried out with the support of ultrasound or magnetic resonance imaging (MRI); in some cases (thermal ablation, for example) procedures are carried out with a strong electromagnetic field. FOSs are made with a glass or plastic dielectric, immune to radiofrequency or microwave fields, whereas traditional electromechanical technologies can fail due to the excess of disturbance.
- Performance: FOSs exhibit good performance in terms of accuracy, repeatability, insensitivity to external effects, and linearity. Compared to their electromechanical counterparts, they usually exhibit higher performance and are well suited for precision measurement.
- Multiplexing and distribution: A key advantage of FOS, unmatched by any other technology so far, is the possibility to develop multiple sensors on the same fiber (multiplexing). This solution allows having multiple sensors stacked in a short space, which can also be employed to detect a plurality of parameters, in a concept often referred to as lab in a fiber. This concept can even extend from a discrete set of sensors to a continuous set of sensing points, where we can measure the variation of one physical parameter (strain or temperature, typically) in every point of an optical fiber with millimeter or lower spatial resolution. Thus, FOSs allow not only the measurement of biophysical parameters, but also their spatial distribution.
- Biocompatibility: FOS probes are made with silica or polymer optical fibers, using biocompatible materials. This makes sensors compliant with medical standards that evaluate the toxicity of materials. It is also possible to fabricate epoxy-free mounts for probes that do not alter the fiber materials.
- Remote sensing: The connection between the fiber-optic probe and the optical source and detector is performed with an optical fiber cable; fibers have very small attenuation, making it possible to position the probe(s) remotely at a long distance from the interrogation hardware. The cabled interconnection also allows an easy guiding of light from the source and to the detectors without external devices.

In contrast, FOSs are often an expensive technology and are not simple to develop, catheterize, and commercialize. The gap between the technology state of the art and the amount and impact of applications is still large, but the property of FOSs makes them the right technology to measure what is measurable within, or on, our body. The effort, from a scientific point of view, should be addressed on this action: use FOS to enable unprecedented applications, inaccessible to other sensing technologies, leveraging on the small size and multiplexing features. This is the operative framework that we should look for: bridging design of sensors and their medical applications. A sensors specialist should consider the requirements and protocols for each application and understand how the sensors behave within the body as opposite as in ideal conditions. However, clinicians should identify how to dramatically improve treatments and diagnostic using sensors and providing engineers with guidelines and standards for development. We have often observed that the technology currently employed in everyday clinical practices is often far from the state of the art, while latest research output is not finding any applications and remains a technology with good potential but no practical users.

Applications for FOS are substantially different from those involving MEMS and piezoelectric or mechanical sensors. FOS succeed to deliver the largest impact wherever traditional technologies fail due to their insufficient number of sensing points, size/weight issues, lack of performance or biocompatibility. Figure I.1 sketches some of the main applications of FOS technology to medical sensing, highlighting the technology readiness level. The expectations for FOS in the near future are to establish as the key technology to measure what is not measurable, with traditional methods. To date, we can roughly divide biomedical FOS into 4 categories:

1. Sensors for supporting treatments and surgery: The probe, packaged in a medical catheter or needle, is inserted percutaneously (insertion through a needle puncture of the skin), endoscopically (examination and treatment of an inner organic tissue), or in other ways within the body. Sensor data are detected in real time, returning the data to the clinician or using it to automatically handle the treatment. Applications include thermotherapies, robotic microsurgery, angioplasty, and intra-aortic balloon pumping.
2. Sensors for diagnostics: These are used to identify and diagnose specific conditions or diseases. Sensors are introduced with medical catheters and diagnostics are partially or totally based on measurement data. Applications

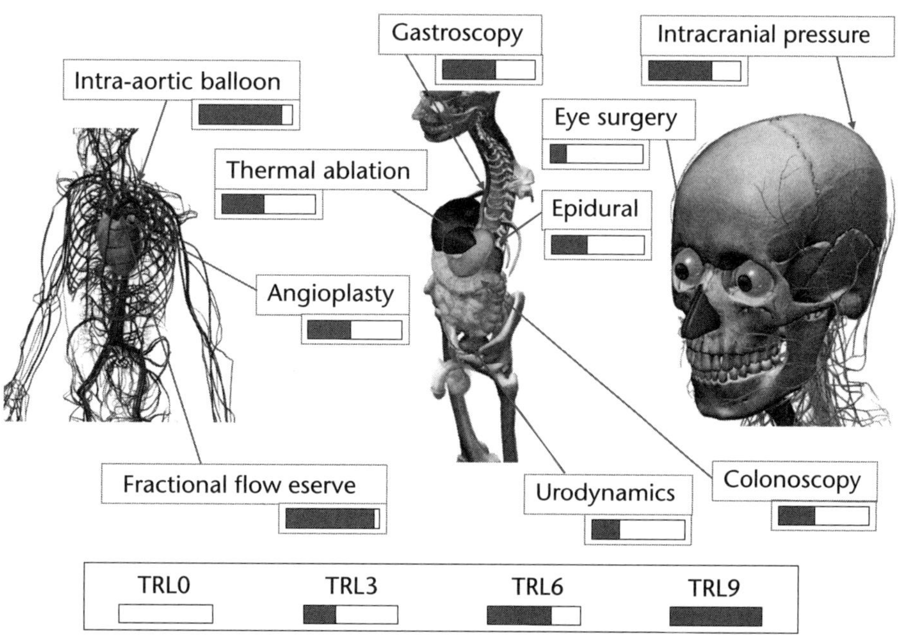

Figure I.1 Sketch of some of the major applications of FOS to medical measurement; the bar shows the level of technological maturity of each application. The legend shows the technology readiness level (TRL).

include blood vessel obstruction detection, urodynamics, gastroscopy, and intracranial pressure measurement.

3. Fiber-optic biosensors: Probes can be functionalized to detect the concentration of specific biological constituents and therefore search for the presence and localization of biomarkers of diseases. Detection can be performed on samples obtained through biopsy or on body fluids, but the future trend is to incorporate the device in a catheter and perform the measurements in vivo. Applications include cancer diagnostics, detection of thrombin, and diagnostics of autoimmune disorders.
4. External or wearable sensors: In this case, the sensing system is designed to be worn by a patient or carried through a portable device such as a smartphone. Sensors operate externally to the body, yet detect parameters of medical interest. Applications include respiratory analysis, body temperature detection through smart textiles, and gait investigation.

Within this book, we will try to merge the principles and practice of FOS and the biomedical applications, present and future, of FOS. For this reason, this book will follow the structure illustrated in Figure I.2. In Chapters 1 to 9, we will illustrate the principle of operation of FOS. An initial introduction on the principles of optical fibers and fiber-optic devices will help readers to understand the basics of fiber-optics and the building blocks of any FOS; this is followed by a discussion on sensors and measurements, understanding how to evaluate the performance of sensors and to define medical devices. These initial chapters are intended for a broad

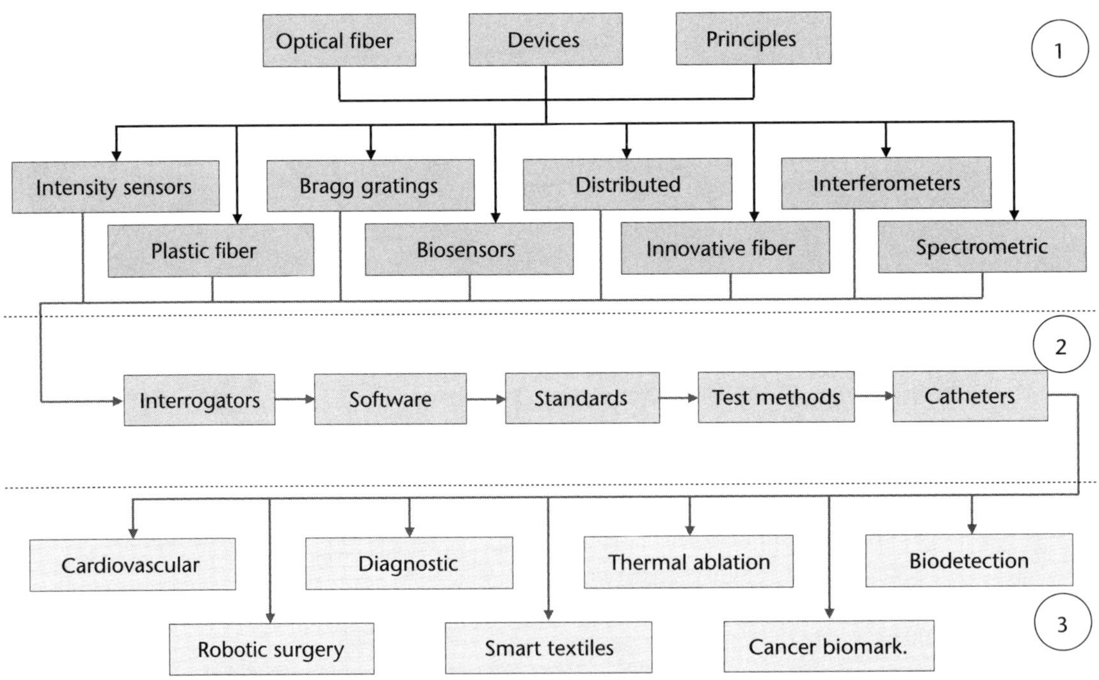

Figure I.2 Sketch of the organization of the book: the first part describes fiber-optic devices and FOS sensing principles; the second part will help the engineer to move from the principle to the realization of a medical device and its validation; the third part describes the main applications, from the most popular to the emerging fields.

audience, and are a tool for students and early-stage engineers with little experience in electromagnetics and electronics to get introduced to the FOS background.

Sensing principles of FOS rely to the wave nature of light. As any wave confined in a waveguide, we can observe the properties of light: amplitude, phase, modes, polarization, and their dependence on wavelength. In sensors, the measurand quantity modulates one or more of these parameters. Chapters 4 to 9 provide the several approaches to sensors, explaining the working principles of the most popular and the rapidly emerging architectures, providing references and practical cases for each sensor type. Intensity-based sensors (Chapter 4) are conceptually the simplest approach to sensing, and fiber Bragg gratings (Chapter 5) are the most popular type of commercial FOS sensor for medical applications. Next to these techniques, distributed sensors (Chapter 6) are considered the next big thing in biomedical engineering. Interferometric sensors (Chapter 7) are at the base of modern optical pressure sensors, and broadly used in medicine. Surface plasmon resonance (Chapter 8) sensors meet the growing demand of optical biosensors and can open in vivo biosensing in real time in the near future. Fiber-optic spectrometry (Chapter 9) is also briefly explained.

In Chapters 10 to 14, we aim at connecting the working principles of sensing with their applicative scenario; through this part, we will explain how to assemble an interrogation unit for sensors and on-board software and how to design and test medical devices. Sensor networks and interrogators (Chapter 10) are the optical hardware of each sensing unit and are required for each sensing principle. Detection software (Chapter 11) is required to read the spectrum or amplitude of sensors and, from them, estimate the measurands. Biomedical standards (Chapter 12) describe the certifications currently in use for medical devices embodying fiber-optic sensors, highlighting the relevant standards for the United States and Europe. Research protocols (Chapter 13) are the set of procedures and tools that engineers should implement when testing and validating sensors; we will provide a description, adapted to FOS technology and a set of useful tools. Catheters and packaging (Chapter 14) are requirements for medical devices requiring percutaneous or endoscopic insertion; the design of catheters with embedded FOS is hereby described.

Finally, Chapters 15 to 21 describe the applications of FOS to medicine and biomedical engineering, presenting ongoing and commercial cases as well as the emerging scenarios. Cardiovascular (CV) sensors (Chapter 15) operate within the cardiac apparatus or in blood vessels and are one of the most established applications of FOS pressure sensors. Sensors for diagnostics (Chapter 16) are a key asset in urology, gastroscopy, and neurology, as the diagnosis of conditions is based on the measurement of biophysical quantities. Thermal ablation (Chapter 17) is a rapidly emerging medical interventional procedure, particularly in cancer care; recently, FOS technologies are playing a key role in monitoring and controlling thermotherapies. Biosensors (Chapter 18) based on optical fibers are potentially the next big thing; we will discuss the most recent works and their impact. Minimally invasive robotic surgery (Chapter 19) is progressively replacing manual interventions with robots; FOSs are the key element to provide tactile perceptions and force estimates. Smart textiles and wearable sensors (Chapter 20) are used externally to the human body, embedded in fabric or located in portable devices, and support continuous monitoring of medical parameters. A brief outline of spectroscopy for cancer care (Chapter 21) shows how this fundamental application can be integrated in the next FOS technologies.

Finally, Chapter 22 will draw conclusions of the book. In Appendixes A and B, we will provide simple code (programmed in MATLAB™) that enables readers to simulate the most important sensing structures, and to implement the software detection algorithms described in the book.

CHAPTER 1

Fundamentals of Wave Optics and Optical Fibers

1.1 Introduction

Optical waves, or waves at optical frequencies, are electromagnetic waves that occupy a portion of the spectrum between infrared (IR) and ultraviolet (UV). Usually, when thinking of optical waves, people refer to the visible (VIS) range, the wavelength corresponding to colors from red (about 700 nm) to violet (about 400 nm). However, researchers in the field include also a group of wavelengths in the 800-nm to 2,000-nm (or sometimes even longer wavelengths) range, the near-infrared (NIR), which are in the infrared region. Figure 1.1 shows a schematic representation of the subdivision of the electromagnetic spectrum in the optical range, which indicated the most relevant wavelengths. The NIR region, specifically around 800 nm, 1,300 nm, and 1,500 nm, is included into the common definition of optical waves because very important in optical communications. For instance, all the long-haul communications take place in the 1,500-nm spectral region, the third optical communications window, whereas fiber-to-the-home (FTTH) applications use also the 1,300-nm spectral region. NIR plays a key role also for fiber-optic sensing, as most of the components used for sensing applications, including the optical fibers, were originally developed for telecommunications.

This chapter is devoted to an introduction of fundamentals of fiber-optics. As this subject is very broad and it is presented in a very large number of books, in the following we will not indicate bibliographical references close to specific topics; instead, we will report at the end of the chapter a list of useful books.

1.2 Electromagnetic Waves

We mentioned that optical waves are electromagnetic waves, so it is worth briefly recalling some basic properties of such waves.

Electromagnetic waves can have any time dependence, but the most important is the time-harmonic regime in which the field components are sinusoidal functions of time at a given frequency. In this case, any field component can be represented

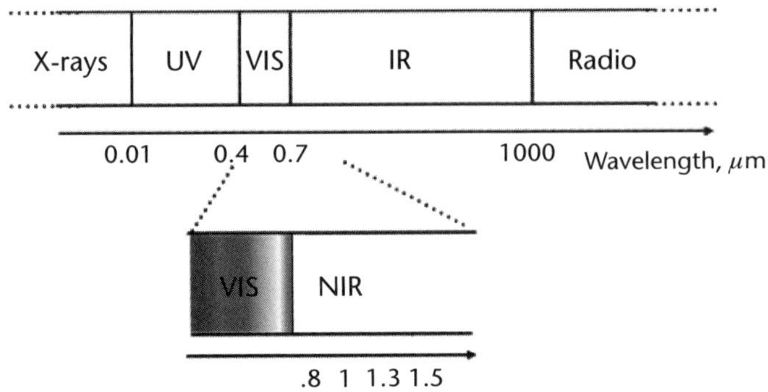

Figure 1.1 Schematic representation of the portion of the electromagnetic spectrum with evidenced what it is usually called the optical wavelength region, which, in turn, includes the two main intervals called the visible (VIS) and the near infrared (NIR).

in the frequency domain by a complex number known as the phasor, which is related to the Fourier transform (FT) of its time-domain expression. Consider as an example a generic electric field component $e(t)$ at a frequency f with sinusoidal dependence:

$$e(t) = A\cos(\omega_0 t + \phi)$$

where A is the (constant) amplitude or magnitude, $\omega_0 = 2\pi f$ is the angular frequency, and ϕ is an initial phase value; the corresponding phasor is:

$$E = Ae^{j\phi}$$

defined so that:

$$e(t) = \mathrm{Re}\left\{Ee^{j\omega_0 t}\right\} = \mathrm{Re}\left\{Ae^{j(\omega_0 t + \phi)}\right\}$$

where j is the imaginary unit (often indicated also with i) and Re is the real-part operator. The phasor notation is particularly convenient since it shares the same properties of the FT (for instance, to transform a differential equation into an algebraic equation), but, differently from FT, it maintains the same physical dimensions of the time domain quantity. For example, the phasor of an electric field will be a complex number with a physical dimension [V/m], as the time-domain field.

The phenomena described in this book can be understood considering the simplest form of electromagnetic waves, the plane waves, a solution of Maxwell equation in Cartesian coordinates, for a homogeneous and isotropic unbounded medium, in absence of sources. In this case, the electric field vector $\overline{E}$ in a generic point of space $\overline{r}$ has the phasor form:

$$\overline{E}(\overline{r}) = \overline{E}_0 e^{-j\overline{k}\cdot\overline{r}} \tag{1.1}$$

where $\overline{E}_0$ is the phasor of the electric field vector in the origin of the coordinate system, $\overline{k}$ is the wave vector, and "·" is the dot-product operator between vectors. In turn,

$$\overline{E}_0 = E_0\hat{e} \tag{1.2}$$

where E_0 is the electric field amplitude and $\hat{e}$ represents a unitary vector indicating its direction. Similarly, $\overline{k} = k\hat{s}$, where $\hat{s}$ represents a unitary vector in space indicating the direction of propagation of the wave and $k = k_0 n$, being k the wave number in the medium characterized by the refractive index n and

$$k_0 = \frac{\omega_0}{c} = \frac{2\pi f}{c} = \frac{2\pi}{\lambda} \tag{1.3}$$

is the vacuum wave-number, c is the speed of light in vacuum, and $\lambda = c/f$ is the wavelength. The refractive index, or index of refraction, of a material is a dimensionless number that describes how light propagates through that medium. It is defined as the ratio between the speed of light in vacuum c and that in the medium v:

$$n = \frac{c}{v}$$

As for ordinary materials $n > 1$, it turns out that the phase velocity (i.e., the speed at which the crests or the phase of the wave moves) in a material is smaller than in the vacuum, so in a material light is slower. The value of the refractive index changes with the wavelength, the chromatic dispersion; this change explains, for example, the separation of colors from white light operated by prisms (or more in general, by what are called dispersive elements). The refractive index of materials used in optics ranges from about 1.3 to 4, with the typical values being around 1.5, which is the value of silica and silicate glasses, by far the most common materials.

For usual optical materials that have no magnetic properties, the refractive index is the square root of the relative permittivity[1].

The wavelength in (1.3), measured in meters (or, actually in its submultiples, micrometer and nanometer, in the frequency range of interest), represents the spatial period of the wave. Given the (inverse) proportionality between frequency and wavelength, they can be used interchangeably to identify the same property of the wave; however, since in optics frequencies are very high (hundreds of terahertz), wavelengths are most commonly used.

For isotropic, homogeneous, and lossless[2] materials as considered throughout this book, n is a scalar real quantity and thus the term $\overline{k} \bullet \overline{r} = k_0 n(\overline{s} \bullet \overline{r})$ in the

1. The relative permittivity of a material is the ratio between the permittivity of the considered material and the vacuum permittivity. The (electrical) permittivity expresses the relation between the electric displacement field $\overline{D}$ and the electric field $\overline{E}$; it is given by the product of ε_0, the vacuum permittivity, which is a constant having dimensions of [F/m], and a dimensionless quantity ε_r, the relative permittivity, which is characteristic of the individual materials.
2. Isotropic means that propagation in the considered material is not dependent on the light polarization (a concept that will be introduced later in this section); in other words, that the material refractive index is not dependent on the light polarization. Homogenous means that the material refractive index is not dependent of the spatial position considered, so it is the same everywhere. Lossless means that light is not

exponential in (1.1) is a number that represents a phase term that changes as the wave propagates. Actually, in the frequency domain, propagation means change of phase, because a change of phase in frequency corresponds to a translation of the signal in the time domain.

The wave expression in (1.1) is said to represent a plane wave because the surfaces with constant phase are planes, perpendicular to the direction of propagation represented by $\bar{k}$ or $\hat{s}$. An expression similar to (1.1) holds for the magnetic field $\bar{H}$:

$$\bar{H}(\bar{r}) = \bar{H}_0 e^{-j\bar{k}\cdot\bar{r}}$$

where, again, $\bar{H}_0$ is the phasor of the magnetic field vector in the origin of the coordinate system. Moreover, it is possible to prove that electric and magnetic fields are orthogonal to each other and related through an impedance relation:

$$\bar{H} = \frac{1}{\omega\mu}\bar{k} \times \bar{E}$$

where μ is the magnetic permeability[3] and $\times$ denotes the vector cross-product.

As the wave propagates, the instantaneous electric field (i.e., the time-domain expression) oscillates in a plane perpendicular to $\bar{k}$; for example, if the wave is propagating in the z-direction, then $\bar{k} = k\hat{z}$ and the electric field can oscillate in any direction that lies in the x-y plane.

The plane geometrical figure described by the electric field in time is known as the polarization of the wave. The most general case is that of an ellipse, so the wave is said to be elliptically polarized; when semi-axes are perpendicular and of equal length, the ellipse becomes a circle and the wave is said to be circularly polarized, either clockwise or counterclockwise depending on how the circle is described in time. Finally, the electric field vector can oscillate along a straight line (a sort of one-dimensional ellipse) and the wave is said to be linearly polarized. With reference to (1.1) and (1.2), the polarization is given by $\bar{E}_0$, or more precisely $\hat{e}$. Assuming for simplicity to be in the x-y plane and writing $\bar{E}_0 = E_{0x}\hat{x} + E_{0y}\hat{y}$, a circular polarization requires that E_{0x} and E_{0y} have the same magnitude and a phase difference of , whereas a linear polarization requires that E_{0x} and E_{0y} are in phase, regardless of their magnitudes, or that one of the two is zero. Figure 1.2 shows some examples of polarizations in the x-y plane: case (a) is for $\hat{e} = \hat{x}$, case (b) is for $\hat{e} = \frac{\hat{x} + \hat{y}}{\sqrt{2}}$, and case (c) is for $\hat{e} = \frac{\hat{x} - j\hat{y}}{\sqrt{2}}$.

attenuated during propagation, so there is no energy dissipation. Do not confuse dissipation (which involves losing power during propagation) with dispersion (which refers to the dependency of the refractive index with frequency, but no loss of power is implied).

3. The magnetic permeability expresses the relation between the magnetic induction and magnetic field and is given by the product of μ_0, the vacuum magnetic permeability in [H/m], and a dimensionless quantity μ_r, the relative permeability. Usually, as in optics the medium of propagation has no magnetic properties, $\mu_r = 1$, so $\mu = \mu_0$.

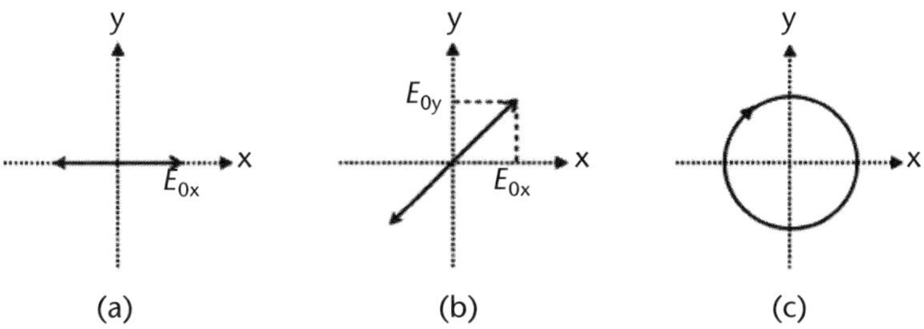

Figure 1.2 Examples of polarizations in the x-y plane: (a) linear horizontal, (b) linear at 45°, and (c) circular clockwise.

Any polarization can be written as the combination of two orthogonal linear polarizations (usually for convenience identified as horizontal and vertical) or two counter-rotating circular polarizations.

By further inspecting (1.1), it is evident that a wave is characterized by four parameters: frequency (or wavelength), amplitude, phase, and polarization. A quantity that affects any of these parameters will therefore somehow change the characteristics of the wave and this, while being detrimental in optical communications, can be efficiently exploited to fabricate optical sensors. For example, a physical stimulus, say, a strain, that changes E_0 as the wave propagates through a certain region of space will modify the field amplitude; this can be used to fabricate an intensity-based sensor (or intensiometric sensor) that allows evaluating the variations of the quantity of interest by measuring the variations in the power associated to the received wave. More on this will be discussed in the following chapters.

1.3 Reflection and Refraction of Plane Waves

It is a common experience that when an optical wave is incident on the interface between two dielectrics, part is reflected (the reflected wave) and part is transmitted (the refracted wave). Equally well known is that the refracted wave changes its direction of propagation as it happens, for example, in the famous experiment of a pen partially immersed in water that appears to be broken at the water-air interface.

We can simplify our discussion by assuming to have chosen a proper reference system that allows us to deal with a two-dimensional problem in the x-z plane, as in Figure 1.3, in which there are no variations along the y-axis and a smooth planar optical interface is defined between two linear, homogenous, and isotropic materials with refractive index values n_1 and n_2, respectively.

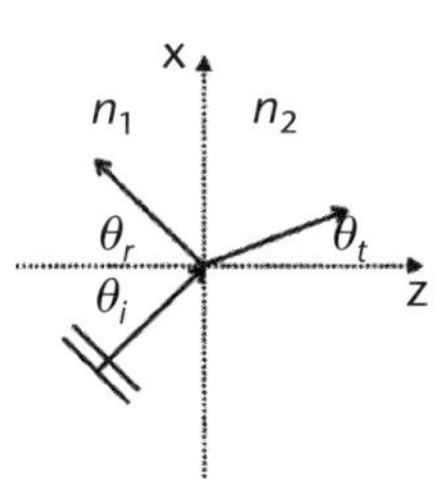

Figure 1.3 Geometry considered to define reflected and refracted plane waves at a dielectric interface.

It can be shown that in these conditions Maxwell equations decouple in two independent sets:

- One that involves only the electric field component along $\hat{y}$, and the magnetic field components along $\hat{x}$ and $\hat{z}$; since the electric field has component transverse to the interface only, the plane wave solutions of this set of equations are called transverse electric (TE) plane waves at microwaves or s-polarized (or perpendicularly polarized) waves in optics.
- One that involves only the magnetic field component along $\hat{y}$, and the electric field components along $\hat{x}$ and $\hat{z}$; since the magnetic field has component transverse to the interface only, the solutions of this set of equations are called transverse magnetic (TM) plane waves at microwaves or p-polarized (or parallel polarized) wave in optics.

The letter "s" to indicate the perpendicular polarization comes from the German term "senkrecht," which means perpendicular ("p" was already taken to indicate parallel also using German terms).

The TE and TM fields represents two orthogonal linear polarizations and thus decomposing a generic plane wave into the sum of s-polarized and p-polarized waves corresponds to decomposing a generic polarization into the sum of two specific linear polarizations that form a basis.

Boundary conditions at the interface between the two dielectrics impose the continuity of the tangential components of the electric and magnetic fields and thus of the normal components of the electric displacement $\overline{D}$ and of the magnetic induction $\overline{B}$ fields. As a consequence, it can be proved that at the interface no change of polarization occurs: an incident s-polarized wave corresponds to reflected and transmitted s-polarized waves, and similarly for p-polarized waves. However, the reflection and transmission coefficients (i.e., the terms that relate the reflected and transmitted field amplitudes) will be dependent on the s- or p-polarization of the wave, except for the case of normal incidence (so for $\hat{s} = \hat{z}$ in Figure 1.3) for which the values are coincident for both polarizations. This introduces a polarization dependent behavior at the interface between the two dielectrics, so an anisotropy, even if the two materials are isotropic.

Moreover, from boundary conditions we can derive the well-known Snell laws: for a plane wave with a given angle of incidence θ_i, the reflected wave will be at an angle $\theta_r = \theta_i$, whereas the angle formed by the transmitted wave θ_t is found from

$$n_1 \sin\theta_{\mathrm{i}} = n_2 \sin\theta_{\mathrm{t}} \tag{1.4}$$

If the medium with n_1 is denser that the medium with n_2, $n_1 > n_2$ and for angles of incidence larger than a certain value θ_c, the critical angle, (1.4) cannot be satisfied by real transmitted angles. We say that we are in total internal reflection (TIR) and the wave is totally reflected back in medium with n_1. Actually, in these conditions, the field in medium with n_2 is not abruptly going to zero, but it has an exponential decay (albeit quite fast, in the order of the wavelength scale), so for a thin region around the interface, the field cannot be neglected. We say that the wave is evanescent in medium with n_2 and this phenomenon is exploited in certain types of sensors, especially for chemicals, as detailed in Chapters 8 and 18. In TIR the magnitude of the reflection coefficient is 1 both for s- and p-polarized waves, but the phase is different for the two polarizations.

1.4 Dielectric Waveguides

A waveguide is a structure used to confine electromagnetic waves in a defined region of space; it means a hollow metal pipe (usually with rectangular, circular or elliptical section) at radio frequency (the metallic waveguides) and an all-dielectric miniaturized rod in optics (the dielectric waveguides) because metals introduce too high losses at these frequencies. The simplest form of dielectric waveguide is the slab waveguide, where the confinement occurs in one direction only [depth in Figure 1.4(a), being breadth ideally unlimited, in practice much larger that the wavelength] and thus it has almost no practical use. It is, however, a good educational example because easier to analyze; moreover, more complex waveguide structures can be studied in an approximate form in terms of equivalent slab waveguides.

More realistic waveguides confine light in the entire transverse section, such as the planar channel waveguides [Figure 1.4(b)], which have an almost rectangular cross-section and are used in integrated optical circuits (millimeter-long devices), or optical fibers [Figure 1.4(c)], which have a circular cross-section and are used in longer range light transmission (from centimeters up to hundreds of kilometers).

The confinement of light in the depicted dielectric waveguides exploits the total internal reflection; hence, it occurs in a region with slightly higher refractive index (the darker regions in Figure 1.4) than the surrounding materials. The TIR condition is necessary but not sufficient to guarantee that a field distribution is guided, meaning that it is confined in the transverse direction and propagates along the longitudinal direction (typically assumed to be the z-axis) so that its energy is delivered at the output, without losses in an ideal dielectric waveguide. A field distribution that maintains the same transverse distribution at all locations along the waveguide axis (i.e., at each z value) and it only changes its phase as it propagates is defined a guided mode (or simply, mode). Besides for being in TIR condition, guided modes must satisfy also a phase-matching condition and this makes them be discrete in number.

For instance, in the slab waveguide case depicted in Figure 1.5, to have confinement in the central medium with refractive index n_1, the phase matching condition requires that the round-trip phase along the transverse direction (the x-axis) is an

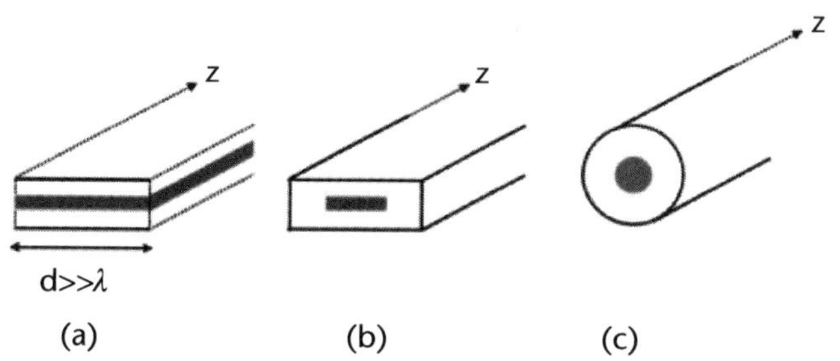

Figure 1.4 Examples of dielectric waveguides: (a) slab; (b) channel; and (c) optical fiber.

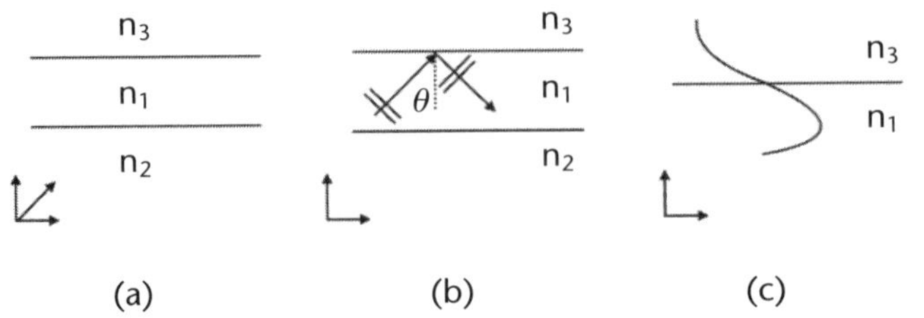

Figure 1.5 Representation of (a) the cross-section of a slab waveguide; (b) a mode in terms of two plane waves upward and downward propagating at an angle θ; and (c) the oscillating and exponential field behavior at the core-cladding interface.

integer multiple of 2π. This means that the sum of the phase shift acquired along the up-down going in the x-axis plus the phase shift acquired due to TIR at the two boundaries[4] must be an integer multiple of 2π.

Field distributions that are not guided are called radiation (or radiated) modes; these are infinite in number and loose power as they travel along the waveguide, so they contribute only for short propagation lengths. Radiation modes can be excited by many causes, such as poor overlapping between the source radiation pattern and the ensemble of guided mode (the guided mode spectrum) of the waveguide, discontinuities (misalignments or inhomogeneities), and too-sharp bends. Stray light that may emerge from the sides of a waveguide (for example, when we bend a fiber, we see it illuminates) is due to excitation of radiation modes, as guided modes are confined inside the waveguide and thus do not leak power as they propagate; hence, they cannot be seen.

Depending on the working frequency, the geometry, and the materials, in general, a waveguide can have zero, one, or a plurality of guided modes. If the number of guided modes is zero, the waveguide is useless; if there is only one guided mode, we say that the waveguide is single mode (SM), whereas if there are more guided modes, the waveguide is multimode (MM).

A generic input field distribution can be written as a combination of guided and radiation modes: simplifying, if the waveguide is long enough so that all the power associated to radiated modes is lost, the output power will be given by the sum of the power carried by each individual guided mode. It is therefore evident that MM waveguides are easier to handle because it is much easier to couple light into as there are many possible field distributions that are guided, providing a better overlap with the radiation pattern of any source. However, the intrinsic working principle of some sensors will require using SM waveguides. The number of guided modes increases as the frequency increases or, equivalently, as the wavelength decreases.

Consider for simplicity a slab waveguide, as in Figure 1.5(a), formed by a layer of refractive index n_1, called the core (or film) in which the confinement takes place, surrounded by a layer of refractive index n_2, called the substrate (or bulk), and a layer of refractive index n_3, called the cover (or cladding). As already mentioned, necessary for the confinement is the possibility of TIR; hence, $n_1 > n_2 \geq n_3$, where $n_2 = n_3$ for symmetric waveguides. A mode can be seen as the result of the

4. Recall that in TIR the magnitude of the reflection coefficient is one, but its phase is nonzero; hence, the reflected field will experience a phase shift.

interference between two plane waves upward and downward propagating at an angle θ with respect to the normal to the dielectric layer surface [Figure 1.5(b)]; for a guided mode, this interference produces inside the core a standing-wave pattern with one or more oscillation peaks depending on the mode order. It is possible to show that the mode having one peak only is the mode that starts to appear first as the frequency increases and is therefore called the fundamental mode. The mode with two oscillations will be the second-order mode, and so on for other higher-order modes (HOMs). When a waveguide is in the single-mode regime, it supports the fundamental mode only, whereas in the multimode operation, it supports the fundamental and a plurality of higher-order modes. Given the wavelength range of usage, the refractive index of the various layers and the thickness of the core can be designed to have just one or more guided modes; however, a waveguide that is single mode at certain wavelength becomes multimode at shorter wavelengths.

In photonics, multimode waveguides may guide even several hundreds of HOMs.

As for the field distribution of a guided mode in the cover and the substrate, being in TIR, it must have an exponential decay, as pictured in Figure 1.5(c).

Figure 1.6 shows a pictorial representation of guided and radiated modes in terms of the rays, which here are basically the direction of propagation of the plane waves whose interference forms the mode field distribution. Although this representation is quite famous because it is usually employed in any text that aims at providing a basic explanation of how it is possible to confine light in a dielectric structure, it should be handled with care, considering that this is a simplified representation. Indeed, it has to be understood that the upward and downward propagating waves are both present in each section along the propagation direction z as it is their combination that must give the same standing-wave pattern in any coordinate z for a mode to be guided. In other words, with reference to Figure 1.6(a), it has not to be thought that regions exist in which there is only the individual up-going or down-going wave, but they are always simultaneously present.

As the phase shift acquired upon TIR depends on the wave polarization, the phase matching condition that defines a guided mode will give different results for s- and p-polarizations, therefore leading to two categories of modes. It is possible to show that the fundamental mode of a slab waveguide is s-polarized.

Each mode is characterized by a propagation constant, typically indicated by the Greek letter β, which represents the phase change rate as it propagates in the waveguide; the propagation constant of the ith mode can be written in terms of the modal effective (or equivalent) refractive index $n_{\text{eff,i}}$ as $\beta_i = k_0\, n_{\text{eff,i}}$ where $n_2 < n_{\text{eff,i}} <$

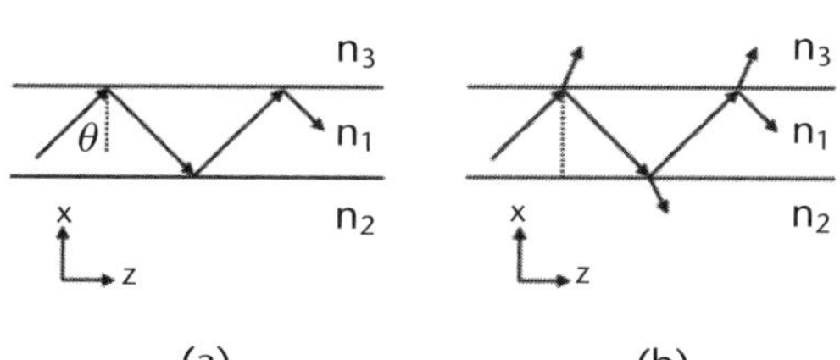

Figure 1.6 A simplified representation in terms of rays of: (a) a guided mode; and (b) a radiation mode that irradiates both in the cladding and in the substrate, as it corresponds in terms of plane-wave decomposition to waves that are not in total internal reflection at these boundaries.

n_1. The modes are ordered with the respect to their effective index, the fundamental being that with the highest possible value.

Figure 1.7 shows a sketch of some guided and radiated mode profiles: note that guided modes are confined thanks to the exponential decay outside the core region, whereas radiated modes do not.

Consider a symmetric waveguide for simplicity. We have already mentioned that, to have confinement, one the conditions requires that the core-cladding incidence angle is larger than the critical angle; with reference to Figure 1.8, through the Snell law, this implies that the entrance angle into the slab from air must be less than θ_a. The sine of θ_a is called the numerical aperture (NA):

$$\mathrm{NA} = \sin\theta_a = \sqrt{n_1^2 - n_2^2} \tag{1.5}$$

which, being a small value, is often approximated as NA $\approx \theta_a$ and relates the acceptance angle with core-cladding refractive index variation (also known as refractive index contrast).

1.5 Optical Fibers

Optical fibers are cylindrical dielectric light guides, whose basic geometry is that sketched in Figure 1.9, in which light is guided in the core by reflecting off at the core-cladding interface. Core and cladding have slightly different refractive index values, typically of few units in 10^{-3}, with $n_{\text{core}} > n_{\text{cladding}}$ to allow TIR. In principle, a core-only fiber could be enough to guide light, but cladding is necessary to avoid perturbations due to the fiber support or surface contamination (recall that the field is not going abruptly to zero at the core interface), besides for allowing the control of the core-cladding refractive index contrast and through that, together with the core diameter, the number of guided modes.

Core and cladding are usually made of ultrapure glass, although some fibers are all plastic and some have glass core and plastic cladding. All-glass fibers are typically made of high-purity fused silica, with the core slightly doped with germanium or phosphorous to increase the refractive index. This is by far the most common type of fibers as it is the fiber used in optical communication networks.

Fibers, as the other dielectric waveguides, can be a single-mode fiber (SMF) or a multimode fiber (MMF). For the fibers used in optical communications, which are intended to work primarily around 1,300 nm and 1,550 nm (the second and third optical communication windows), SMFs have a core diameter of about 9 to 10 μm and a cladding diameter of 125 μm; they are therefore indicated as 9/125 fibers. As for MMF, the most common have a core diameter of 50 μm or 62.5 μm

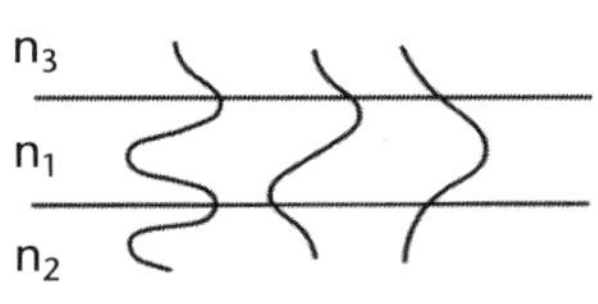

Figure 1.7 Sketch of mode profiles for a slab waveguide: from right to left, fundamental mode, first higher-order mode, and substrate radiated mode.

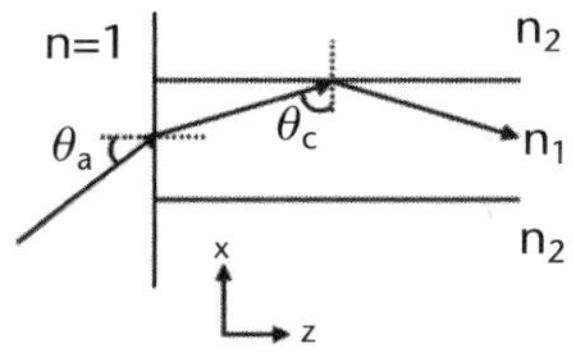

Figure 1.8 Definition of the numerical aperture of a symmetric slab waveguide.

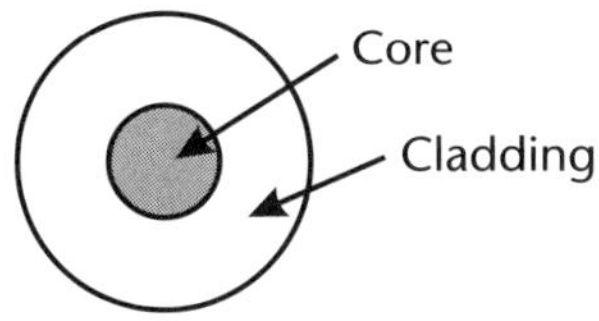

Figure 1.9 Cross-section of basic optical fiber structure.

and a cladding diameter, again, of 125 μm; they are therefore indicated as 50/125 or 62.5/125 fibers. Both these MMFs guide several hundred modes at 1,550 nm.

It is worthy highlighting that the number of modes increases as the wavelength shortens: hence, an SMF at 1,550 nm will not be single mode in the visible range anymore, but will guide some higher-order modes as well. For a single mode in the visible, say, at 633 nm, one should look for a fiber with either an even smaller refractive index contrast or, most commonly, a smaller core diameter, in the order of 4 to 5 μm.

The propagation loss of all-glass fibers is very low: for SMFs, it can go down to the incredible value of 0.17 dB/km (the decibel is a logarithmic unit expressing the ratio between two power levels[5]), meaning that the power is reduced by a factor of 2 only after about 18 km. From the sensor point of view, this implies that such optical fibers can be used route the measurement signals is remote locations without appreciable degradations.

In biomedical sensing fibers with larger core diameters, for example from 200 μm up to 600 μm and more, hence largely multimode, are also used, especially for intensity-based system, as discussed in Chapter 4. This choice is partly due to the necessity of higher light collection capabilities, and partly to improve the mechanical strength. The latter aspect is particularly relevant to fabricate applicators that can be inserted as probes. Moreover, dirt on the fiber ends has a smaller impact than with smaller-core fibers. Another category of applications that requires large fiber is imaging: in this case core diameters of some hundreds of micrometers are employed, possibly obtained through a bundle of smaller fibers not to worsen the flexibility. A typical example is in endoscopes.

5. The decibel, dB, is defined as 10 times the logarithm base 10 of the ratio between two power values dB = 10 $\log_{10} P_1/P_2$, so it expresses a relative variation. It can have positive or negative values depending if the power ratio is larger or smaller than 1. A ratio of 2 corresponds to 3 dB; a ratio of 1/2 to –3 dB, a ratio of 10 to 10 dB, and a ratio of 100 to 20 dB. It has not to be confused with the dBm, another highly utilized way to express power levels in photonics: the dBm expresses an absolute value of power compared to the reference level of 1 mW: dBm = 10 $\log_{10} P/1\text{mW}$. For instance, 0 dBm corresponds to 1 mW, and 30 dBm to 1W. The maximum emitted power or the maximum accepted power for instruments is often expressed in the datasheets in dBm.

Besides for all-glass fibers, biomedical sensing also makes use of fibers having a glass core and a polymeric cladding and of all-polymer fibers. The latter are also known as plastic optical fibers (POFs), and are characterized by a core made of polymethyl methacrylate (PMMA), surrounded by lower-index materials, such as fluorine containing polymers. Compared to telecom fibers, they usually have a much larger core, from 250 μm to 1,000 μm and above, which makes them supporting thousands of modes, so that coupling with any source is really facilitated. The tolerance in coupling allows using very simple tools for cutting and polishing the fiber and of inexpensive connectors (i.e., connectors with quite low mechanical characteristics). Moreover, as already mentioned, the combination of large core diameter and high NA makes the fiber more tolerant to dirt. Another important advantage of POFs for biomedical sensing is their high mechanical resilience, which is intrinsic of plastic materials, and that makes possible to severely bend or stress the fiber without risking to break it. As for the drawbacks, POFs have much higher propagation loss reducing the usable length to few tens of meters only. This may not represent a key limitation in biomedical sensing, at least not as much as the uniformity of behavior and the time stability that for POFs are worse than those of corresponding glass fibers. Therefore, given their characteristics, POFs are most suitable for low-cost, intensity-based sensing systems.

Another distinction among the different type of fibers is based on the nature of the core-cladding boundary: step index (SI), if the change of the refractive index (and thus of the material composition) between core and cladding is abrupt; graded index (GRIN) if the refractive index gradually decreases from the value at center of the core to that of the cladding. In the latter case, different profiles are possible, the most common being a close to parabolic one. Usually, SMFs have an SI structure, whereas glass telecom grade MMFs have a GRIN one (this is to reduce a value known as modal dispersion, particularly relevant in optical communications).

A parameter always provided by fiber manufacturers, besides for core and cladding diameters, is the numerical aperture (NA), which has the same definition and meaning already introduced for the slab waveguide. With reference to Figure 1.10, for an SI fiber it is defined as the sine of the angle θ_a, which, geometrically, represents the angular aperture of a cone, called the acceptance cone.

In practice, the NA describes how easy it is to couple light into the fiber. A typical SMF has NA ~ 0.1 (i.e., difficult to couple light into; it requires a laser diode with suitable emission pattern), while for an MMF it ranges from about 0.3 of telecom fibers to 0.5 of POFs (i.e., very easy to couple light into; LEDs can be used as source). For multimode fibers, given that the large number of guided modes makes the angle of incidence corresponding to a guided mode a continuous function from

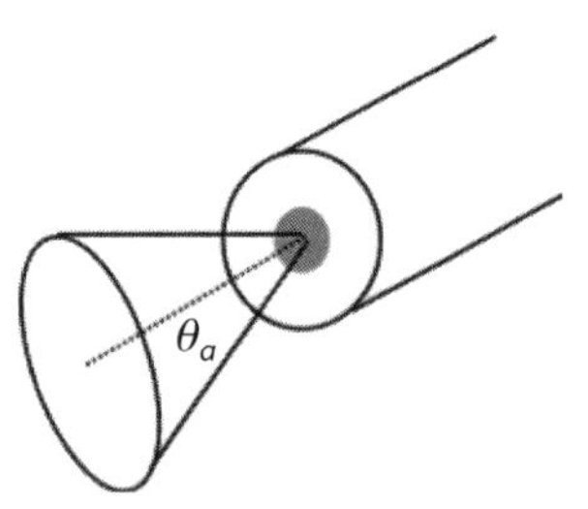

Figure 1.10 Definition of the numerical aperture for fibers.

the practical point of view[6], we can interpret the NA as the aperture of the cone that limits accepted rays, which are incident rays that originate rays confined into the fiber core.

However, picturing the modes in terms of angles formed by rays, as done for slab waveguides, is difficult for fibers since there are two categories of rays (Figure 1.11): meridional rays, which are rays that bounce at the core-cladding interface while remaining in a plane that contains the z-axis and are similar to those of the slab waveguides, and skew rays, which rotates around the z-axis describing a helical path, never crossing the axis.

The two types of rays reflect in a more difficult solution of the rigorous mode computation problem because in addition to TE and TM modes (corresponding to meridional rays, as in slab waveguides), a new set of modes, the hybrid modes (corresponding to skew rays), appears. However, for the scope of this book, we can consider the weakly guiding fibers (i.e., very low contrast), for which modes can be approximated as linearly polarized transverse only fields, the LP modes (or, more correctly, pseudo-modes since they are modes only for the limit case of $\Delta n \rightarrow 0$). Within this notation, the fundamental mode, the only mode of an SMF, can be approximated by a Gaussian distribution (a sort of a hat-shape as in Figure 1.10), while HOM have peaks and valleys in the core along the radial and azimuthal direction. Again, the field is evanescent in the cladding.

Each LP mode exists simultaneously in two orthogonal polarization states given the symmetry of the fiber; we say that LP modes are degenerate in polarization. Therefore, common fibers are not able to maintain the polarization: this means that even if you launch, at the input of a fiber span, a light distribution with a well-defined polarization state, at the output the polarization state will be unpredictable because it will change depending on how microstrains, which are unavoidable, couple light between the polarization degenerate modes along the propagation length. For example, let us suppose that we excite the vertically polarized fundamental mode of an SMF; ideally, at the output we should obtain a linear vertically polarized signal, as the horizontal polarization state of the fundamental mode that we can use as the basis to express any polarization state is orthogonal, hence it will not couple with the input field. However, perturbations, due to small-scale inhomogeneity along the fiber, to microbending or torsions in the cable cause

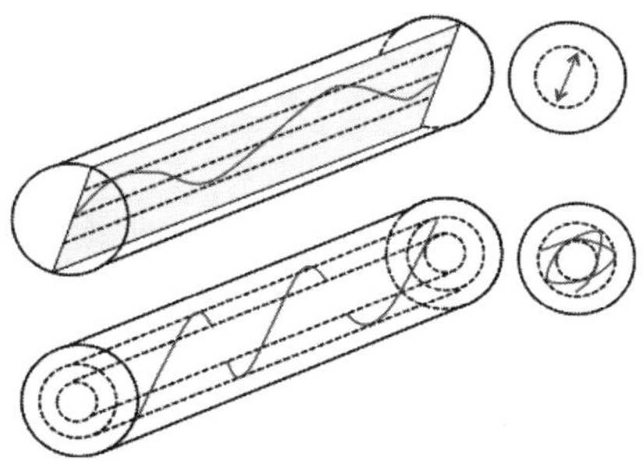

Figure 1.11 Schematic representation of meridional (above) and skew (below) rays in optical fibers.

6. Recall that MMFs usually guide hundreds or millions of modes so there are hundreds or millions of possible incidence angles corresponding to a guided mode in a restricted angular range above the critical angle.

the two fundamental modes to exchange power and since they are degenerate (i.e., they have the same propagation constant), this exchange is effective, leading to a mixture of the two fundamental-mode states at the output. If polarization is a relevant issue for a specific sensing system, then suitable polarization maintaining SMFs (PMFs) should be used. These fibers introduce a certain degree of asymmetry in the refractive index distribution so as to avoid the polarization degeneracy.

Figure 1.12 shows a picture of the fundamental mode of SI fiber.

Given the large number of modes in MMFs, we can assume that peaks and valleys of the various modes average out, so to give an almost constant distribution over the core area. In this representation, the emission from an MMF is a uniform illumination that spread out of a cone defined by the fiber NA.

Other types of fibers with more complex geometries (such as double cladding fibers) or exploiting other guiding principles than TIR (such as photonic crystal fibers) can be used for biomedical sensing. They will be briefly described in the following chapters when needed.

1.6 Practical Aspects in Using Optical Fibers

Most common fibers in sensing are those made in silicate glasses and used in optical communications:

- SI SMFs with core diameter of about 9 to 10 μm and cladding diameter of 125 μm, compliant with geometrical, optical, transmission, and mechanical parameters described in the Recommendation ITU-T G.652 issued by the ITU-T (the International Telecommunication Union, a United Nations specialized agency in the field of telecommunications);

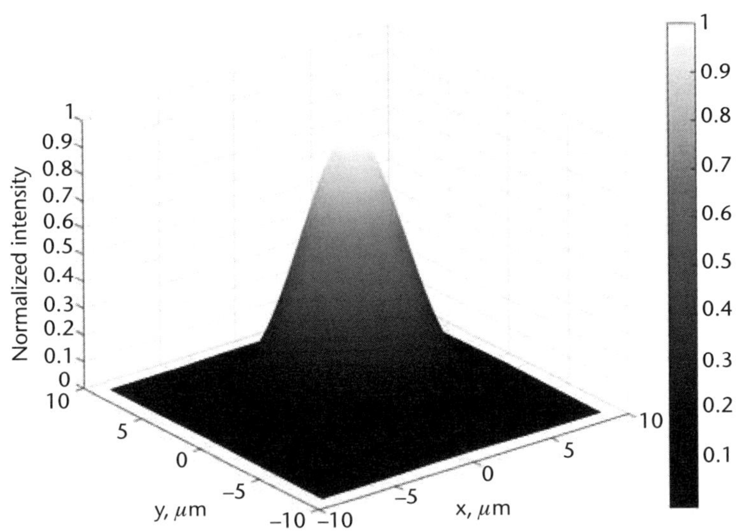

Figure 1.12 Picture of the fundamental mode of an SMF.

- GI MMFs with core diameter of 50 or 62.5 μm and cladding diameter of 125 μm, compliant with the parameters described in the Recommendation ITU-T G.651.

Commercial fibers have additional protective coatings besides for the cladding. The minimum size is for bare fibers, which have a plastic coating (the primary coating) covering the cladding, for an overall diameter of 250 μm. Bare fibers are used for component pigtails (pigtails are fibers entering and exiting a fiber-coupled device, used to route signals to and from the component) intended to be placed inside protective enclosures. The most common types of fiber coating found in sensors are the 900-μm fibers, which have a secondary protective buffer of 900-μm diameter, and plastic jacketed fibers, which further surround the 900-μm coating with high-strength aramid yarn (e.g., Kevlar) encapsulated in an about-3-mm protective plastic tube. The latter is the preferred cabling when fibers are supposed to be exposed to repeated handling. Additional protective jacketing, including metallic tubes, is used for exposure to harsh environment of for very high-power applications, but usually are not so relevant in biomedical sensing.

Plastic jacketed fibers are normally colored yellow for single-mode fiber and orange for multimode fiber.

Figure 1.13 shows examples of some types of fibers for sensing applications.

Sooner or later, any fiber sensor will have to be connected to an interrogator or, more in general, to another fiber-optic system. This can be done through connectors or splices. A connector is a device that can be mated and unmated repeatedly with similar devices to transfer the light signal, while a splice is a permanent junction between two fibers.

Optical connectors essentially include a ferrule (often ceramic), which holds the bare fiber, surrounded by metallic or plastic parts that ensure a reproducible mechanical coupling with low insertion loss and back-reflections [Figure 1.14(a)]. Optical connectors lack the male-female polarity common in electrical connectors and require specific mating adapters that fit between two connectors. The connector of Figure 1.14(a) is called FC/PC and is particularly used in measurement equipment because of its very good characteristics; it can be found both on SMFs and in MMFs.

For applications that require SMFs and very low back-reflection, the ferrule end-face can be polished at about 8° to prevent light that reflects from the interface from being guided back in the fiber, as this angle is larger than that corresponding to the fiber NA. Such connectors, which are identified by a green strain relief boot,

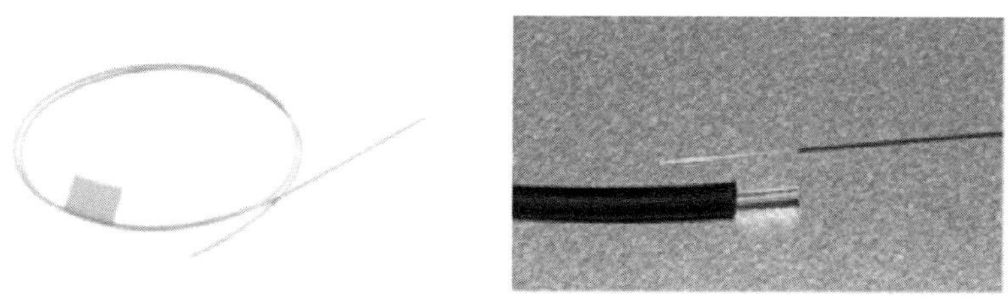

Figure 1.13 Picture of some optical fibers: 900-μm fiber (left) and comparison between a POF and SMF for telecom (right; POF is the larger one).

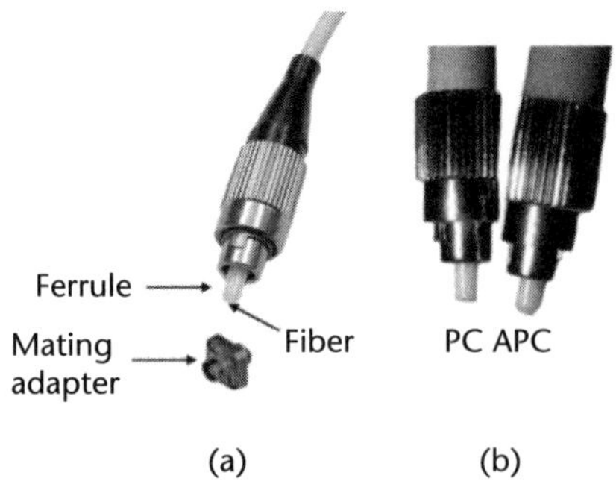

Figure 1.14 Picture of: (a) a typical high-quality optical connector with its mating adapter; and (b) detail of PC and APC connectors.

are called APC; for instance, Figure 1.14(b) compares an FC/PC and an FC/APC connector. Angle-polished connectors should be mated only to other angle-polished connectors to avoid damaging the ferrule and very high insertion loss.

Different types of standard connectors have been proposed over the years, depending on which specific characteristics (including cost) one wants to privilege. Figure 1.15 shows a picture of the most likely to be encountered in sensors; they are all typical of glass fiber, with the exception of the SMA, which is also used for POFs.

Splices are made through fusion splicers, a specialized equipment in which an electrical discharge raises the temperature while fibers are pushed against each other; number and intensity of pulses must be adjusted depending on the fibers. Before splicing, fibers must be properly cut using specific cleavers to ensure a flat and neat end-face. Cleavers and splicers are quite expensive instruments; moreover, they require some training and a relatively clean environment for the best results.

However, they are typically used in the development of the sensor only and not for its normal use, unless you buy a sensor not already equipped with connectorized fiber pigtails.

What we have seen so far applies basically to glass fibers only, as plastic optical fibers, due to their larger diameters and higher NA guide several millions of modes and this dramatically lowers the mechanical requirements of junctions, and thus of connectors.

Indeed, POFs can be cut using conventional scissors and then the end-face can be just polished using extra-fine sandpaper. Besides for using SMA connector, junctions can be made also by clamping the fiber to a receptacle. Moreover, as POFs

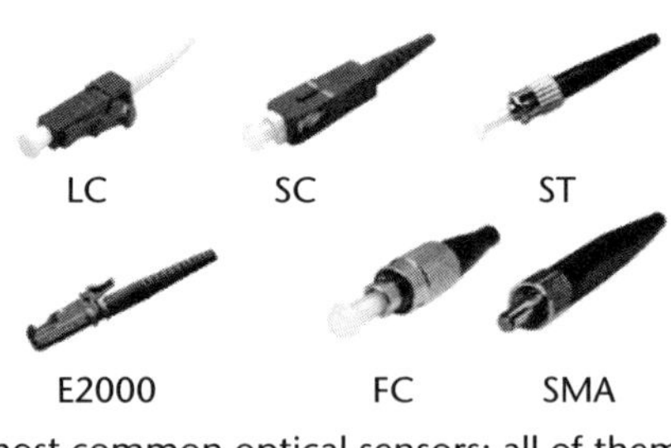

Figure 1.15 Picture of the most common optical sensors; all of them, except the SMA, exist in the PC and APC version.

are mainly used with visible wavelengths only (typically read at about 650 nm), seeing the light helps to have a quick evaluation of the alignment and behavior of the connection.

Selected Bibliography

Crisp, J., and B. Elliott, *Introduction to Fiber Optics*, 3rd ed., Amsterdam: Newnes, 2005.

Ghatak, A., and K. Thyangarajan, *Introduction to Fiber Optics*, Cambridge, U.K.: Cambridge University Press, 1998.

Hecht, J., *Understanding Fiber Optics*, 5th ed., Englewood Cliffs: Laser Light Press, 2015.

Kuzyk, M. G., *Polymer Fiber Optics*, Boca Raton, FL: CRC Press, 2007.

Marcuse, D., *Theory of Dielectric Optical Waveguides*, 2nd ed., New York: Academic Press, 1991.

Rogers, A., *Essential of Photonics*, Boca Raton, FL: CRC Press, 2009.

Shotwell, R. A., *An Introduction to Fiber Optics*, Upper Saddle River, NJ: Prentice Hall, 1997

Snyder, W., and J. D. Love, *Optical Waveguide Theory*, London, U.K.: Chapman and Hall, 1983.

CHAPTER 2

Devices for Fiber-Optic Sensing Applications

2.1 Introduction

Fiber-sensing systems are fiber-optic links in which the propagation in part of the fiber is somehow affected by the quantity under measurement.

A basic fiber-optic link includes a light source and a receiver, connected through an optical fiber using connectors or fusion splices, as sketched in Figure 2.1. The practical implementation of actual fiber-optic sensing systems, however, requires additional components: indeed, besides for the connectors, which have been discussed in the previous chapter, it is highly probable that the layout will include other devices, such as an isolator to protect the source from back-reflected light, a coupler or a circulator to properly route the optical signal from the source to the sensor and from the sensor to the receiver, filters to allow passing or reflecting only specific wavelengths, and polarizers to select only specific light polarization states. The main characteristics of all these devices will be analyzed in this chapter; if you are interested in further information, you can easily find many textbooks about optoelectronics and photonics devices, some of which are listed at the end of the chapter, and also white papers and tutorials on the internet.

2.2 Light Sources

Light sources for biomedical sensing systems are basically made out of a semiconductor device, such as a light-emitting diode (LED), often of the superluminescent type (SLED), or a laser diode (LD), because of their miniature size and low cost; therefore, only this kind of sources will be considered in this book. In the following, we suppose that the reader has a very basic knowledge of the physics of semiconductors; accurate descriptions of semiconductors for optoelectronics and of the working principles of LEDs and LDs can be found in [1–6].

Both LEDs and LDs are based on forward-biased PN junctions, as diodes used in electronic circuits, and emit an output optical power that is directly propor-

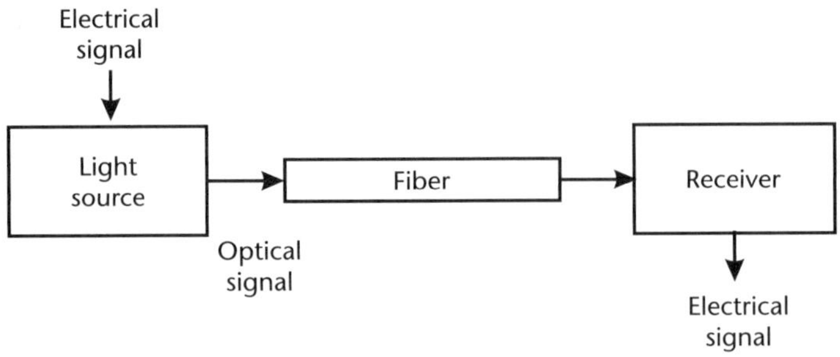

Figure 2.1 Schematic representation of a basic fiber optic link.

tional to the supplied electric current; their difference lays in the light emission mechanism, which then reflects on the properties of the emitted light.

The driving current can be constant in time or can be constituted by a constant value (the bias) with a superimposed varying signal (the modulation); in the first case, the output optical power will be constant (albeit neglecting noise induced fluctuations), whereas in the latter case it will follow the input current oscillations. A modulated optical signal can be used, for example, to implement the lock-in amplifier scheme, in which the receiver is able to extract a useful signal in very noisy environment by detecting only the received signal that is synchronous with the transmitted signal [7].

For both LDs and LEDs, light is emitted due to the recombination of electrons and electron holes, releasing energy in the form of photons (Figure 2.2).

The emission wavelength depends on the semiconductors used to form the P-N junction (and in particular on its bandgap energy): as most of the photonic-sensing approaches take advantage of the devices originally developed for the optical communication applications, the most important semiconductors for the realization of sources are those based on gallium arsenide (GaAs) alloys (such as AlGaAs) for emission in the red and infrared (IR) up to about 1,000 nm, and on indium phosphide (InP) alloys (such as InGaAsP) for emission in the 1,300 to 1,500-nm range. Other relevant semiconductors include the gallium nitride (GaN) alloys for emission in the ultraviolet (UV) to blue.

2.2.1 Light-Emitting Diodes

The light output of LEDs is due to the spontaneous emission, which is a randomly occurring recombination of electron-hole pairs, therefore generating photons with independent frequencies, phases, and direction of propagation. This results in incoherent light with a quite large emission spectrum, usually in the order of several

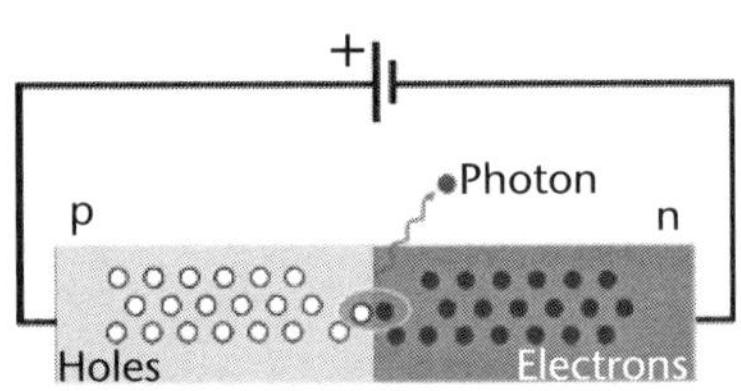

Figure 2.2 Schematic representation of the electron-hole recombination at a P-N junction with emission of a photon (light).

tens of nanometers. Emitted power is linearly dependent on the driving current, at least for currents lower than a certain saturation value, with the proportionality factor that changes with emitted light color (actually, the specific semiconductor alloy used to have emission at that wavelength) and structure design. In general, we can write that the emitted power is given by the number of recombinations per unit time, multiplied by the number of photons per recombination, multiplied by the energy of each photon:

$$P = \left[\frac{\textit{recombinations}}{\textit{unit time}}\right]\left[\frac{\textit{photons}}{\textit{recombinations}}\right]\left[\frac{\textit{energy}}{\textit{photon}}\right]$$

where the number of recombinations per unit time is given by the number of electrons entering the junction (I/e, with e as the charge of the electron), the number of photons per recombinations is given by a parameter characteristic of the called internal efficiency η_i that depends on the semiconductor material, and the energy of photons is related to their frequency n through the Planck's constant. In the formula:

$$P = \left[\frac{I}{e}\right][\eta_i][h\nu] = \text{const} \bullet I \tag{2.1}$$

The power given by (2.1) is that generated at the junction; the useful power due to photon that escape from the LED is much lower for different reasons, so an additional proportionality term, the external efficiency h_e (lower than 1), must be taken into account. In the end, usual LEDs emit only hundreds of microwatts for currents in the order of about 20 mA, although higher-power LEDs can emit up to 10 mW at larger currents, say, 50 mA or above (here we are not considering LEDs specifically designed for illumination).

In principle, it is very easy to drive an LED: as an example, Figure 2.3 shows how it is possible to drive a LED from a digital output (e.g., from a microprocessor) using an NPN (n-doped, p-doped, n-doped) transistor. For more demanding applications, such as in long-term measurements in which drifts and fluctuations are more critical, more sophisticated current sources are advisable. In the market, there

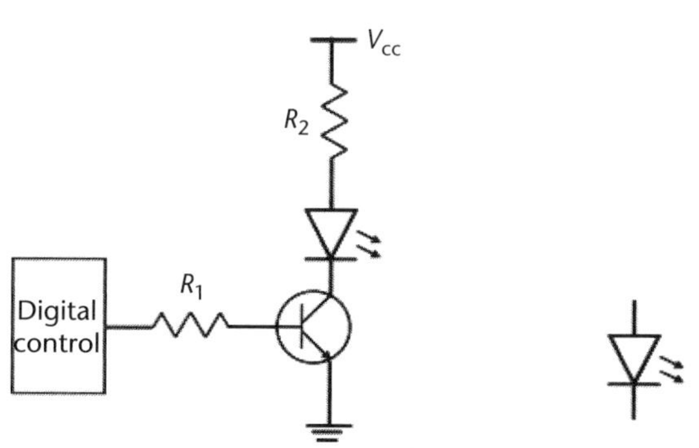

Figure 2.3 Basic scheme to drive an LED (left) from a digital output using an NPN transistor and symbol of LEDs (right).

are a lot of possible ready-made solutions, from bench-top equipment to electronic chips specifically designed to be integrated in custom boards.

LEDs are characterized by broad spatial and spectral emissions (from 30 nm to 100 nm), so that they are practically limited to be used with largely multimode fibers, such as plastic optical fibers (otherwise, the collected power by an SMF would be too low). Actually, LEDs and POFs, where applicable, are the cheapest combination for fiber-optic links and fiber-sensing applications.

Figure 2.4 summarizes the main spectral and power versus current characteristics of LEDs and compares them with those of lasers.

Fiber coupled broadband source for sensing, specifically with single-mode fiber pigtail, is typical based on SLEDs. It is a special type of LEDs in which the spontaneous emission is amplified with a mechanism similar to that occurring in lasers, but without the feedback that forms the cavity; we could think of an SLED as a sort of an LED followed by an optical amplifier. The emitted power versus current of SLEDs is similar to that of LDs, but with a softer transition between a regime dominated by spontaneous emission (typical of standard LEDs) and one that is dominated by the amplification of the spontaneous emission (the superluminescence). This results in higher emitted power (but at higher operating currents than usual LEDs, up to several hundreds of milliamperes) over a slightly narrower spectral region, but, most importantly, over a much narrower spatial region so that coupling with single-mode fiber is possible.

2.2.2 Laser Diodes

A laser is the optical equivalent of electrical oscillators. In electronics, an amplifier, such as a transistor or an operational amplifier, is connected with its output partially fed back into its input through a frequency selective network (i.e., a filter) to provide positive feedback [Figure 2.5(a)]. When power is supplied to the amplifier by the bias, electronic noise in the circuit provides a nonzero signal that is successively amplified to get oscillations started. Stable oscillation occurs when the magnitude of the loop gain ($A\beta$, which is the product of the amplifier gain A and the losses introduced by the feedback network β) is unity and phase shift around the loop is a multiple of 2π. These are known as the Barkhausen conditions. There is a similar situation for lasers: the equivalent of the oscillator layout is an optical cavity formed by an active medium (i.e., the amplifier in which an optical signal can be

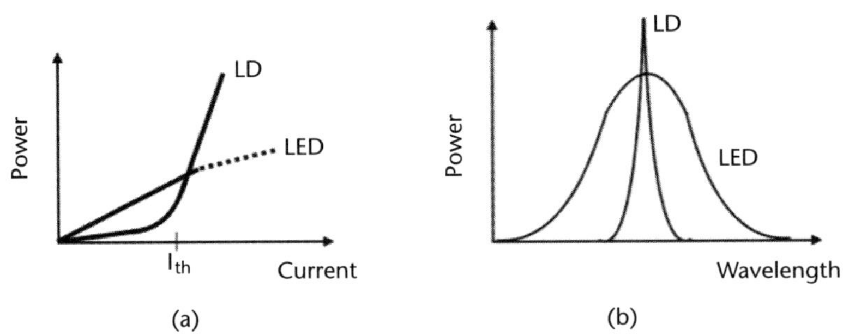

Figure 2.4 Qualitative comparison between LEDs and LDs in terms of power versus current characteristics (left) and spectral emission (right); I_{th} identifies the threshold current of laser diodes.

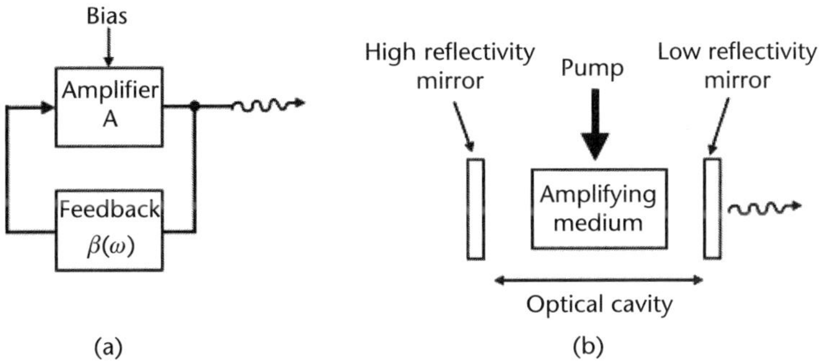

Figure 2.5 Sketch of: (a) the working principle of an oscillator and (b) the generic configuration of a laser.

amplified thanks to the energy delivered by the mechanism called the pump acting as the bias for electronic amplifiers) inserted in between two mirrors that provide a positive feedback.

One mirror (the HR mirror from the initials of high reflectivity) has a high reflectivity, as we want emission from a single side of the cavity only, whereas the other mirror (the OC mirror from the initials of output coupler) has a low reflectivity to allow extracting enough power from the cavity. Oscillations occur at wavelengths for which the gain provided by the active medium compensates propagation and mirror losses (including the power exiting the cavity) and the round-trip phase condition is met.

In a laser diode the amplifier is a forward-biased heavily-doped P-N junction, similar to that used for LEDs but working in a different regime as detailed later, properly injected by the bias current that acts as the pump. This semiconductor amplifier is also used as a stand-alone device to amplify optical signals in communications, the semiconductor optical amplifier (SOA). Mirrors can be realized with different techniques: in the most simplified structure, the optical feedback is given by the reflectivity of the surface obtained by cleaving the semiconductor material along its crystal planes [Figure 2.6(a)], possibly increased or decreased to form the HR and OC by depositing alternate layers of suitable materials until the desired reflectivity is obtained.

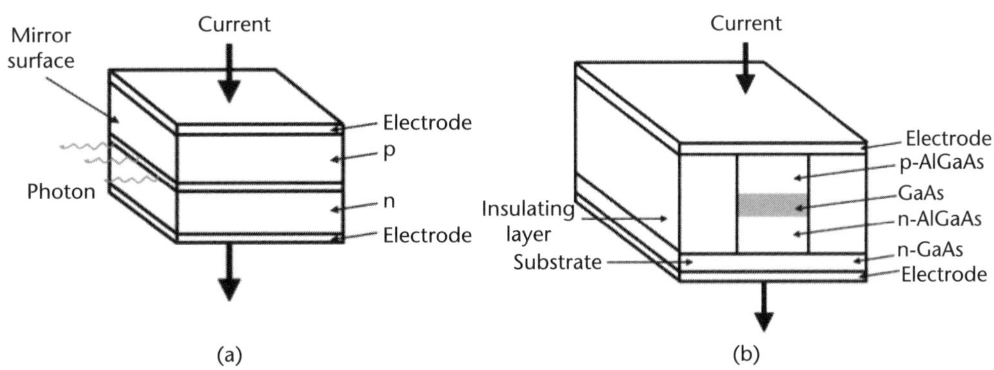

Figure 2.6 Sketch (not in scale) of (a) the basic structure of a semiconductor laser diode and (b) a more realistic structure of a semiconductor laser diode.

Actual LD devices have more complex structure that require the combination of semiconductor alloys of different composition to add potential barriers and form optical waveguides to confine both the injected electrons and the emitted photons, so to have a stronger interaction and increase the efficiency. An example is sketched in Figure 2.6(b).

In LDs light is essentially generated by stimulated emission, which the key mechanism underlying amplification of light: an electron in the conduction band drops to a lower energy level following the interaction with an electromagnetic wave of suitable wavelength (hence, the stimulated name) transferring its energy to that field. Another way to see this phenomenon is to say that the electron-hole recombination with subsequent release of a photon does not occur randomly, but it is triggered by an incident photon (of suitable energy equal to the bandgap energy). The created photon has the same phase, frequency, polarization, and direction of travel as the photons of the incident wave. These peculiar characteristics of the emitted photons are responsible for the unique property of laser light of being coherent, which expresses the possibility of having a stationary interference. The injected current is the pump mechanism of the laser system and represents the source of energy at expenses of which the light amplification occurs.

Laser action starts when the gain provided by stimulated emission compensates all the losses, which include absorption, scattering, and transmission through the output mirror. The gain equal loss condition defines a threshold current (I_{th}) above which the stimulated emission is capable of sustaining lasing. In these conditions, the stimulated emission largely prevails the spontaneous emission and the slope of the curve emitted-power versus injected current sharply changes as outlined in Figure 2.4(a).

It can be shown that the emitted optical power above the threshold is a linear function of the current as for LEDs (at least up to safe working conditions) and that is given by:

$$P_{\text{out}} = \beta_{\text{s}} \left(I - I_{\text{th}} \right) \tag{2.2}$$

where β_s is a parameter known as the *slope efficiency*. The slope efficiency depends on the laser type: for the most common lasers using in sensing, it is about 0.2 to 0.5 mW/mA. Semiconductor lasers with output power ranging from few milliwatts to about 10W to 15W and different forms of packages can be found as off-the-shelf products; for fiber-sensing applications, the most common are those fiber-coupled, with emitted power in the order of 1 to 5 mW for current of 20 to 50 mA and with threshold around 5 to 10 mA. Current is supplied to laser through specific laser drivers, which are precise constant current generators, often controlled in closed loop by monitoring the power delivered from the laser through a photodiode, equipped with circuitry for a soft start and to avoid overshoots.

The slope efficiency should not be confused with the wall-plug efficiency, which is the ratio between the emitted optical power and the supplied electrical power. The wall-plug efficiency takes into account the voltage drop across the LD, which ranges from 1.5V to 2V, much larger than that of a rectifying diode. Roughly speaking, we could therefore assume that the wall-plug efficiency of LDs is about 50% of its slope efficiency. This difference between absorbed and emitted power is converted into heat that raises the temperature of the laser chip. A too-high temperature leads the laser chip to a rapid death, but even smaller changes of temperature can be detrimental as they cause an increase of the threshold current and a variation of the emitted wavelength. Typically, wavelength shifts by 0.3nm/°C, quite a large value (for comparison, consider that fiber Bragg grating sensors, discussed in Chapter 5, shift their spectral response by about 10 pm/°C). Therefore, especially for applications where the stability of the emitted power (e.g., in amplitude-base sensors) or of the wavelength (e.g., in interferometric or in wavelength-based sensors) significantly contributes to the accuracy of the measure, the laser chip must be accurately cooled. In the low-power versions usually employed in sensing this is usually accomplished through Peltier cells, which may already be included in the laser diode package, especially for those of the highest quality like the butterfly package used for optical communications (Figure 2.7). The butterfly package has quite a complex optical layout because, besides for the laser diode chip and the Peltier cells, it includes a thermistor to measure the chip temperature and thus to control the Peltier cells in a closed loop, an isolator, and miniaturized collimating and focusing lenses to couple the emitted beam to the fiber pigtail. Often, a monitoring photodiode is also integrated into the laser diode package.

From the cavity viewpoint, the simplest type depicted in Figure 2.6 and characterized by cleaving of the semiconductor facets so as to introduce a uniform reflectivity over a broad wavelength range is known as the Fabry-Perot (FP) laser diode. As already mentioned additional optical coatings can also be deposited on the cleaved facets to optimize the laser performance by creating low-reflectivity front and high-reflectivity back mirrors; nevertheless, though the spectral width of emission is much narrower than that of LEDs [Figure 2.4(b)], FP lasers have quite a broad spectrum, in the order of 2 nm, usually characterized by many closely spaced peaks. The spectral linewidth can be narrowed using a distributed feedback (DFB) laser structure. In this case, a wavelength selective device (a Bragg grating extending over the longitudinal direction of the laser chip) is integrated into the semiconductor structure to provide the required amount of feedback to have laser action only at a specific wavelength. DFB lasers are those used also in high performance optical communication systems and have very narrow linewidth, on the order of 1 nm or less. Packages like that depicted in Figure 2.7 are quite expensive and thus reserved only for high-performance laser diodes, such the DFB type. Usually, FP laser diodes make use of cheaper TO-can packages in which the components for controlling the temperature are not included. Figure 2.8 shows two examples of the most common fiber-pigtailed LD packages: a TO-can on the left and a butterfly package on the right.

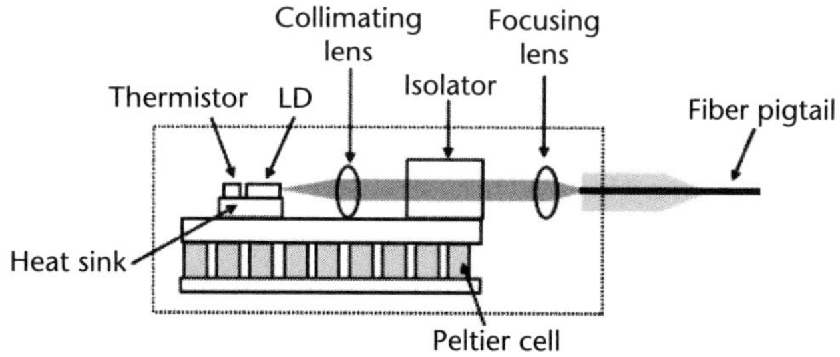

Figure 2.7 Sketch (not in scale) of the basic structure of a butterfly laser package.

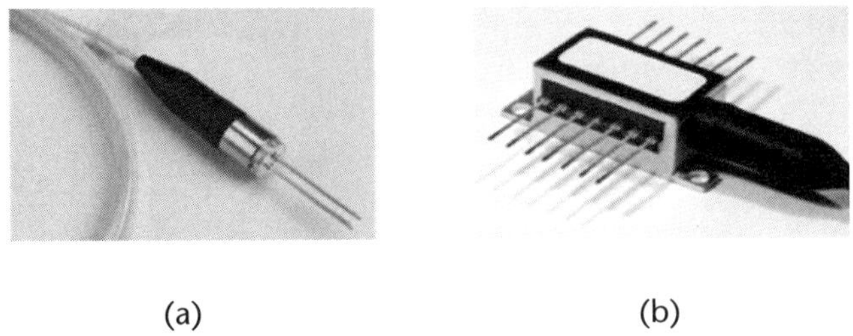

Figure 2.8 Picture of typical packages for fiber pigtailed laser diodes: (a) TO-can mainly for FP and (b) butterfly mainly for DFB lasers.

The beam emitted by an FP or DFB LD is astigmatic (Figure 2.9), with an elliptical cross-section: the vertical direction, which has the minimum size on the diode facet but spreads very quickly and originates the larger axis far from the diode, is called the fast axis; the perpendicular direction has the opposite characteristics and it is called the slow axis. Depending on the architecture of the package, given these characteristics of the emitted laser beam, sometimes only the fast axis needs to be collimated before focusing on a collecting fiber; other times, an aspherical lens or a couple of different cylindrical lenses, known as the fast axis collimator (FAC) and the slow axis collimator (SAC), respectively, are used to collimate the beam in both directions.

Both the FP and the DFB lasers considered so far have a horizontal optical cavity and side emission; therefore, they are said to be of the edge-emitting family. This is the layout first proposed at the beginning of the laser diode era, in the 1960s. Alternatively, in the late 1980s lasers were proposed with a vertical optical cavity and beam emission perpendicular from the top surface, the vertical-cavity surface-emitting laser (VCSEL) [8, 9]. Although the structure of VCSELs is somehow more complicated, there are production advantages that make the yield more controllable and therefore the chip overall less expensive. VCSELs are made by growing alternated layers of different materials, first to realize the back (lower) mirror, then the active region, and finally the front (upper) mirror. The most important advantage from the production viewpoint is that they can be tested and characterized directly after growth, at the wafer stage, without requiring labor-intensive (thus expensive) dicing and mounting processes. VCSELs have also many optical advantages over FP-type LD, such as higher wavelength stability and uniformity, lower sensitivity to temperature, and emission of an almost perfectly circular beam (at least for small mode area and low power). Although some fiber-coupled VCSELs emitting in the third optical communication window can be found on the market, most common devices emit in the 700 to 800-nm range and are widely used in con-

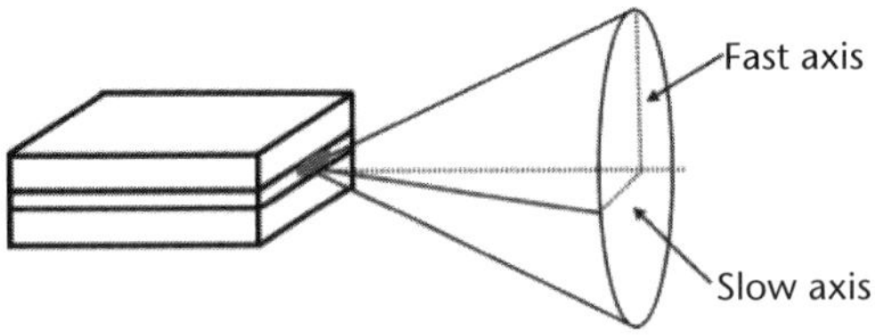

Figure 2.9 Sketch of the astigmatic far-field output beam characteristic of a FP or DFB laser diode.

sumer applications (e.g., computer mice) and for short-range data communications and local area networks.

Other interesting, albeit very expensive, semiconductor laser sources are the tunable laser diodes, which can be implemented both using the edge-emitting and the vertical cavity approaches [10]. Tunable LD means that their emission wavelength can be changed by applying an external control signal. In sensing, they can find application for the measurement of gas concentrations by measuring the relative absorption at two wavelengths or in one of the possible implementations of interrogation systems for wavelength encoded sensor, such as the fiber Bragg grating (FBG) sensors.

The simplest approach for tuning the emission wavelength is to exploit its naturally dependency on the temperature (thermal tuning), but this is slow, difficult to precisely control (even using thermistors and Peltier cells) and in practice limited to few nanometers as the temperature can be changed only in a narrow interval to maintain the diode in safety operation conditions. For example, such approach would be practically impossible to employ for precise demodulation of FBG sensors.

In edge-emitting lasers, tunability can be obtained more effectively with distributed Bragg reflector (DBR) structures in which one of the mirrors is replaced by a passive grating section integrated in the semiconductor base material and whose reflection peak position is determined by the decrease or increase of the refractive index contrast through the control of injected currents. Alternative structures make use of external cavities in which a movable discrete external mirror or grating can change the resonance condition and thus the emitted wavelength.

VCSELs can be made tunable by adjusting the position of the upper mirror using thermal expansion or piezoelectric miniaturized movements.

In practice, as the industrial research on optoelectronic devices is mainly driven by telecommunication applications, commercial tunable devices exists only for emission around 1,500 nm (i.e., in the third optical communication windows).

2.3 Photodiodes

A photodiode (PD) is a semiconductor P-N junction used to convert light into a current, a behavior opposite to that of a LED [2, 4]. Actually, in practice, the structure is often of the P-I-N type (p-doped, intrinsic, n-doped), but for simplicity in the following, we will just consider the P-N case.

When a p-type semiconductor (i.e., doped to produce excess holes) comes into contact with an n-type semiconductor (i.e., doped to produce excess electrons), the holes and the electrons begin to flow across the junction as they both experience a lower potential on the opposite side of the junction. This establishes an electric field that opposes to the flow and a steady state condition is achieved with the creation of a depletion region (Figure 2.10).

An incident photon with energy greater than the semiconductor bandgap is absorbed, which means that the associated energy is used to promote an electron from the valence to the conduction band with the creation of an electron-hole pair. If the absorption occurs in the depletion region (or close by), the carriers are swept across by the electric field, so that holes move toward the anode and electrons

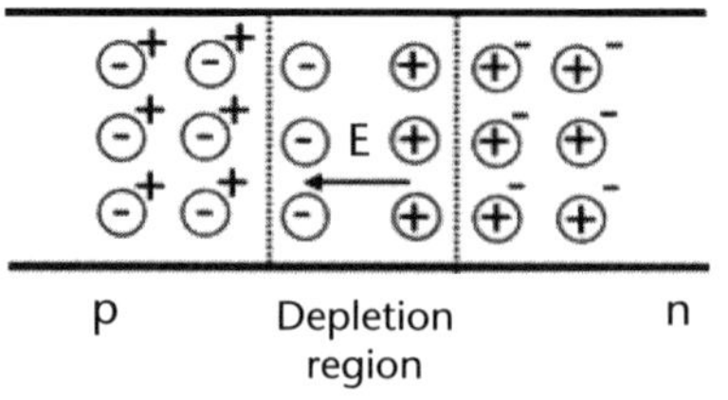

Figure 2.10 P-N junction and creation of the depletion region.

toward the cathode, therefore producing a current (photocurrent). This is called the photovoltaic mode of operation. However, as in these conditions a forward bias that opposes to the current flow is generated, PDs are usually operated in the photoconductive mode by applying a reverse bias (i.e., with the cathode positive with respect to the anode).

Some photodiodes, the avalanche photodiodes, are optimized to work under high reverse bias to allow each photo-generated carrier to be multiplied by avalanche breakdown resulting in higher sensitivity.

The semiconductor used in photodiode determines its properties, starting from the wavelength range of usability, as it will be able to absorb only photons with energy larger than its bandgap. Most common materials are silicon, which can be used from UV (~190 nm) to NIR up to 1,100 nm, germanium, from 400 to 1,700 nm, and a ternary alloy of indium gallium arsenide, from 800 to 2,600 nm. The latter is therefore the most popular for applications that use wavelengths in the third optical communications window (~1,550 nm).

The most important parameters that qualify the performance of photodiodes in biosensing applications are:

- Responsivity or photosensitivity, function of the wavelength, which is the ratio of the generated photocurrent to incident light power and is expressed in A/W;
- Spectral response characteristic, which is the relation between the photoelectric sensitivity and wavelength;
- Dark current, which is the current through the photodiode in the absence of light, constituting therefore a source of noise for low incident power applications.

Besides for these quantities, there are also the response time and noise-equivalent power (NEP), which is basically the minimum detectable power.

It should be also considered that all the parameters are dependent on the temperature. In particular, the dark current increases with temperature because at higher temperatures more electrons in the valence band can be promoted in the conduction band. Sensitivity increases with increasing temperature too particularly for longer wavelengths.

The generated photocurrent is a linear function of incident power through the responsivity; so from a practical point of view, PDs in receivers are usually connected to transimpedance amplifiers (TIAs), which are a current-to-voltage converter, to transform optical power changes into voltage changes. In its simplest form a TIA can be implemented with an operational amplifier with a large-valued feedback resistor, R_f, which defines the current to voltage conversion gain (Figure 2.11). Some commercial photodiodes incorporate a preamplifier in their package to help achieving a high current to voltage gain while keeping the noise low and optimizing the frequency response (which characterize the dynamic response and has not to be confused with the spectral response). Other photodiodes are actually phototransistors, which are bipolar transistors with a transparent package so that light can reach the base-collector junction and generate a photocurrent that is amplified by the transistor current gain.

LDs and PDs are used as the key elements of the optical source and the receiver, respectively, used to interrogate sensors. Figure 2.12 shows a schematic of the simplest layout, which can be used for sensors working in transmission as sensors working in reflection require additional components as detailed in the following.

2.4 Isolators and Circulators

Isolators and circulators are passive nonreciprocal devices; this means that they are devices having the property that the behavior for forward propagating light is not the same as that for backward propagating light.

An isolator is a two-port device that transmits optical signals in one direction only, similar to the one-way sign for road traffic. Ideally, with reference to Figure 2.13(a), a signal entering from port 1 is transmitted to port 2 without attenuation (in practice, low insertion loss, in the order of fractions of decibels), whereas a signal entering from port 2 is totally absorbed (in practice, it experiences high insertion loss, called the isolation, on the order of 20 to 40 dB). It is used to shield equipment on its input side, from the effects of conditions on its output side; for example, to prevent unwanted backreflections from being fed back to a laser causing perturbations in its oscillations. Figure 2.13(a) shows the symbol of an isolator. Isolators are already included in the package of high quality laser diodes, as we have seen for the butterfly package sketched in Figure 2.7.

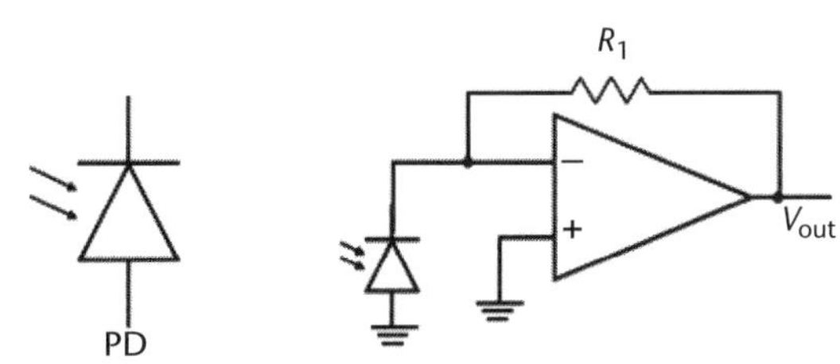

Figure 2.11 Basic optical receiver circuit (right), with indicated the symbol of photodiode (left).

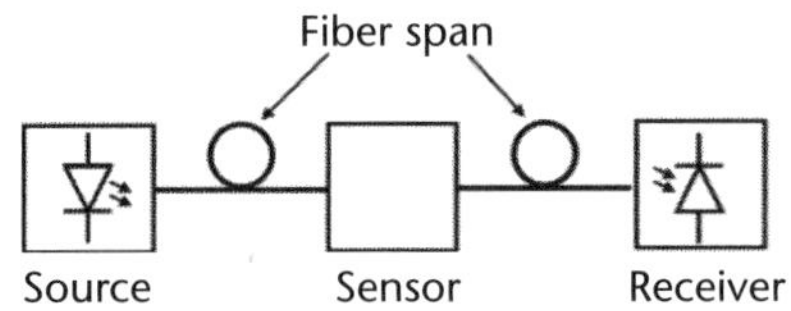

Figure 2.12 Combination of a source and of a receiver to interrogate a sensor working in transmission.

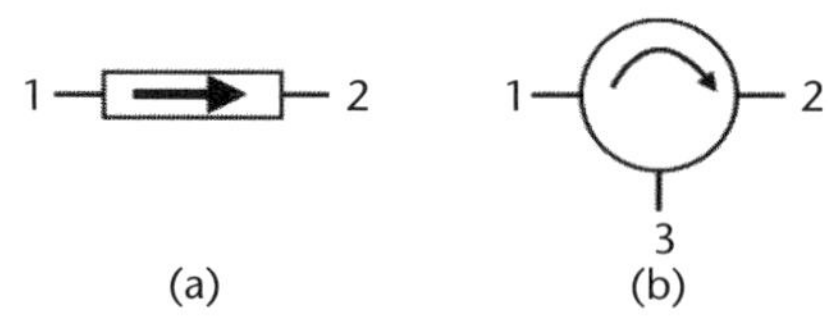

Figure 2.13 Symbol of an isolator (left) and of a three-port circulator (right).

A circulator is a three-port or four-port device used to separate light traveling in different directions, with a behavior similar to roundabouts for road traffic. More specifically, with reference to Figure 2.13(b) where a three-port device is represented, light entering from port 1 is routed with little attenuation (ideally, zero) to port 2 only while nothing exits from port 3; similarly, light entering from port 2 is routed to port 3 only and nothing exits from port 1.

As devices mainly developed for telecommunication applications (as almost all the devices using in fiber-optic sensing), the most common commercial circulators are designed to work in the 1,200-nm to 1,600-nm wavelength range.

Their typical use in sensing is to separate incident and backreflected signals from a sensor supposed to work in reflection. For example, fiber Bragg grating (FBG) sensors, one of the most widespread fiber sensors, as discussed in later chapters, are often interrogated in the reflection mode, so for them we may use the circuit layout sketched in Figure 2.14.

More detail on the working principle of isolators and circulators and on their applications can be found in [11] or in white papers and tutorials that can be easily found on the internet. These devices are traditionally made with discrete micro-optic components; a review of current efforts for developing integrated version can be found in [12].

2.5 Couplers

A fiber-optic coupler is a multiport device in which light signals entering from one or more input ports are routed to one or more output ports so that the total input power is distributed over the outputs, potentially depending on the wavelength and the polarization.

Couplers can be fabricated using fiber optic technology by fusing together a certain number of fibers or using planar lightwave circuits by exploiting coupling

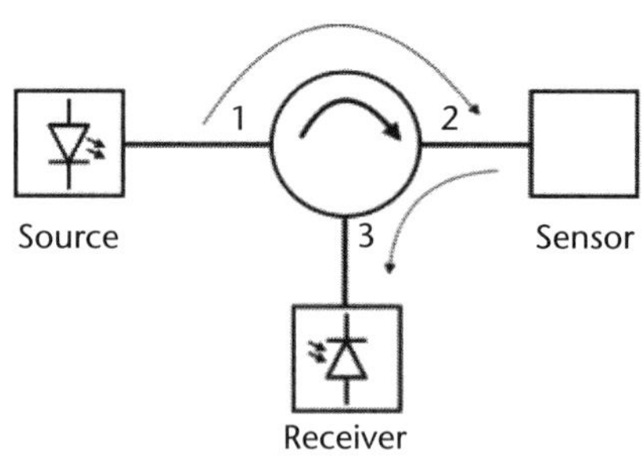

Figure 2.14 Schematic representation of the use of a circulator to interrogate a sensor working in reflection.

between different mode orders or between modes guided by different waveguides in proximity.

Form the functional viewpoint, couplers can be subdivided in wavelength insensitive or wavelength depending, as in the following.

2.5.1 Wavelength-Insensitive Couplers

Wavelength-insensitive couplers (WINCs) are devices that can be used as splitters to distribute the input signal to a plurality of outputs or as directional couplers (DCs) to separate forward and backward propagating signals, independently from their wavelength (at least over the operating range). Most common examples of splitters are 50:50 1×2 (read 1 to 2) single-mode couplers (also known as 3-dB splitters), which are used to divided an input signal equally between the two output ports, and 90:10 or 99:1 1×2 single-mode couplers (also known as 10-dB or 20-dB splitters, respectively), which are used to divide unevenly the input between the outputs, for example for sampling the signal flowing along a certain path. Another example is a 1×4 splitter, which can be used to divide evenly the input power among four outputs. Examples are shown in Figure 2.15. A 3-dB 1×2 splitter is also called y-junction (as it can be practically implemented dividing a single input waveguide in two output waveguides with a geometry that resembles that of the letter Y), while a 1×4 divider can be implemented with the cascade of two y-junctions (Figure 2.16). Splitters, particularly 3-dB 1×2 devices, can be used also as combiners to merge two signals (amplitude and phase combination).

The most important parameters that qualify a splitter are the insertion loss for the each input-output path (this is the ratio expressed in decibels between the power at the input port and the power at a specific output port and it includes both an intrinsic contribution due to the splitting of the input power and possible additional contributions due to extrinsic loss mechanisms) and the uniformity of splitting among the various output ports. The 3-dB 1×2 splitters can be used also to interrogate sensors working in reflection as sketched in Figure 2.17.

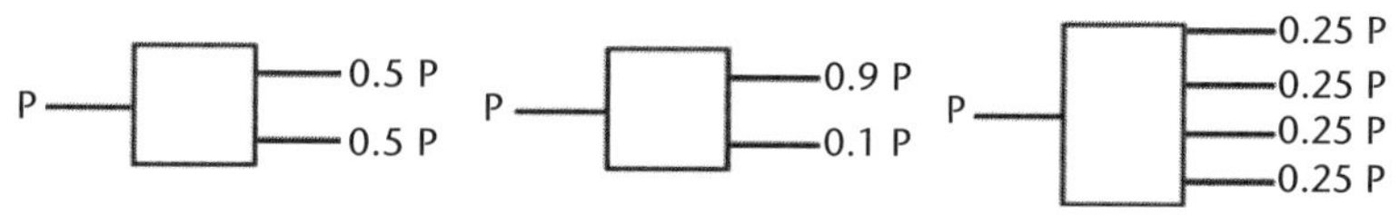

Figure 2.15 Schematic representation of power splitters: (left to right) 3-dB 1×2, 10-dB 1×2 and 1×4.

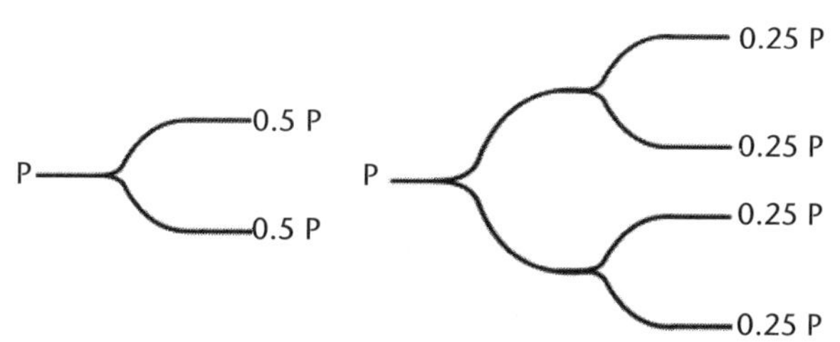

Figure 2.16 Implementation of a 3-dB 1×2 splitter as a y-junction (left) and of a 1×4 splitter as the cascade of y-junctions (right).

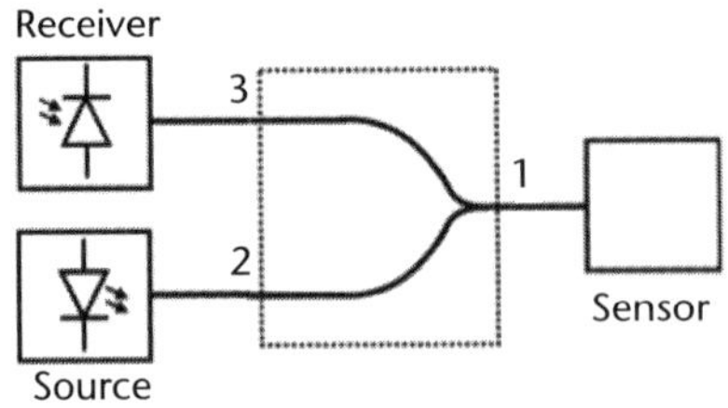

Figure 2.17 Schematic representation of the use of a 3-dB 1 × 2 splitter to interrogate a sensors working in reflection.

As for the DCs, it is a four-port device, as sketched in Figure 2.18(left). Ideal devices have the ports on the same side uncoupled: for instance, a signal entering at port 1 cannot exit at port 4 and a signal entering from port 2 cannot exit from port 3. The representation in Figure 2.18(right) shows that the transfer of power between ports that are at the opposite ends of the diagonals is characterized by coefficient C (called the coupling of the DC), whereas that between ports that are on the same row by (1-C); as already mentioned ports along the same column are uncoupled, so the transfer of power is zero (the corresponding parameter is called isolation). The most common DCs used in sensing have a coupling of 3 dB, so the input power is split evenly among the output ports; isolation is usually at least 20 to 30 dB, meaning the signal exiting from the wrong port carried between 100 to 1,000 times less power than the incident one.

Figure 2.19 shows an example of how a DC can be used to interrogate a sensor working in reflection as an alternative to the circulator. The ratio between the power at the receiver and the input power is C(1-C), which is clearly maximized when C = 1/2 (i.e., for a 3 dB coupler). Splitters/combiners and DCs are also the basic components of interferential sensor schemes: for example, a Mach-Zehnder interferometer is built using two 1 × 2 splitters (one actually to split and one to recombine the beams) or two couplers.

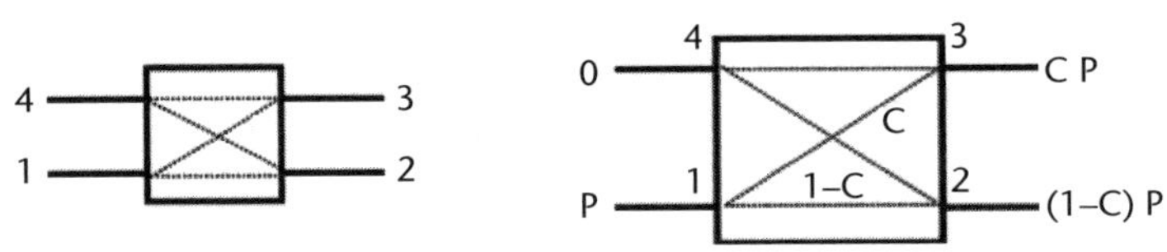

Figure 2.18 Four port device block indicating a directional coupler (left) and representation of the relation among the various ports when a power P enters in port 1 (right).

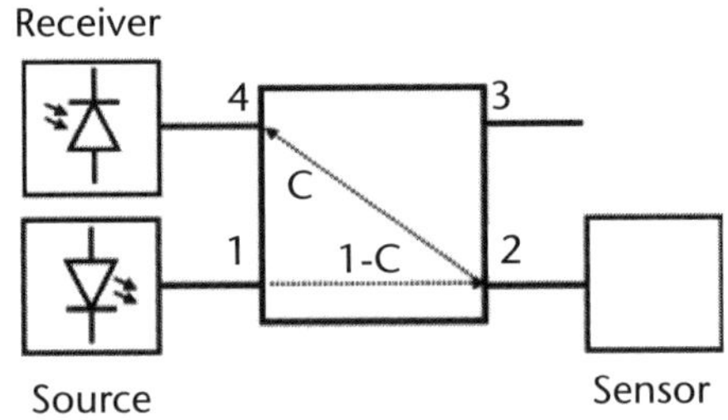

Figure 2.19 Schematic representation of the use of a directional coupler to interrogate a sensors working in reflection. Port 3 can be used to sample the incident signal, for example, to compensate for power fluctuations.

2.5.2 Wavelength-Sensitive Couplers

These are couplers designed to have a strong dependence of the coupling coefficient on the wavelength and are therefore used as wavelength-division multiplexers (WDM) to combine several input channels with different wavelengths or to route different wavelengths to separate outputs (Figure 2.20). In biosensing, they can be used, for example, to combine and separate the excitation signal and the produced fluorescence (Figure 2.21).

More details on the working principle of couplers and on their applications can be found in [13, 14] or in white papers and tutorials that can be easily found on the Internet.

2.6 Polarization Optics

In this category, we find all the components that change the polarization state, the most relevant being polarizers and waveplates.

A polarizer is a two-port device that lets light waves with a specific polarization only pass, while attenuating light waves with other polarizations up to ideally blocking waves that are orthogonally polarized.

It can be used either to generate light with a well-defined polarization state or as a polarization analyzer to extract only signals with a given polarization. Common are linear and circular polarizers, which, as their name states, allow passing linearly or circularly polarized light only.

Linear polarizers can be further subdivided into two general categories: absorptive polarizers, in which one of the two basic polarization states experiences higher propagation loss in passing through the device, and beam-splitting polarizers, in which the two basic linear polarization states composing the input beam are separated at the output (Figure 2.22). Most of us are familiar with absorptive polarizers (first developed by E. H. Land and commercialized under the brand

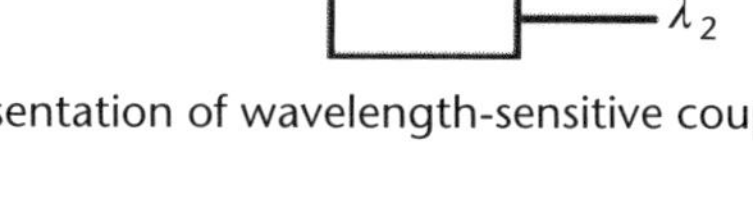

Figure 2.20 Schematic representation of wavelength-sensitive coupler used as a WDM device.

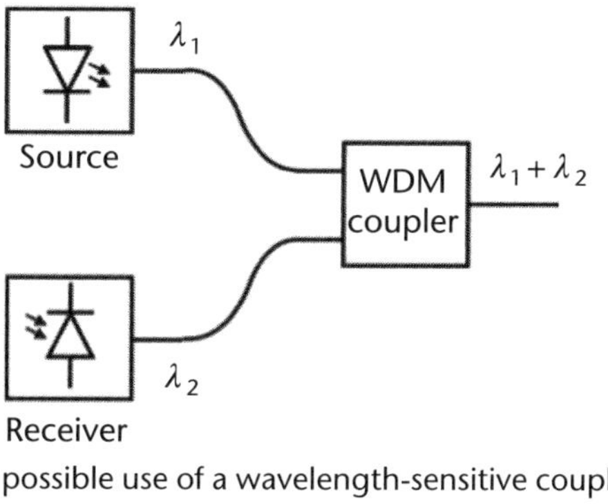

Figure 2.21 Schematic of a possible use of a wavelength-sensitive coupler for separating excitation and fluorescence signals.

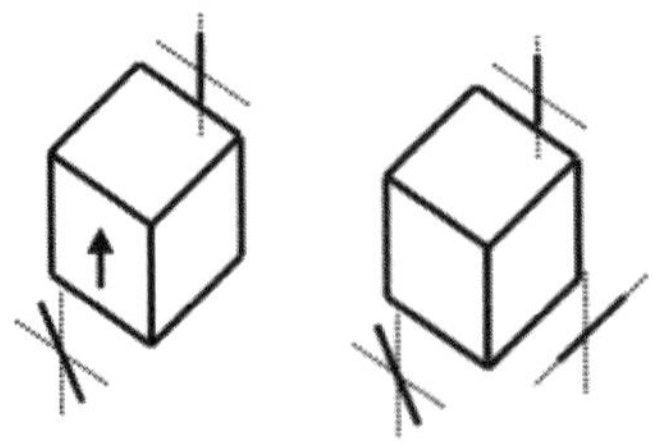

(a) (b)

Figure 2.22 Schematic representation of: (a) the behavior of an absorptive linear polarizer oriented to pass the vertical polarization only and (b) a polarization beam splitter that separates the input polarization in two orthogonal linear polarization states.

Polaroid) because this mechanism is used in sunglasses with polarizing lenses. Given the exploited working principle, this type of polarizers is limited to low-power applications, but this should not be a concern for sensing applications.

Besides the insertion loss, the most important parameter that qualifies a polarizer is the extinction ratio, which is the ratio between the powers associated to transmitted and rejected polarizations.

Waveplates (also called retarders) are devices used to change the polarization state of light wave traveling through them. They are fabricated using birefringent materials and, although in principle they could be built for any type of polarization conversion, the only practical devices that are commonly used are the half-wave plate and the quarter-wave plate. A half-wave plate is built in such a way to decompose an input linear polarization into the two elements of the basis (say, horizontal and vertical polarizations) and recombine them at the output while inserting a phase shift of π between them. In practice, this converts a linear input polarization into another linear output polarization with orientation that is 2θ if θ is the angle formed by the input polarization with a reference axis of the waveplate. The typical use is to insert this waveplate at 22.5° so that the output polarization is rotated by 45° (for example, to convert a horizontal polarization into a 45° polarization) or at 45° so that the output polarization is rotated by 90° (for example, to convert a horizontal polarization into a vertical polarization without having to rotate the source).

A quarter-wave plate is built in such a way to decompose an input linear polarization into the two elements of the basis and recombine them at the output while inserting a phase shift of $\pi/2$ between them. In general, this converts a linear input polarization into an elliptical output polarization and vice versa; however, if the input polarization is at ±45°, the output beam will be circularly polarized (and vice versa, a circular polarization is converted into a linear polarization at ±45°).

A behavior similar to that of the waveplates can also be obtained by stressing on purpose a piece of SMF and exploiting the induced birefringence. This can be done with automatic bench-top instruments or with manual polarization controllers made of two or three independently coiled optical fiber sections that can be rotated using paddles (Figure 2.23).

In practice, each coil behaves like a waveplate, with the amount of induced birefringence that is a function of the fiber cladding and spool diameters, the number of fiber loops per spool, and the wavelength of light. By rotating the paddles, we change the relative orientation between the input polarization and the equivalent

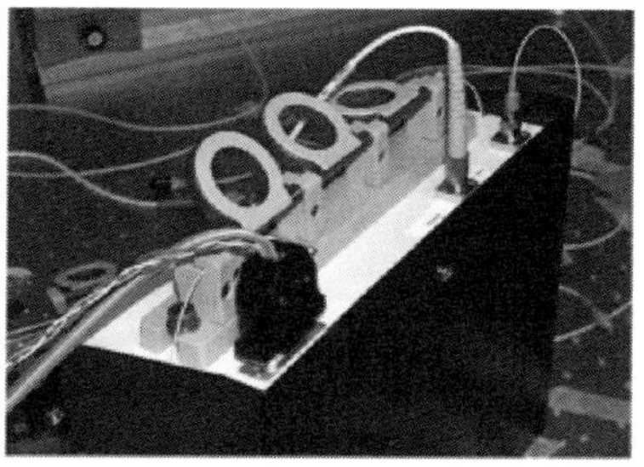

Figure 2.23 Picture of a prototype of a fiber-optic sensing system developed by the authors exploiting the stress induced birefringence in fibers. On the top of the box, it is evident the manual polarization controller made of three independently coiled optical fiber sections.

waveplate reference axis (much like we do when we rotate a discrete waveplate); it should be evidenced that the waveplate effect is due to the birefringence induced by the loop in the fiber, not by stresses due to the twisting of the paddles. Most common are two-paddle (designed to behave like two quarter-waveplates) or three-paddle (designed to behave like a quarter-waveplate, a half-waveplate, and a quarter-waveplate again) combinations, which form device that allow transforming an arbitrary input polarization state into any output polarization state by rotating the paddles.

More details on the working principle of polarization optics and on possible applications can be found in [15, 16] or in white papers and tutorials that can be easily found on the internet.

2.7 Optical Spectrum Analyzers and Spectrometers

Optical spectrum analyzers (OSAs) and spectrometers are instruments used to analyze the spectral content of an optical signal. In other words, they are used to represent plots of the power versus wavelength. Both terms indicates similar type of apparatus, although typically the name OSA is reserved for lab-type instruments, usually capable of superior performance in terms of wavelength range (the most common instruments are used for telecommunications applications and cover a range that spans from the red, 600 nm, to the past the third optical communication window, about 1,700 nm), dynamic range (over 70 dB) and resolution bandwidth (down to 1 pm). OSAs are built around single-pass or double-pass monochromators (basically tunable filters that allow passing light only in a very narrow band), which are, in turn, based on movable gratings or scanning Fabry-Pérot interferometers. The latter will be discussed in one of the following chapters because it is the basis of many interferometric sensors, whereas a grating is an optical component with a periodic structure, thanks to which it is capable of diffracting light into different directions depending on the wavelength. This behavior is similar to that in the well-known experiment of the prism that decomposes a beam of white light into the constituent colors. Figure 2.24 illustrates the working principle of a grating.

In OSAs, only part of the diffracted beam from a rotating grating hits a high-sensitivity photodiode through a narrow slit; thus, as the grating in the monochromator rotates, bands of different wavelengths illuminate the detector, which then returns a signal proportional to the power in that band (Figure 2.25). The

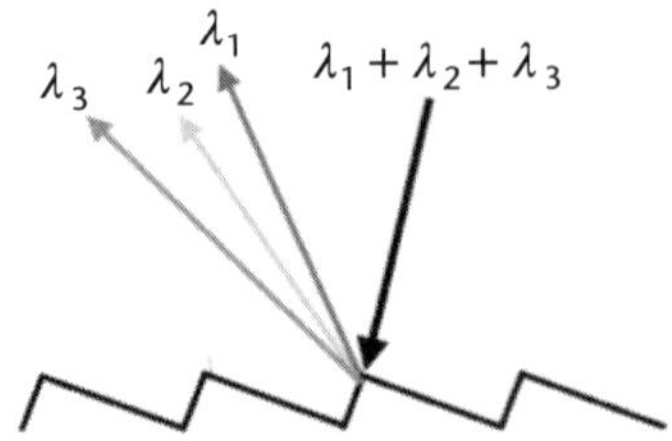

Figure 2.24 Schematic representation of the behavior of grating for an incident wave composed of several wavelengths.

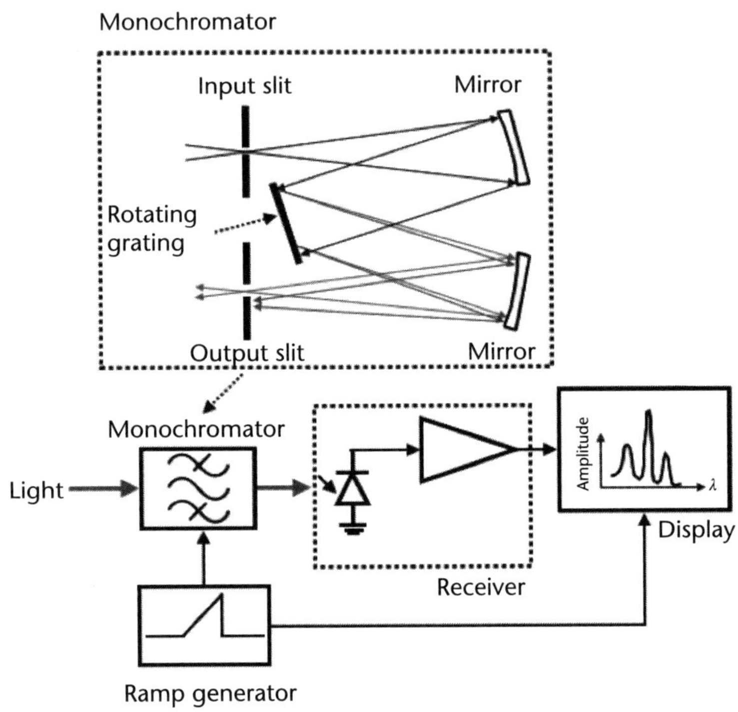

Figure 2.25 Block diagram of an OSA (below) and possible implementation of a monochromator using a grating (above).

resolution of the OSA are controlled by the grating dispersion and by the output slit size, which, together with the input slit, also affect the quantity of light arriving at the photodetector and thus should be designed also taking into account the minimum sensitivity and the acceptable noise level.

However, a spectrometer, at least in the world of fiber sensing, usually indicates a compact, portable instrument, with more limited measurement characteristics with respect to OSAs. Often they exploit a fixed diffractive element, such as a grating or a prism, and image the diffracted beam on a charge-coupled device (CCD) array so that each wavelength corresponds to a given position. For example, you may imagine using a CCD to intercept the diffracted beams from the grating in Figure 2.24.

More details on the working principle of optical spectrum analyzers can be found in [17] or in white papers and tutorials that can be easily found on the internet.

References

[1] Sands, D., *Diode Lasers*, Boca Raton, FL: CRC Press, 2004.

[2] Quimby, R. S., *Photonics and Lasers*, New York: Wiley Interscience, 2006.

[3] Held, G., *Introduction to Light Emitting Diode Technology and Applications*, Boca Raton, FL: CRC Press, 2008.

[4] Agrawal, G. P., *Fiber-Optic Communication Systems*, 4th ed., New York: Wiley, 2011.

[5] Sun, H., *Laser Diode Beam Basics, Manipulations and Characterizations*, New York: Springer, 2012.

[6] Coldren, L. A., S. W. Corzine, and M. L. Mashanovitch, *Diode Lasers and Photonic Integrated Circuits*, 2nd ed., New York: Wiley, 2012.

[7] Meade, M. L., *Lock-In Amplifier: Principles and Applications*, London, U.K.: Peter Peregrinus, 1983.

[8] Choquette, K. D., and H. Q. Hou, "Vertical-Cavity Surface Emitting Lasers: Moving from Research to Manufacturing," *Proceedings of the IEEE*, Vol. 85, 1997, pp. 1730–1739.

[9] Michalzik, R., (ed.), *VCSELs: Fundamentals, Technology and Applications of Vertical-Cavity Surface Emitting Lasers*, New York: Springer, 2012.

[10] Buus, J., M. C. Amann, and D. J. Blumenthal, *Tunable Laser Diodes and Related Optical Sources*, 2nd ed., New York: SPIE Press and Wiley-IEEE Press, 2005.

[11] Liu, J. -M., *Photonic Devices*, Cambridge, U.K.: Cambridge University Press, 2005.

[12] Stadler, B. J. H., and T. Mizumoto, "Integrated Magneto-Optical Materials and Isolators: A Review," *IEEE Photonics Journal*, Vol. 6, 2014.

[13] Hecht, J., *Understanding Fiber Optics*, 5th ed., Auburndale, MA: Laser Light Press, 2015.

[14] Kanamori, H., "Passive Optical Components and Their Applications to FTTH Networks," *SEI Technical Review*, Vol. 73, 2011, pp. 14–21.

[15] Hecht, E., *Optics*, 4th ed., Boston, MA: Pearson, 2012.

[16] Iizuka, K., *Elements of Photonics: Volume I, In Free Space and Special Media*, New York: Wiley, 2002.

[17] Hui, R., and M. O'Sullivan, *Fiber Optic Measurement Techniques*, New York: Elsevier Academic Press, 2009.

CHAPTER 3

Principles of Fiber-Optic Sensing

3.1 Definitions

In the previous sections, we described the light propagation in optical fibers, and the main devices that can be used to emit, detect, and process light in a fiber. Now we can use such devices to create measurement instruments based on such concepts.

A measurement instrument is a device that detects changes of a physical, thermal, electrical, chemical, or biological parameter (called measurand, in measurement science), and returns an output that is related to such parameter. A transducer is a device that converts a signal in one form, to a signal in a different form. Typically, we aim at transducing a signal that is hard to measure, into one that can be easily detected: an example is pressure sensors, which transduce pressure into displacement that can be easily measured. Sensors, particularly fiber-optic sensors (FOSs), typically assume a physical separation between the place where sensing mechanism happens (typically called the active part), where the transducing takes place, and the interrogator, which is responsible to send energy to the active area and collect the output energy. The combination of a sensor, its transducing effect, and its interrogator constitute the working principle of a sensing system.

FOSs or optical fiber sensors (OFS) are a subclass of sensors, which make use of light as a sensing principle, and optical fibers either as a medium to transport light from the source to the active area, or as a medium to modulate light as a function of the measurand, or both. At a minimum complexity, the skeleton of a fiber-optic sensor is composed of: (1) a light source; (2) an optical fiber; (3) one or more active areas that can be devices embedded inside the fiber or externally placed; (4) a detector, that detects the sensor output; and (5) a technique (usually implemented in software) that estimates the measurand from the sensor output. FOSs enable remote sensing, as the active area can be positioned far away from the light source and detector. This description is valid for most sensing techniques; however, distributed sensors do not have an active region, but rather the whole fiber (element 2) acts as the active region.

In the intuitive and broadly accepted definition, a medical device has 3 characteristics: (1) it is used on humans; (2) it has therapeutic benefits; and (3) it has a physical or mechanical effect on the body, or it is used to measure functions of the body. The majority of biomedical sensors belong to the group of medical devices, which is also extremely restrictive for standards and certifications [1]; most of the sensors described in Chapters 15 to 19 belong to this group. There is an emerging class of biomedical sensors that are not medical devices (such as wearable sensors, shown in Chapter 20) that is not operating in humans but is operating externally.

3.2 Classification

The modern advances, particularly after 2010, have pushed for a new classification of FOS: an efficient standardization divides sensing architectures into 6 categories, which are sketched in Figure 3.1 [2]:

1. *Single-point extrinsic sensors:* Single-point sensors measure one parameter in one active point. Extrinsic sensors use the optical fiber just to propagate light to the active point and back (or forward) to the detector. In the figure, we show an example of sensor, in which an optical fiber emits light against a target, positioned at a certain distance and externally to the fiber, and the same fiber collects the backreflection.
2. *Single-point intrinsic sensors:* Intrinsic sensors include the sensing mechanism inside the fiber, in a well-confined region: typically, an in-fiber device is used to modulate light emitted from the fiber, backreflecting part of it. In the figure, a fiber Bragg grating (FBG) fabricated inside the fiber core reflects a portion of the input light, letting the remaining light pass through.
3. *Multiplexed sensors:* FOS gives the possibility of integrating several sensors inside a single fiber. This architecture allows multiple sensing, with a discrete number of active areas, spatially separated from each other. By using a suitable interrogation, it is then possible to retrieve each individual sensor without ambiguity. An array of FBG is an example of multiplexed sensing network, having multiple sensors (each reflecting a different wavelength) aligned on the same fiber.
4. *Multiparameter sensors:* This nomenclature refers to the detection of multiple measurands, obtained with sensors having a different working principle. For example, an FBG can be used for temperature sensing, while an extrinsic Fabry-Perot interferometer (FPI) can be used for pressure sensing. In biomedical sensors, we inherently consider that the active areas of each of element are overlapping, or in proximity, such that the detection of multiple parameters happens in a single sensing point.

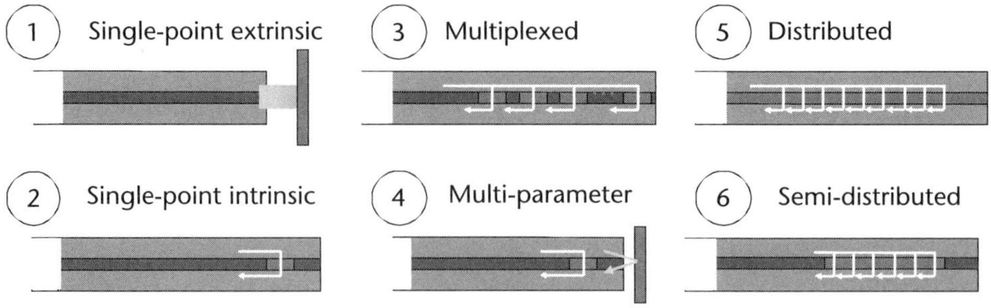

Figure 3.1 Definition of fiber-optic sensing schemes.

5. *Distributed sensors:* In all previous categories, we have a discrete number of sensing points; distributed sensors extend this concept to a continuous flow of sensing point. Distributed sensors turn an entire optical fiber to a sensor exploiting internal phenomena (such as Rayleigh scattering, as in the figure) that allow a spatially resolved detection of each point, estimating the measurand in each measurement point along the whole fiber length. The common denominator of all distributed sensing architectures is that when coherent light is emitted inside an optical fiber, a small portion of it is backreflected: in the time, or frequency, domain, it is possible then to separate each of these small reflection contributions, and each of them is dependent upon variations of the measurand. In this case, the sensor does not have a separate active area, but rather the whole fiber is actively sensing.
6. *Semidistributed sensors:* This class of sensors lies in between multiplexed sensors and distributed sensors: we identify a well-defined active region formed by a fiber-optic device, and within this length, it is possible to continuously detect the measurand, as in distributed sensors. An example of a semidistributed sensor is a chirped FBG (CFBG), as in the picture: it is a device whose reflection spectrum is dependent upon the variations of the measurand in each point of the grating.

Single-point sensors operate similarly to other sensing technologies, such as microelectromechanical systems (MEMS), screen-printed sensors, transistors, piezoelectric, and electrochemical. The key advantage of FOS is given by the possibility of using a single fiber to host a plurality of sensing points, and detecting multiple parameters in the same fiber. This is the unique feature of FOS, unmatched by any other technology.

When a system detects light passing through the active area(s), it operates in transmission. Most biomedical sensors, however, operate in reflection mode, by detecting and processing the light reflected by the active area(s). In this case, the same fiber is used as input, delivering light to the active point, and as output, guiding the light back to the detector. This configuration minimizes also the cabling size, and is highly recommend particularly for sensors operating in vivo. Reflection-mode sensors are often referred as probes. From the hardware point of view, directional couplers (or circulators) allow routing light to the sensor(s) and back to the detector. Figure 3.2 shows the difference between transmission and reflection sensors.

3.3 Working Principles

In order to achieve a sensing structure, with any of the schematics of Figure 3.1, it is necessary to unambiguously encode a variation of a measurand with a variation of one or more of light parameters; here the description of light as a wave (lightwave) comes in support [3]. The simplest form of sensing is given by intensity-based sensors (IBSs), which modulate the amplitude of light as a function of variations of the measurand. IBSs usually are single-point, both extrinsic and intrinsic, and often operate on extended lengths of fiber. Nearly any of the devices used for sensing can be turned into an IBS by either operating on a single wavelength or using an adequate

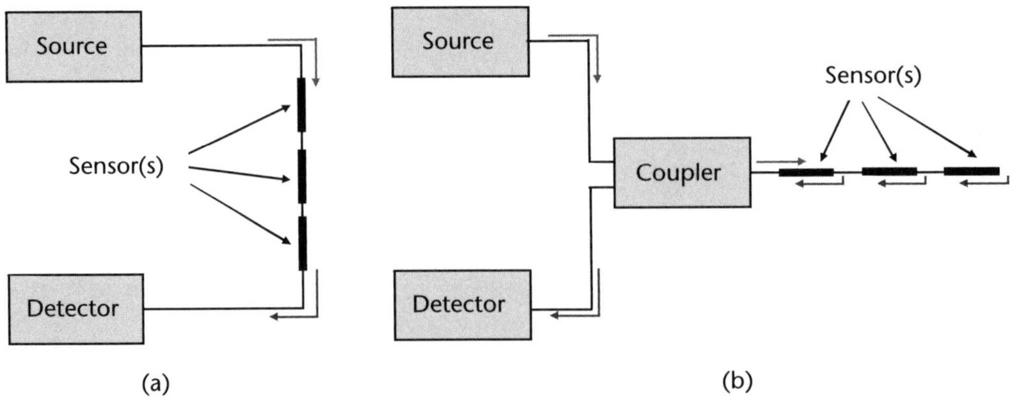

Figure 3.2 Typical layout of (a) transmission and (b) reflection-based sensors. Light gray arrows show the path of incident light, dark gray arrows show the light path from sensors to detectors, either transmitted through the sensor(s) (a) or reflected by each sensor (b).

setup. IBS represents the simplest form of sensing, but also the most vulnerable to external parameters, as will be detailed in the next section.

A more refined and elegant technique for sensing encodes variations of the measurand into variations of the optical spectrum, which can be recorded with optical spectrum analyzers, scanning-wavelength devices, or spectrometers that are well available on the market. Measurand variations cause changes in the reflection (or transmission) spectrum, which are detected by performing a signal analysis on the optical spectrum. This class of sensors represents, to date, the majority of both established and emerging techniques for biomedical sensing and allows operating with multiplexed, multiparameters and quasi-distributed sensors. Figure 3.3 shows examples of sensors based on IBS and optical spectrum working principles.

It is also possible to encode measurand variations in the phase of the light phasor. This approach leads to most interferometric sensors, which usually exploit light interference between two or more light beams travelling in different spaces. The light phase is detectable in the time domain, and this is often performed in distributed sensing (particularly on short-length distributed sensors described in

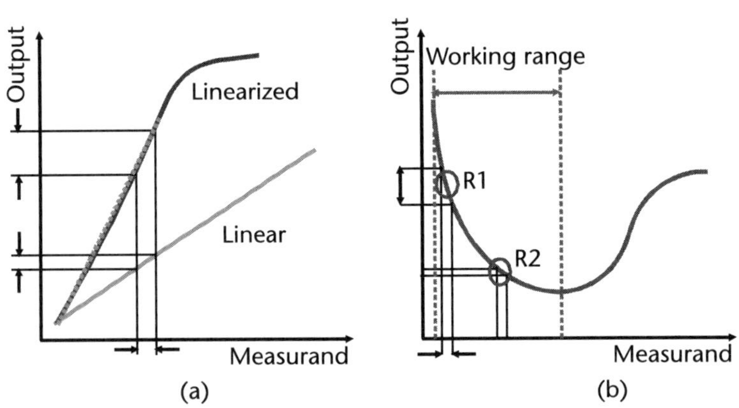

Figure 3.3 Example of calibration function of a generic sensor, operating in (a) linear or quasi-linear mode or (b) nonlinear mode.

Chapter 6). However, most architectures use artifacts that convert the phase term into a spectral or amplitude term, which is easier to detect.

Finally, polarization is another way to encode light variations, and it was popular particularly at the end of the last decade; polarization-based sensors make use of variations in the incident wave polarization due to external effects due to measurand (for example, strain acting on the fiber), which can often be detected with the support of polarization filters. Nowadays, its application in biomedical sensors is more marginal.

When designing a sensor, it is always necessary to have a firm grasp of how to transduce the measurand(s), into a quantity that is easy to measure and provides a clear and unambiguous interaction with light: a good sensor design aims at transducing a parameter that is hard to measure into a parameter that can be more easily detected by a probe.

In biomedical sensors, we can substantially divide between physical, biophysical, and biological parameters. Physical parameters refer to the measurement of physical quantities, that include temperature, strain, pressure, acoustics, vibrations, displacement, time, and level; when such parameters, and the sensors to measure them, are located within a living organ, or are closely related to physical activities of the body (e.g., respiration, cardiac activity, urologic variations), they are typically labeled as biophysical. Temperature measurements are relatively easy with FOS, because we can exploit the thermal expansion coefficient of glass (or plastic) used to manufacture fibers and devices. Displacement is also a quantity that can be easily measured, using a plurality of sensing principles. In contrast, pressure is a parameter that can be hardly measured, as FOS devices have very small response to pressure variations: usually, we transduce pressure to strain or to displacement, by means of mechanical structures based on flexible diaphragms. Similar considerations can be often made for flow or level sensors.

Biological measurements instead refer to the detection of variations of biological organisms, such as cells, blood, and biomarkers. The detection can be performed by simply monitoring variations of the refractive index and, in this case, is often referred to as label-free. More advanced sensors adapt the active part to the detection of individual and well-selected organisms that are usually captured on the sensing part: in this case, we often label the sensor as functionalized to a specific analyte.

We can draw, in Table 3.1, a design and selection tool for engineers and biomedical specialists, as a guideline for developing and evaluating sensors. The matrix has on the rows the principal measurands, to be detected in medical sensors and devices, and on the columns the sensor classes. FBGs are, by far, the most popular technology for biophysical sensing, due to their ease of implementation in single-point and multiplexed sensing schemes, and because of their dual sensitivity to strain and temperature; however, FBGs can also detect pressure and flow (transducing them to strain). FPI is also an extremely popular approach to detect temperature, displacement, and is a relatively unique device for its pressure sensing (with small footprint) capability. IBS techniques, such as fiber bundles and macrobending, are often used for low-cost, single-point sensing, and only rarely for multiplexed sensing (limited to dynamic strain in frequency domain). Distributed sensors use a different approach: optical backscattering meter (OBR) is used for thermal and strain sensing, and the most recent optical frequency-domain reflectometry (OFDR) systems can

Table 3.1 Most Popular Working Principles for Biomedical Sensors

	Single-Point	*Multiplexed*	*Distributed*	*Semidistributed*
Temperature	FBG, FPI, interferom	FBG, FPI	OBR	CFBG, MWPF, AGF
Strain (static)	FBG, FPI, macrobend	FBG	OBR	CFBG, MWPF. AGF
Strain (dynamic)	FBG, FPI, macrobend	FBG, macrobend		
Displacement	Bundle, FPI, SMI	Bundle, FPI		
Pressure	FPI, bundle, FBG	FBG		
Flow	FPI, FBG	FBG		
Shape			OFDR	AGF
Label-free biosensing	SPR, TFBG, LoF	SPR, LPG		
Functional biosensing	SPR, LPG, TFBG, LoF	SPR, LPG		

access also shape sensing in multicore fibers. Semidistributed sensing is a recent trend and is usually accomplished through chirped FBG (CFBG), all-grating fibers (AGF) and microwave photonic filters (MWPF). However, biosensors use different approaches: surface plasmon resonance (SPR), long-period grating (LPG), tilted FBG (TFBG), typically combined with SPR, and emerging lab-on-fiber (LoF) structures. Most of these sensing techniques can be integrated in a multiparameter sensor, but can be hardly multiplexed only to few sensing points. Distributed sensing for biological and biochemical sensors has not been demonstrated yet.

3.4 Sensor Performance Analysis

Metrology, or measurement science, has a rigorous standard for evaluating the performance of sensors and measurement instrumentation [4]; in this section, we focus on the fundamental criteria for evaluating the working principles of a sensor, which can be easily accessed and mastered by most engineers and are in line with the National Instruments white paper [5].

The calibration function is the relationship between the input and the output of a sensor. This is the most important characteristic of a sensing device, as it relates the measurand (x axis) with the output returned by the sensor to the interrogator (y axis). Figure 3.3 shows an example of calibration functions. In the left chart, the light gray curve draws an ideal sensor, in which the output is related by the input according to a linear function. Since we read the output and aim at estimating the measurand, if a linear relationship between the two is found, the estimation process is straightforward. The dark gray curve represents a different case, in which the calibration function is still monotonic, but not linear. The monotonic trend allows one to unambiguously relate any value of the output to an input value; but as a linear relationship is not guaranteed, the estimation of the measurand relates to a nonlinear relationship. In several applications, however, we are interested only in a restricted range of measurand values: in this case, for example, it is possible to consider a wide portion of the calibration curve (dotted line in the figure), and approximate it as a linear function. This operation introduces an estimation error, which can often be negligible. The right chart shows a nonlinear, and nonmonotonic calibration function. In this case, sensing can only be limited to a restricted range of values where the curve is monotonic, otherwise detection would be ambiguous (several input values corresponding to the same output value).

The sensitivity is typically defined as the variation of the output corresponding to a small variation of the input. The sensitivity corresponds to the derivative of the calibration function, and its measurement unit is the ratio between the output units and the input units. For a linear sensor, the sensitivity is constant; in quasi-linear sensors, usually the sensitivity is approximated as constant within a certain range. In the case of Figure 3.3(b), the sensitivity changes in every point of the calibration curve: in the region R1 the sensitivity is high, as a broad variation of output is obtained for a small input variation; however, in R2 the same input variation leads to a smaller output change.

The calibration process is the estimation of the input/output characteristic and usually is performed comparing the sensor under test with a reference instrument, which is required to have a much greater accuracy than the sensor under test. In biomedical sensors, calibration is typically performed in vitro, making use of accurate instrumentation, and afterwards operate the sensor in situ. Calibration is typically performed in a measurement chamber that applies a measurable variation of temperature, strain, pressure, or any other measurand and compares the fiber-optic sensor output with a reference instrument. There is a large body of literature in metrology related to the calibration of individual sensors; however, a hard task in optical fibers is the calibration of distributed and quasi-distributed sensors. This occurs because of the absence of alternative technologies for distributed sensors, which makes it impossible to apply and detect an arbitrary variation of measurands throughout the entire length of the sensor. In this case, sensors can often be calibrated in few points and artificially extend the calibration to the whole fiber length.

Few more properties are typically required to complete the characterization of a sensor, and are shown in Figure 3.4. The resolution is the minimum increment or decrement of the input that can be clearly, and unambiguously, detected by the output. The accuracy is the often defined as the difference between the sensor output and the true value of the measurand. The estimation of the accuracy is not straightforward and is often performed in a statistical way; accuracy is defined either as an absolute value or in percentage of the output. The accuracy depends on the accuracy of the active sensor, the accuracy of the transduction process, the output accuracy, and the overall effect of the sensing scheme and its intereferences (e.g., fiber twists, ambient light). Engineers typically estimate the accuracy during the calibration process, calculating the standard deviation between reference and sensor output as a main accuracy metric. The repeatability is a figure related to the sensor reliability and can either refer to several measurements performed with the same sensing devices, or to the fabrication of multiple sensors having the same nominal response. Hysteresis is the deviation of the response from the calibration function, whether the measurand is increasing or decreasing from a static value. Often, hysteresis is significant for sensors having a slow response time.

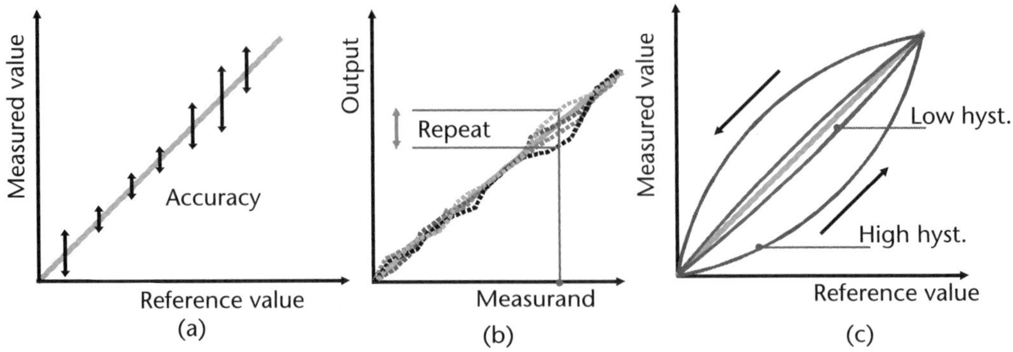

Figure 3.4 Graphic representation of parameters for sensor performance evaluation: (a) accuracy, (b) repeatability, and (c) hysteresis.

3.5 Application-Integrated Design

Throughout the last decade, the FOS design has been often separated from real applications. This has resulted in a neat separation between the requirements for which each application demands and the actual technology; as a consequence, scientists have often investigated solutions with little or no commercial viability, failing to make impact. This gap has been significant for biomedical engineering, due to its higher standards for safety and operations and the high barriers towards applications. A modern and efficient structure for sensor design is shown in Figure 3.5 and is application-driven. The model becomes effective when a single research group embeds the whole three building blocks: (1) top part: understanding the physical principles of fiber-optic sensing and transduction; (2) left part: design of a sensing system, its interrogation, and its capability to estimate measurands; and (3) right part: biomedical applications and their requirements and standards. Only by drawing the dark gray arrows, moving from the application to the design, it is possible to have a successful path towards biomedical sensing, and meet the huge demand for medical devices. The reverse process (i.e., design a sensing system and imposing its applications) has so far constrained FOS in small niches, with little

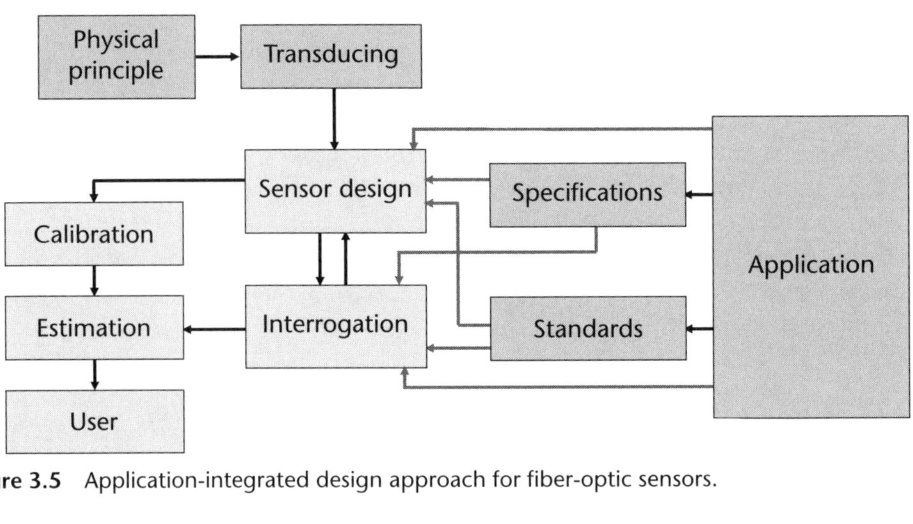

Figure 3.5 Application-integrated design approach for fiber-optic sensors.

market values. The scope of this book is to provide engineers with the right information at all levels of this chart, so that they can develop and iterate sensing systems but also translate the demand coming from each biomedical application into an addressable set of specifications (size, performance, budget) and identify the right standards and way to prototype systems.

References

[1] Peterson, J. I., and G. G. Vurek, "Fiber-Optic Sensors for Biomedical Applications," *Science*, Vol. 224, April 13, 1984, pp. 123–128.

[2] Udd, E., and W. B. Spillman, Jr., (eds.), *Fiber Optic Sensors: An Introduction for Engineers and Scientists*, New York: John Wiley & Sons, 2011.

[3] Agrawal, G. P., *Lightwave Technology: Telecommunication Systems*, New York: John Wiley & Sons, 2005.

[4] Potts, P. J., "Glossary of Analytical and Metrological Terms from the International Vocabulary of Metrology (2008)," *Geostandards and Geoanalytical Research*, Vol. 36, No. 3, 2012, pp. 231–246.

[5] National Instruments, "Sensors Terminology," white paper, September 23, 2013, http://www.ni.com/white-paper/14860/en/.

CHAPTER 4

Intensity-Based Sensors

4.1 Introduction

Intensity-based sensors, also known as intensiometric sensors, are fiber-optic sensors (FOSs), in which, as their name implies, the quantity under measurement affects the intensity of the light propagating in a fiber optic span and thus directly the power at the receiver.

Actually, in the end, all FOSs are somewhat intensity-based because at the receiving side there will be a photodiode or a similar power-sensitive device; however, with the name intensity-based sensors, we commonly refer only to intrinsic sensors in which the quantity under measurement directly modulates the received power.

Intensiometric sensors are the simplest and lowest-cost among the fiber sensors because they just require a source, a receiver, and a connecting fiber-optic span that acts both as the transducer and as the waveguide for the measurement signal. Figure 4.1 shows the schematic representation of an intensity-based fiber-optic sensing system in which the sensor works in transmission, that is the transmitter and the receiver are at the opposite ends of the sensing fiber span. Another possibility is to have a sensor working in reflection; in this case, the signal is reflected at one end of the fiber through a mirror and both the transmitter and the receiver are on the same side of the fiber-optic span. Clearly, the latter layout is more effective for building probes, but requires a circulator or a coupler must to separate incident and reflected signals (Figure 4.2, analogous to Figures 2.14 and 2.17).

Most notable examples of intensity-based sensors are those used to measure physical quantities, such as distances, vibrations, and accelerations, although intensity-based detection principles can be exploited also to detect chemicals or changes in their concentration.

Given that the working principle does not require a tight control of the propagating modes, this type of sensor is mainly implemented using multimode fibers to take advantage of their lower requirements in terms of coupling with sources and of connector alignments. Particularly suited is the use of the multimode plastic optical fibers (POFs) introduced in Chapter 1 because with type of fibers it is possible to deploy very inexpensive all-fiber sensing systems as they do not require com-

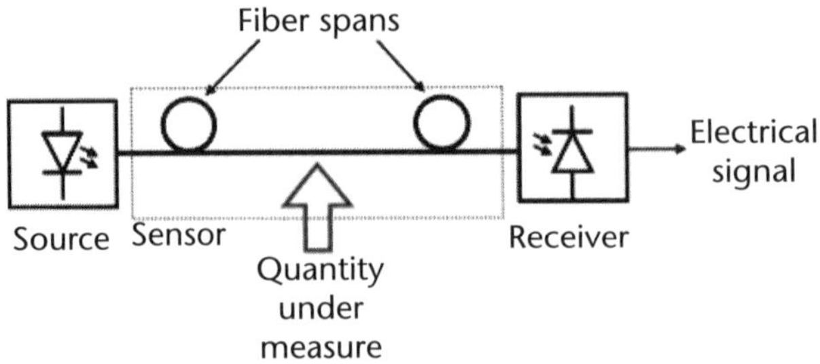

Figure 4.1 Schematic representation of an intensity-based fiber optic sensing system in which the sensor works in transmission.

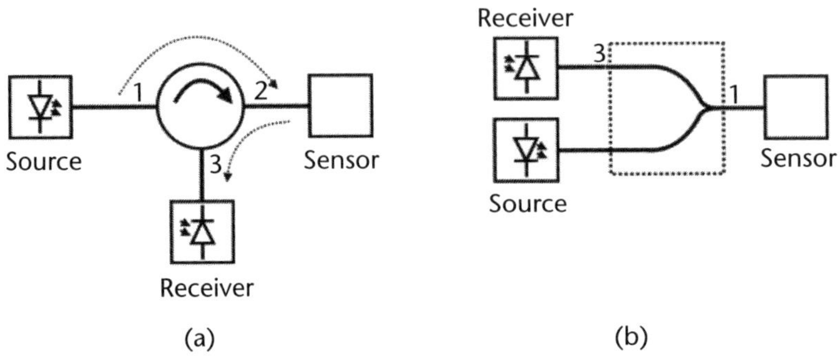

Figure 4.2 Schematic representation of two possible implementations of an intensity-based fiber-optic sensing system in which the sensor works in reflection.

plex machines for splicing and polishing and, moreover, they make use of simpler connectors and of visible LED sources.

The simplicity of intensity-based sensors, however, is also their main drawback since any contribution impacting on the transmitted light amplitude, such as power source fluctuations, curvatures, or mechanical instabilities leading to misalignments, are indistinguishable from actual measurement signals. Power source fluctuations can be compensated by tapping part of the launched power for monitoring, but the other causes of perturbations can be harder to compensate and may require an additional null-sensor or the use of more than one wavelength [1, 2]. Moreover, although specific parts of the fiber can be made more sensitive to the quantity under measurement, the entire fiber-optic span connecting the source and the receiver is actually sensitive to phenomena that can change the received power, making de facto the entire fiber length a sensor. This cross-sensitivity with parasitic effects occurring along the entire fiber length is particularly limiting the use of this approach for sensors that need to be repeatedly repositioned during their use. As an example in the biomedical field, let us consider a sensor in the form of a fiber probe that is inserted through a percutaneous needle or through a sort of gastroscope: the actual reading from the probe will depend on how the fiber is inserted and on the bending along the entire fiber. If the effects of the quantity under measurement largely exceed those due to cross-sensitivity, the probe can still be successfully used for the considered application; nevertheless, the impact of disturbances that may occur along the entire fiber length must be carefully taken into account in the sensor and characterization phases.

For these reasons, intensity-based sensors are best suited for low-cost[1] and not highly demanding applications or for dynamic measurements in which the quantity of interest produces changes in the received power in a specific frequency range, well distinct from disturbances, so they can be easily filtered out.

Another limitation of intensity-based sensors is the difficulty, if not impossibility, to multiplex them along a single fiber to create quasi-distributed sensing systems, unless expensive detection techniques based on light scattering (mainly Rayleigh scattering) are employed [3, 4]. In practice, the only simple solution for a complex monitoring system is to use spatial multiplexing in which sensors can possible share the source through power dividers. An example of a layout that exploits such approach is shown in Figure 4.3.

A somewhat arbitrary classification of intensity-based sensor can be in terms of the exploited specific mechanism: modulation of propagation loss, coupling loss, evanescent field absorption loss, and surface plasmon resonance. This chapter is devoted to the first two approaches, whereas the other two are discussed in Chapter 8.

A review of intensity-modulated sensor (actually, including also some sensors that in this book have other classifications) can be found in [5, 6].

4.2 Sensors Exploiting the Modulation of Losses

One of the well-known causes that can increment propagation loss in optical fibers is bending: in telecommunication applications, for standard fibers, negligible contribution to loss requires bending radii larger than about 5 cm. In sensing, this drawback of communication links can be exploited to sense, for instance, pressures or deformations, as sketched in Figure 4.4, in which the effect of bending is enhanced by the rods that impose tight curvatures.

A setup similar to that in Figure 4.4 introduces macro bending; alternatively microbending can also be exploited, obtaining the curvature sensors [7]. Sensitivity to bending can be locally increased by realizing grooves or micropatterns in the fiber cladding by milling, laser ablation, or plasma treatments [8]. These approaches produce grooves with different shapes and depths and hence lead to different bend-sensitivity characteristics. Usually the outcome of the fiber surface modification is

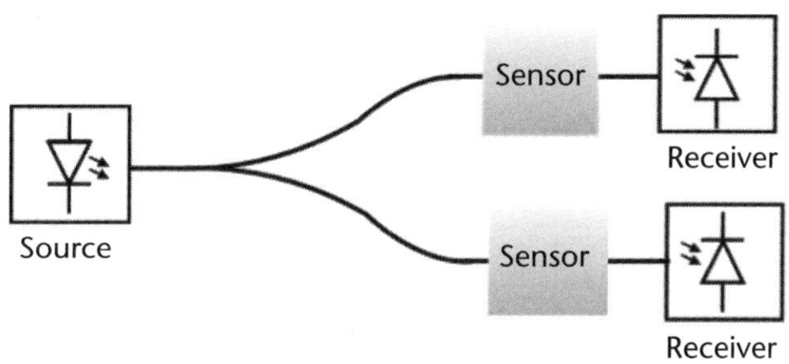

Figure 4.3 Example of spatial multiplexing (limited to two sensors for simplicity) for intensity-based sensors.

1. Here low-cost is relative to fiber-optic sensors, which, nevertheless, can be much more expensive that their electrical counterparts.

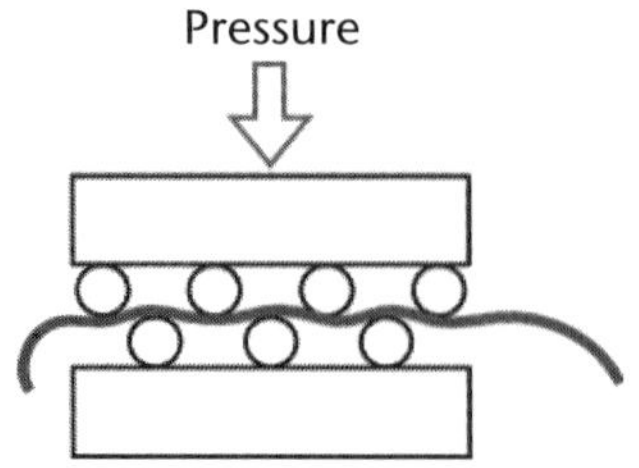

Figure 4.4 Schematic representation of a sensor that exploits the increase in propagation loss due to bending induced by pressure.

quite difficult to predict a priori from simulations, so the optimization of the fiber sensor is typically mainly based on experimental tests.

Clearly, there must be a balance between increasing the sensitivity to bending and maintaining the structural integrity to guarantee the necessary mechanical strength of the fiber.

All these surface-modifying approaches are more effective when using MMFs for their intrinsic higher sensitivity to bending. In particular, POFs are well suited not only because they are less brittle than glass fibers of the same thickness, but also because bend-loss sensitivity increasing techniques are easier to implement in polymers. For instance, for a not-so-demanding (and reproducible) application, a scalpel can be effectively used to form the grooves. Then, as POFs have a very thin cladding[2], the sensitivity to bending can be largely increased compared to their all-glass fiber counterparts because it is very easy to modify the entire cladding thickness if it does not affect the core itself too. An example of surface grooving in POF is reported in Figure 4.5: on the left, the figure shows a sketch of the grooved fiber, while on the right there is a picture in which the region with grooves is clearly evidenced by the brighter region consequence of more irradiation of light. As the curvature radius along the fiber longitudinal axis changes, the grooves are either stretched or compressed (depending on the reduction or increase of the curvature radius) and this modulates the propagation loss, enhancing or reducing the irradiation from the grooves.

Curvature sensors can be made to have a very high sensitivity. In the biomedical field these curvature sensors have been mainly used for thoracic and abdominal circumference measurements in respiratory and heart rate monitoring, for examining

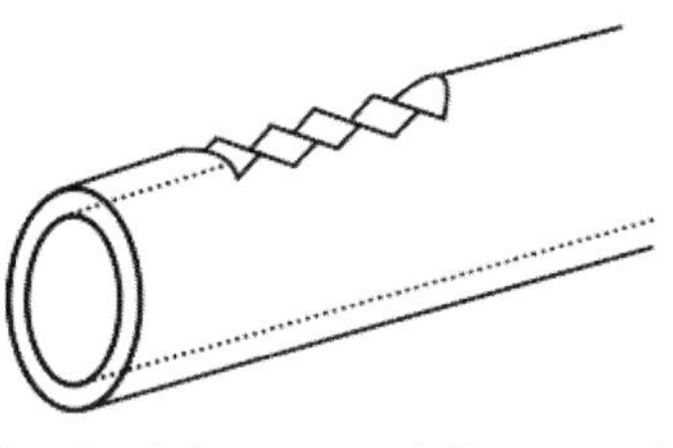

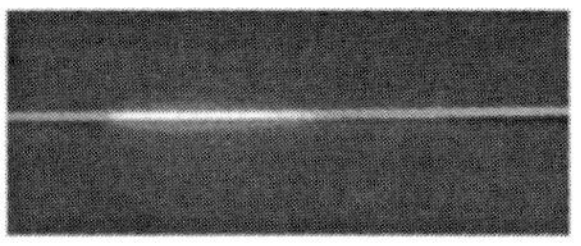

Figure 4.5 Sketch of the grooved fiber to enhance the sensitivity to bending (left) and picture of one of such grooved fibers (right).

2. Most common POFs have 1-mm cladding diameter, of which 0.98 mm are the core; even the smallest commercial POFs have 0.25-mm cladding diameter, of which 0.22 mm are the core. For comparison, recall that standard multimode telecom grade fibers have 0.125-mm cladding diameter, of which only 0.06 mm are the core.

joint movements, and for measuring pressures, possibly embedding the sensing fibers into textiles [4, 9].

The relation between the received power and the deformation is nonlinear and difficult to predict a priori because it is strictly dependent on the sensor implementation (for example, the position and shape of the microgrooves); however, for fixed working conditions and small movements, as in respiratory or heart rate monitoring, after proper calibration, the deformation can be accurately recovered from the measure of the received power.

Figure 4.6 reports an example of application in which a POF-based curvature sensor has been located on the abdomen. Thanks to the high sensitivity, it is possible not only to evaluate the respiratory movements, but also to detect the heart activity during apnea phases [10].

In another example, curvature POF sensors have been employed to monitor knee joint movements [11].

The simplicity of such sensors has triggered the attention of many other research groups; additional examples of applications can be found in many other papers, such as [12–16], to cite a few. In particular, [16] revises different fiber-optic technologies for monitoring patient physiological parameters, with a special focus on the application during magnetic resonance imaging, and compares the characteristics and performance of propagation-loss intensity-modulated sensors with interferometric and fiber Bragg grating-based sensors.

4.3 Sensors Based on Coupling Loss

Another possibility to modulate the received power is to change the coupling conditions between two fibers. The easiest implementation is by changing the separation between two facing fibers or, equivalently, the distance between two parallel fibers and a facing mirror. The first configuration is said to work in transmission, whereas the latter is said to work in reflection. The two setups have different power-versus-distance responses, but overall similar performance and application: contrary to the reflection configuration, the transmission configuration exhibits a monotonic response, but the reflection configuration is more practical because, as already mentioned, both receiving and transmitting fibers are on the same side.

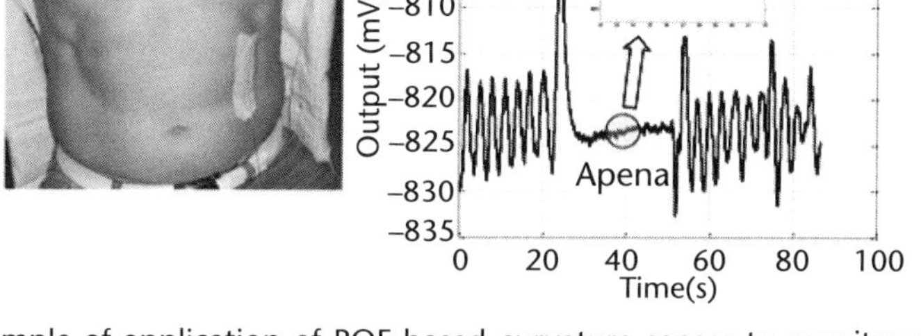

Figure 4.6 Example of application of POF-based curvature sensor to monitor respiratory movements and heart rate.

Although these schemes measure directly only a distance, they can be easily adapted to measure any other quantity (for instance, a pressure), provided that this quantity changes the separation between the fiber end-faces or between the fiber end-faces and the mirror.

Figure 4.7 shows a schematic representation of a sensor working in transmission: light from the transmitting fiber spreads within a cone determined by the numerical aperture and the fraction collected by the receiving fiber, among other parameters (e.g., core size and numerical aperture), depends on the separation *d*, the quantity under measure.

The working range is strongly dependent on the characteristics of the fibers and increases going from the use of SMFs (a few micrometers) to that of MMFs (a few millimeters). Usually, MMFs only are used in practical applications; in this case, the received power versus distance can be written in terms of geometrical optics using Gaussian beams [17], finding a hyperbolical relation. So, also in this case, the relation is highly nonlinear and the sensitivity decreases with distance. However, by properly choosing the operating point, sensors with a good price/performance ratio can be developed [8]. An example of an actual curve of the collected power versus distance for a 1-mm diameter, 0.5-NA POF is reported in Figure 4.8.

A variant of such sensor is sketched in Figure 4.9: the loss between two facing fibers at a fixed distance is modulated by the insertion of an obstacle that controls the amount of transmitted light. This scheme can be used, for example, to sense a pressure by changing the insertion length of the obstacle.

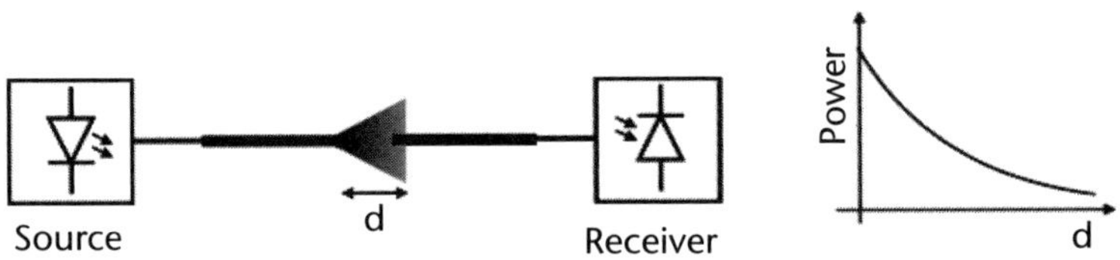

Figure 4.7 Schematic representation of displacement sensor that exploits the separation between two facing fibers (left) and qualitative relation between the separation d and the received power (right).

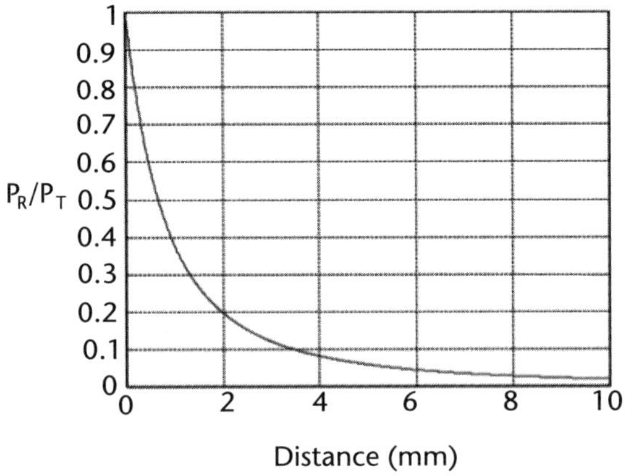

Figure 4.8 Ratio between received (PR) and transmitted (PT) power versus separation for a sensor as in Figure 4.7 implemented using two 1-mm, 0.5-NA, facing POFs.

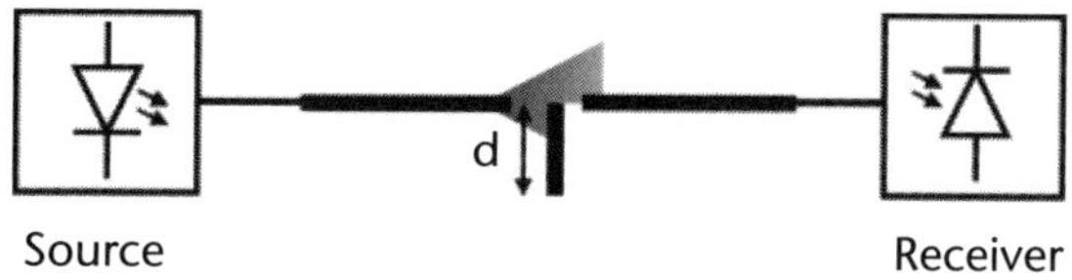

Figure 4.9 Working principle of a transmission-based displacement sensor working by controlling the amount of transmitted light.

The schematic representation of a sensor working in reflection is reported in Figure 4.10; in this case the transmitting and collecting fibers are parallel, often in contact (so their separation is zero), both facing a reflecting target.

An example of an actual curve of the collected power versus distance for a 1-mm-diameter, 0.5-NA POF is reported in Figure 4.11.

Comparing the curve shown in Figure 4.8 with that in Figure 4.11, the latter exhibits two monotonic regions with complementary features: high sensitivity in a short operating range in the first part of the curve and lower sensitivity but larger operating range in the second part. For larger distances, both sensors in Figure 4.7 and in Figure 4.10 have the same performance.

A specific implementation of this type of sensor is shown in Figure 4.12, in which a reflective membrane is used as the mirror to respond to an external effect such as pressure. In other words, the device in Figure 4.12 acts as an optical microphone, so a nonmetallic microphone, totally MRI compatible and particularly low-cost using POFs, can be used in biomedical applications for voice monitoring or heartbeats. As the membrane/mirror position shifts the effective separation between the optical fibers shift with a resultant intensity modulation.

A variant of the configuration in Figure 4.10 uses a single fiber both for transmitting and collecting and makes use of a coupler to separate forward and backward signals. In this case the response curve becomes monotonic again; however, as this configuration is implemented in practice only with high multimode fibers (mainly POFs, which guide some millions of modes) to have reasonable working ranges and couplers for such fibers do not guarantee a good isolation between forward and backward signals, the maximum operating distance can be again severely reduced.

Sensors based on coupling loss can also be implemented using fiber bundles; actually, this was the first implementation of such sensors, especially before the

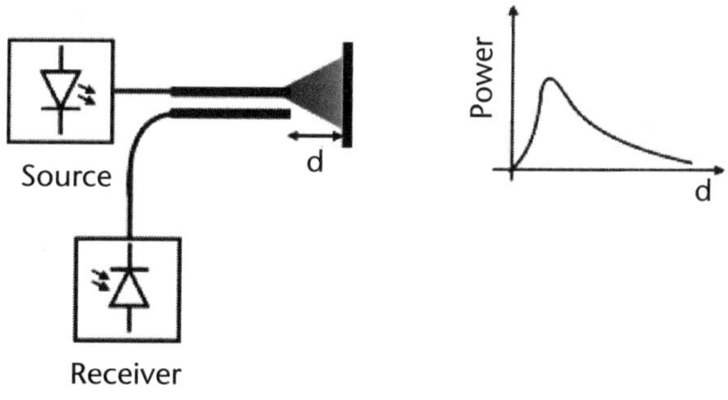

Figure 4.10 Schematic representation of a displacement sensor that exploits the reflection from a target facing transmitting and collecting fibers (left) and qualitative relation between the separation d and the received power (right).

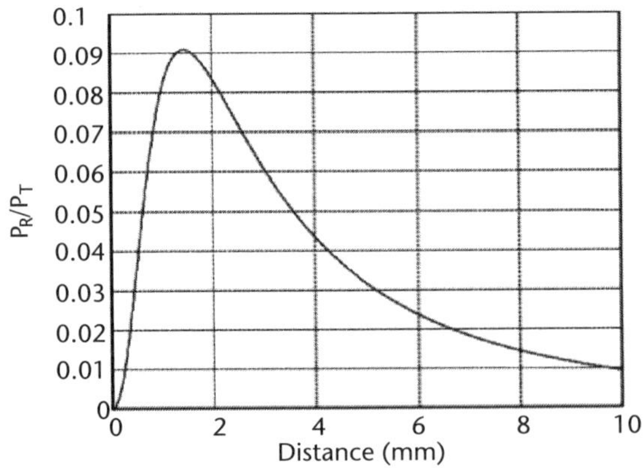

Figure 4.11 Ratio between received (PR) and transmitted (PT) power versus fiber tip to target separation for a sensor as in Figure 4.10 implemented with two 1-mm, 0.5-NA POFs with zero separation in between.

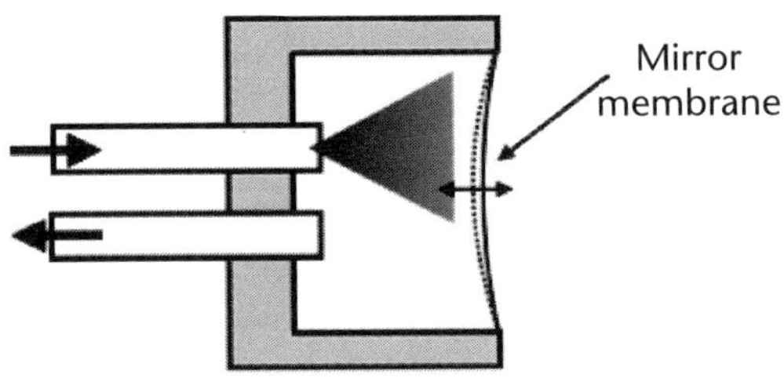

Figure 4.12 Schematic representation of a low-cost, all-optical pressure sensor (microphone) realized modulating the collected light intensity.

advent of large numerical aperture POFs. Still, this configuration adds flexibility in developing a probe with multiple receiving fibers to enhance the sensitivity or improve the working range, as reported in [18–20]. Commercial sensors mainly exploit this fiber bundle approach. Besides, some of the additional receiving fibers can be used to compensate for the influence of unwanted quantities, such as temperature and target reflectivity [21–24].

In general, this family of sensors is used in biomedical applications mainly as a simpler and lower cost alternative to measure pressures or displacements to more performing types of fiber-optic sensors, such as interferometric or those relying on fiber Bragg gratings, especially in the early days of fiber-optic sensors (see, for example, [25, 26] for a review and [27] for comparison that includes also commercial devices).

The conclusion for the sensors based on coupling loss is similar to that already presented for the sensors based on propagation loss: the large sensitivity to parasitic quantities that may affect the intensity of light during propagation severely limits the performance when used in static or quasi-static applications. However, these sensors can be fruitfully used to measure dynamic displacements, such as vibrations, with very excellent price/performance ratio. For example, their resolution can be good enough to detect heart pulsations through the skin [28].

References

[1] Casalicchio, M. L., et al., "Non-Contact Low-Cost Fiber Distance Sensor with Compensation of Target Reflectivity," *I2MTC/2009, Proceedings of the IEEE International Instrumentation and Measurement Technology Conference*, 2009.

[2] Vallan, A., et al., "Assessment of a Dual-Wavelength Compensation Technique for Displacement Sensors Using Plastic Optical Fibers," *IEEE Transactions on Instrumentation and Measurement*, Vol. 61, 2012, pp. 1377–1383.

[3] Liehr, S., et al., "Polymer Optical Fiber Sensors for Distributed Strain Measurement and Application in Structural Health Monitoring," *IEEE Sensors Journal*, Vol. 9, 2009, pp. 1330–1338.

[4] Krebber, K., "Smart Technical Textiles Based on Fiber Optic Sensors," *Intech Current Developments in Optical Fiber Technology*, 2013.

[5] Roriz, P., et al., "Fiber Optic Intensity-Modulated Sensors: A Review in Biomechanics," *Photonic Sensors*, Vol. 2, 2012, pp. 315–330.

[6] Zawawi, M. A., S. O'Keffe, and E. Lewis, "Intensity-Modulated Fiber Optic Sensor for Health Monitoring Applications: A Comparative Review," *Sensor Review*, Vol. 33, 2013, pp. 57–67.

[7] Vallan, A., et al., "Static Characterization of Curvature Sensors Based on Plastic Optical Fibers," *IEEE Transactions on Instrumentation and Measurement*, Vol. 63, 2014, pp. 1293–1300.

[8] Vallan, A., et al., "Two-Dimensional Displacement Sensor Based on Plastic Optical Fibers," *IEEE Transactions on Instrumentation and Measurement*, Vol. 62, 2013, pp. 1233–1240.

[9] Massaroni, C., P. Saccomandi, and E. Schena, "Medical Smart Textiles Based on Fiber Optic Technology: An Overview," *J. Funct. Biomater.*, Vol. 6, 2015, pp. 204–221.

[10] Vallan, A., et al., "A Plasma Modified Fiber Sensor for Breath Rate Monitoring," *IEEE International Symposium on Medical Measurements and Applications (MeMeA)*, 2014, pp. 1–5.

[11] Stupar, D. Z., et al., "Wearable Low-Cost System for Human Joint Movements Monitoring Based on Fiber-Optic Curvature Sensor," *IEEE Sensors Journal*, Vol. 12, 2012, pp. 3424–3431.

[12] Augousti, A. T., F. X. Maletras, and J. Mason, "The Use of a Figure-of-Eight Coil for Fibre Optic Respiratory Plethysmography: Geometrical Analysis and Experimental Characterization," *Optical Fiber Technology*, Vol. 11, 2005, pp. 346–360.

[13] Grillet, A., et al., "Optical Fiber Sensors Embedded into Medical Textiles for Healthcare Monitoring," *IEEE Sens. J.*, Vol. 8, 2008, pp. 1215–1222.

[14] Chen, Z., et al., "Simultaneous Measurement of Breathing Rate and Heart Rate Using a Microbend Multimode Fiber Optic Sensor," *J. Biomed. Optics*, Vol. 19, 2014, pp. 1–11.

[15] Krehel, M., et al., "An Optical Fibre-Based Sensor for Respiratory Monitoring," *Sensors*, Vol. 14, 2014, pp. 13088–13101.

[16] Dziuda, L., "Fiber-Optic Sensors for Monitoring Patient Physiological Parameters: A Review of Applicable Technologies and Relevance to Use During Magnetic Resonance Imaging Procedures," *J. Biomed. Opt.*, Vol. 20, 2015, pp. 1–23.

[17] Faria, J. B., "A Theoretical Analysis of the Bifurcated Fibre Bundle Displacement Sensor," *IEEE Transactions on Instrumentation and Measurement*, Vol. 47, 1998, pp. 742–747.

[18] Shimamoto, A., and K. Tanaka, "Optical Fiber Bundle Displacement Sensor Using an ac-Modulated Light Source with Subnanometer Resolution and Low Thermal Drift," *Appl. Opt.*, Vol. 34, 1995, pp. 5854–5860.

[19] Cao, H., et al., "Theoretical and Experimental Study on the Optical Fiber Bundle Displacement Sensors," *Sensors and Actuators A: Physical*, Vol. 136, 2007, pp. 580–587.

[20] Yasin, M., et al., "The Performance of a Fiber Optic Displacement Sensor for Different Types of Probes and Targets," *Laser Physics Letters*, Vol. 5, 2007, pp. 55–58.

[21] Casalicchio, M. L., et al., "Optical Fiber Sensors for Long- and Short-Term Crack Monitoring," *IEEE Workshop on Environmental, Energy, and Structural Monitoring Systems, EESMS 2009*, 2009, pp. 87–92.

[22] Tosi, D., G. Perrone, and A. Vallan, "Performance Analysis of a Noncontact Plastic Fiber Optical Fiber Displacement Sensor with Compensation of Target Reflectivity," *J. of Sensors*, Vol. 2013, 2013, pp. 1–12.

[23] Wego, A., and G. Geske, "Fiber Optic Displacement Sensor with New Reflectivity Compensation Method," *J. Sensor Technology*, Vol. 3, 2013, pp. 21–24.

[24] Jia, B., et al., "A Differential Reflective Intensity Optical Fiber Angular Displacement Sensor," *Sensors*, Vol. 16, 2016, p. 1508.

[25] Roriz, P., et al., "Review of Fiber-Optic Pressure Sensors for Biomedical and Biomechanical Applications," *J. Biomed. Opt.*, Vol. 18, 2013, pp. 050903-1–18.

[26] Poeggel, S., et al., "Optical Fibre Pressure Sensors in Medical Applications," *Sensors*, Vol. 15, 2015, pp. 17115–17148.

[27] Pinet, E., "Pressure Measurement with Fiber-Optic Sensors: Commercial Technologies and Applications," *Proc. SPIE*, Vol. 7753, 2011, pp. 1–4.

[28] Vallan, A., M. L. Casalicchio, and G. Perrone, "Displacement and Acceleration Measurements in Vibration Tests Using a Fiber Optic Sensor," *IEEE Transactions on Instrumentation and Measurement*, Vol. 59, 2010, pp. 1389–1396.

CHAPTER 5

Fiber Bragg Gratings

5.1 Description

Fiber Bragg gratings (FBGs) are one of the most popular fiber-based sensing techniques, and are strongly established in biomedical sensing [1, 2]. FBGs are substantially the fiber-optic equivalent to a narrowband-reject filter or when operated in reflection narrow bandpass filter, whose pitch frequency changes as a function of physical parameters (strain and temperature). The principle of operation of an FBG is described in Figure 5.1, as in [3]: when a broadband light is incident on the grating device, this reflects a very narrow spectrum centered around its pitch wavelength, also called the Bragg wavelength; the remainder of the spectrum is transmitted through the grating. Using a coupler or a circulator, it is possible to decouple the input and output light and collect the reflected spectrum. The FBG is sensitive to both thermal variations and changes of strain applied to the grating sides; such variations result in a change of the Bragg wavelength of the quantity Δ. Thus, from the system point of view, FBGs behave as narrowband reflection filters, and the task for sensing purpose is the determination of the wavelength shift $\Delta\lambda$.

5.2 Uniform FBGs

In order to analyze in depth the characteristics of the FBG, and its sensing features, the coupled-mode theory (CMT) is the most appropriate tool. Introduced by Erdogan [4], and established in [5] as the leading technique for FBG analysis, the CMT aims at modeling, for each specific optical structure forming the FBG, the wave equations for the propagating wave (transmission) and the counterpropagating wave (reflection). The CMT can provide estimation in the closed form, for several grating structures, or can be approximated to discrete portions of a nonstandard grating as in the layer-peeling analysis by Skaar et al. [6]. As the time of this writing, CMT in its original and discretized form were at the basis of the most popular software for grating modeling and analysis, such as RSoft (BeamPROP™ and GratingMOD™) and OptiGrating, or to several modeling tools developed for MATLAB and partially included in the Appendix A. In this chapter, we will focus on the CMT solutions for uniform FBG and extend it to chirped FBGs and tilted FBGs.

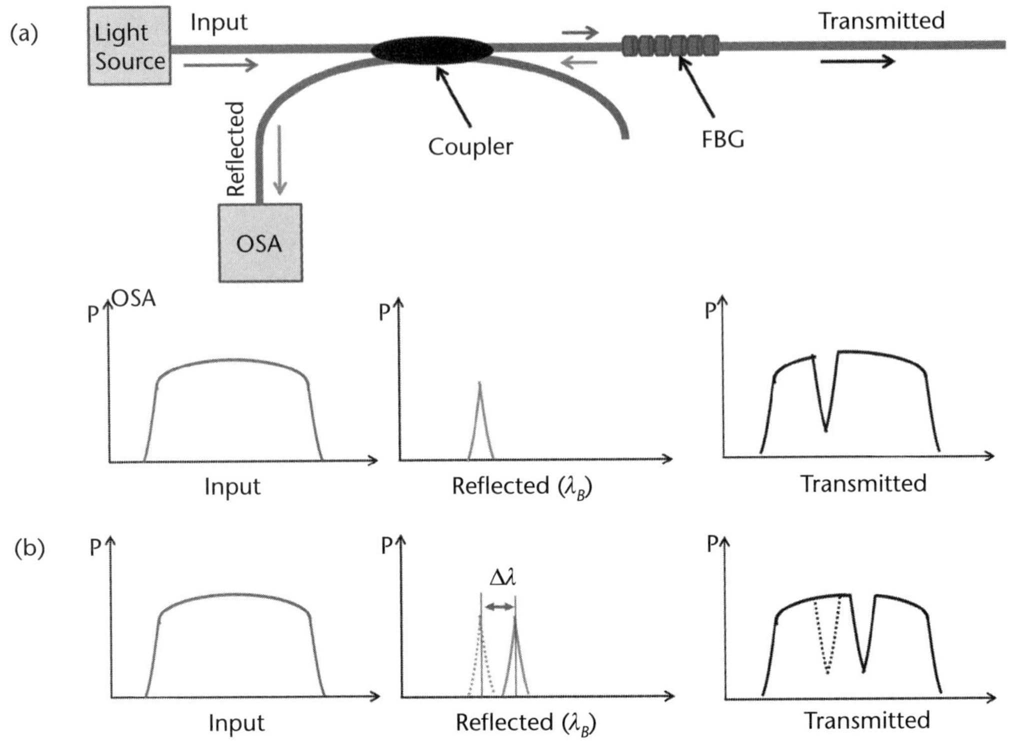

Figure 5.1 Principle of operation of an FBG-based optical sensor, as illustrated in [3].

From the optical perspective, an optical FBG is a periodic modulation of the refractive index of the core of an optical fiber. Exploiting the photosensitivity effect of a standard fiber or of a specifically designed photosensitive fiber, it is possible to produce local variations of the refractive index in each point of the fiber. Calling z the fiber propagation axis, the effective refractive index $n_{eff}(z)$ is not constant, but varies along z. In a uniform FBG, the pattern of $n_{eff}(z)$ has a sine-wave dependence, with constant period Λ, and amplitude δn_{eff}. In these conditions, the Bragg wavelength λ_B can be determined as [5]:

$$\lambda_B = 2n_{eff}\Lambda \tag{5.1}$$

where n_{eff} is the average effective refractive index. For such uniform grating, under the assumption that the refractive index modulation has a sine-wave profile of spatially uniform period Λ and amplitude δn_{eff}, the CMT provides a closed-form expression for the reflection coefficient of the grating:

$$\rho(\lambda) = \frac{-k\sinh\left(L\sqrt{k^2-\sigma^2}\right)}{\sigma\sinh\left(L\sqrt{k^2-\sigma^2}\right)+j\sqrt{k^2-\sigma^2}\cosh\left(L\sqrt{k^2-\sigma^2}\right)} \tag{5.2}$$

where ρ is the complex reflection coefficient, λ is the wavelength, L is the grating length, and kL is a unitless number that defines the grating strength and typically ranges from 0.1 to 4. If δn_{eff} is the amplitude of the refractive index modulation, usually ranging from 10^{-6} to 10^{-4} according to the type of fiber and efficiency of the inscription setup, σ has the following expression:

$$\sigma(\lambda) = \frac{\pi}{\lambda}\delta n_{eff} + 2\pi n_{eff}\left(\frac{1}{\lambda} - \frac{1}{\lambda_B}\right) \tag{5.3}$$

Usually, interrogation setups as in Figure 5.1 detect the reflection spectrum of the FBG, and therefore we are interested in the grating reflectivity R:

$$R(\lambda) = |\rho(\lambda)|^2 = \frac{\sinh^2\left(L\sqrt{k^2 - \sigma^2}\right)}{\cosh^2\left(L\sqrt{k^2 - \sigma^2}\right) - \frac{\sigma^2}{k^2}} \tag{5.4}$$

Equations (5.1) through (5.4) provide a CMT-based model for standard uniform FBGs, which are to date the most popular type of FBG used in biomedical applications. Figure 5.2 shows the spectrum of an FBG having a wavelength of 1,550 nm, simulated with CMT, and having different values of kL ranging from 0.1 to 4. The figure shows that all spectra have a main lobe with a high reflectivity, followed by a series of sidelobes with decreasing reflectivity. For very small values of kL, the grating is poorly reflective (only a few percentage), and therefore is barely visible on a spectrometer or an OSA. When kL ranges within 0.3 and 0.7, the grating approaches the 50% reflectivity at the Bragg wavelength, which allows easily detecting the grating on a spectrometer; also, sidelobes have a low reflectivity. High values of kL result in a grating with a reflectivity close to 100%, and with a flat profile in the central part: the bandwidth increases with kL, and the amplitude of sidelobes also tends to increase. It is possible to see that, for any value of grating strength, the resonance condition of the FBG is very strong, and the FBG behaves as a filter with

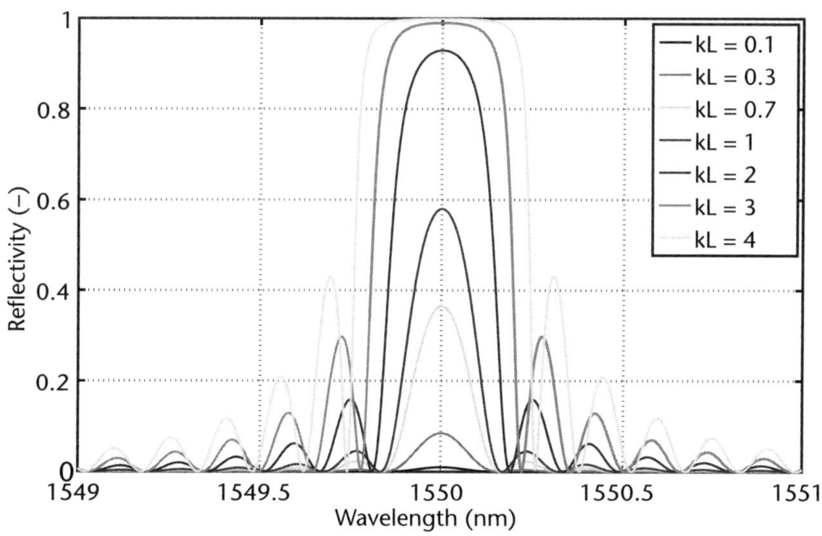

Figure 5.2 Reflection spectrum of a 1,550-nm FBG, modeled with Erdogan's CMT for different values of grating strength kL. Spectra have been simulated with MATLAB model included in Appendix A.

Table 5.1 Characteristics of CMT-Modeled FBGs, Operating Around 1,550 nm, in Terms of Peak Reflectivity, FWHM, and Q Factor, for Different Values of Grating Strength

kL	*Peak Reflection (%)*	*FWHM (pm)*	$Q = \lambda_B/FWHM$
0.1	1.0	142	10,942
0.3	8.5	146	10,641
0.5	21.4	154	10,087
0.7	36.5	166	9,356
1	58.0	189	8,207
1.5	81.9	235	6,587
2	92.9	286	5,425
3	99.0	384	4,034
4	99.9	479	3,239

an extremely small bandwidth compared to the central frequency. As in electronics and electromagnetics disciplines, it is common to observe the Q factor, or quality factor, defined as the pitch frequency divided by the full-width half maximum (FWHM) bandwidth, for an FBG the Q value is typically around 10,000, similar to microwave resonant cavities and much larger than electric filters.

As a design tool, we include in Table 5.1 the nominal performance of FBGs in the third optical window (centered at 1,550 nm), in terms of peak reflectivity, FWHM bandwidth, and Q factor. Also, in Appendix A, we include a MATLAB function *FBGgeneration.m* that returns the FBG reflection spectrum given the input parameters, in accordance to (5.1) through (5.4) CMT. Spectra in Figure 5.2 have been obtained for $L = 5$ mm, $n_{eff} = 1.5$, $\delta n_{eff} = 2 \times 10^{-6}$, $\lambda_B = 1{,}550$ nm, and maximum fringe visibility (=1).

In sensing applications and for the vast majority of the systems described in this book, the grating length ranges from 1 mm to 10 mm, and the reflectivity is within 10% and 98% (0.5 dB to 13 dB, counting in logarithmic units). In this range, the effect of sidelobes is almost negligible considering that their amplitude is very small, and their bandwidth is so narrow that it cannot be discriminated with a spectrometer or an OSA. Also, in this case the FBG spectrum can be approximated, with a good degree of reproduction, with a Gaussian function. Figure 5.3 shows the spectrum of an FBG fabricated at Politecnico di Torino as described in [7, 8]; the spectrum is measured with an OSA with 0.1-nm resolution bandwidth. The maximum amplitude is equal to 13.8 dB, corresponding to 95.8% reflectivity. Sidelobes are too small to be visualized with the OSA. The Gaussian fit well reproduces the FBG bandwidth, which is the most important part in sensing as it is used to estimate the pitch frequency (see Chapter 11), with a slight overestimation of the peak.

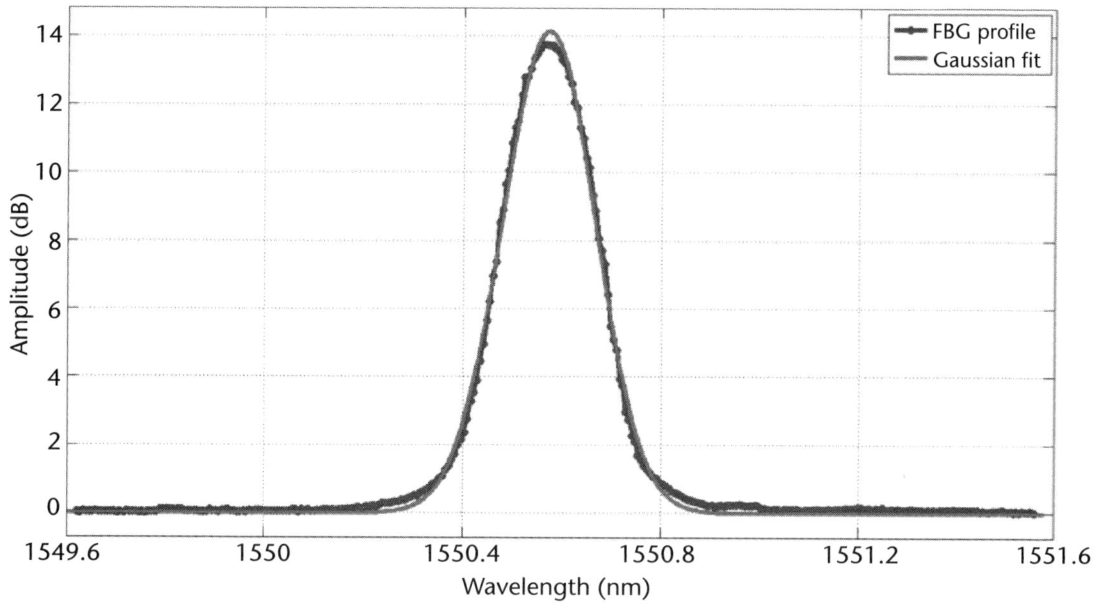

Figure 5.3 Reflection spectrum of an FBG inscribed at Politecnico di Torino using the phase mask technique [7, 8], measured with an optical spectrum analyzer, and its Gaussian fit.

5.3 FBG Thermal and Mechanical Response

The CMT provides an evaluation of the spectrum of an FBG, under static conditions (i.e., when its temperature is maintained constant and it is not subjected to mechanical strain). The underlying principle of FBG sensors relies in the variation of the FBG Bragg wavelength due to these two physical effects. In (5.1), both the modulation period and the refractive index have a dependence on temperature variation ΔT and strain variation $\Delta\varepsilon$. Strain is usually reported in microstrain relative units ($\mu\varepsilon$), whereas +1 $\mu\varepsilon$ corresponds to an elongation of 1 μm applied to a 1-m device length; negative strain units refer instead to compression. We observe a variation of the Bragg wavelength by the quantity λ_B, when a temperature/strain variation is applied, according to [5]:

$$\Delta\lambda_B = 2\left(\frac{\partial n_{eff}}{\partial T}\Lambda + \frac{\partial \Lambda}{\partial T} n_{eff}\right)\Delta T + 2\left(\frac{\partial n_{eff}}{\partial \varepsilon}\Lambda + \frac{\partial \Lambda}{\partial \varepsilon} n_{eff}\right)\Delta\varepsilon \tag{5.5}$$

where T is the temperature. For relatively small values of strain and temperature variations, as typically occurring in biomedical sensors which operate mainly in precision sensing, the variation $\Delta\lambda_B$ has an almost linear dependence on both ΔT and $\Delta\varepsilon$ on the whole sensing range [1]. In this case, it is common as in [9] to rewrite (5.5) highlighting the sensitivity terms:

$$\Delta\lambda_B(\Delta T, \varepsilon) = k_T \Delta T + k_\varepsilon \Delta\varepsilon \tag{5.6}$$

whereas the k_T [pm/°C] is the thermal sensitivity and k_ε [pm/$\mu\varepsilon$] is the strain sensitivity. The two sensitivity terms mainly depend on the wavelength of operation and the fiber structure; at 1,550 nm, typical values are $k_T \approx 10$ pm/°C and $k_\varepsilon \approx 1$ pm/$\mu\varepsilon$.

Figure 5.4 shows the shift of the Bragg wavelength for an FBG fabricated at Politecnico di Torino and subjected to tensile strain, within 0 and 1,500 $\mu\varepsilon$. The FBG has been positioned within two micropositioners, at the distance of 20 cm between each other; the grating is pretensioned, so that it is sensitive also to small values of strain. The micropositioner is tuned with ±0.1-μm accuracy, returning a strain calibration with 0.5-$\mu\varepsilon$ accuracy. As the strain increases, the FBG Bragg wavelength shifts from the initial value of 1,650.5 nm to 1,562.0 nm, according to an almost perfectly linear trend as in (5.6). The linearity of the curve is >0.999, over this range, and the linear fit shows a sensitivity of $k_\varepsilon = 1.08$ pm/$\mu\varepsilon$.

A similar consideration can be drawn for thermal variations. In Figure 5.5, the same FBG is calibrated for thermal variations ranging from –30°C to 20°C. For this characterization, the FBG has been enclosed in a climatic chamber (Angelantoni ACS Thermoclimatic Chamber), which maintains a controlled temperature and a constant relative humidity. Several ramps of temperature reproduce the cooling from ambient temperature (20°C) to low temperature (–30°C) and the reversed heating; each ramp has 1-hour duration, with 1-hour maintenance of temperature within each ramp. Inside the chamber, an LM35 fast-response chip sensor is used to obtain the reference temperature (±0.1°C). Figure 5.5 shows the entire spectrum, recorded with an OSA with a 0.1-nm resolution bandwidth, for each temperature value along the ramp. It is possible to notice that, as for the strain sensitivity, the thermal sensitivity is linear with a very high linearity coefficient, and the thermal sensitivity obtained is $k_T = 10.22$ pm/°C. It is also possible to notice that, as the Bragg wavelength changes, the entire spectrum shifts without distortions. This condition is always verified for small variations of temperature and strain, uniform on

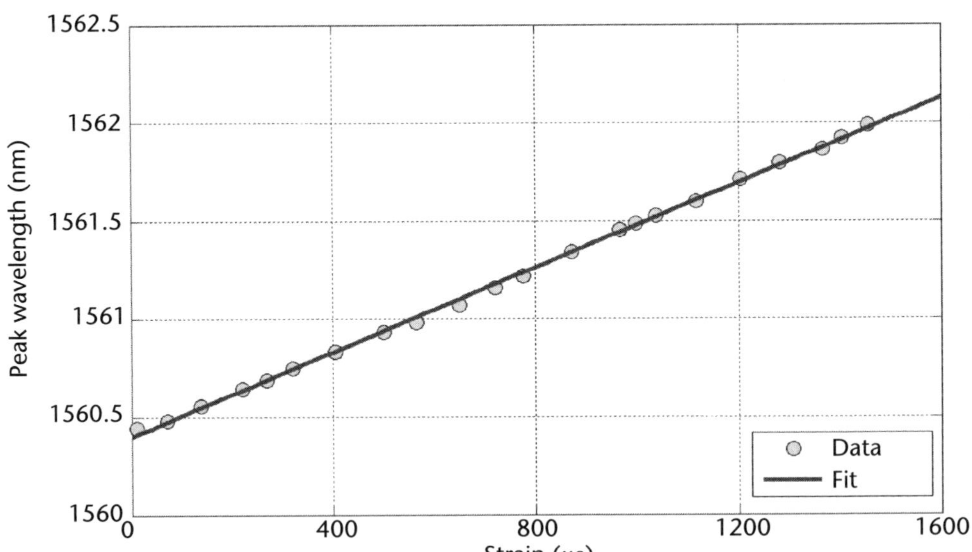

Figure 5.4 Strain sensitivity of an FBG; the chart reports the estimated Bragg wavelength, obtained with an OSA, as a function of the longitudinal strain applied to the FBG with micropositioners. Linear fit is used to estimate the sensitivity as 1.08 pm/$\mu\varepsilon$.

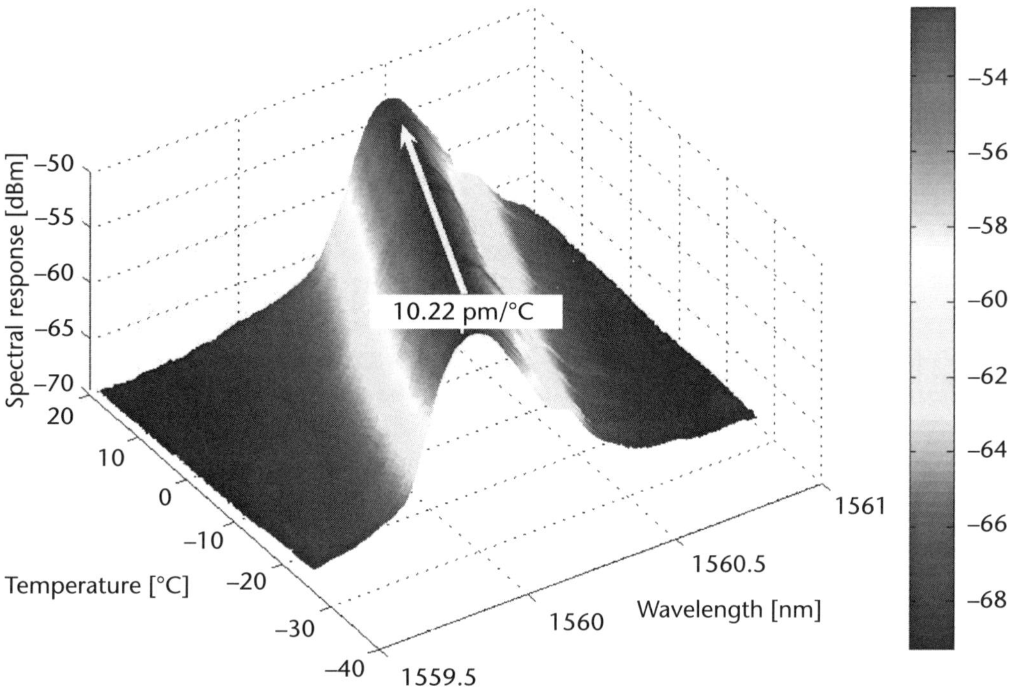

Figure 5.5 Temperature spectral response of an FBG; the chart reports the whole FBG spectrum, obtained with an OSA with 0.1-nm resolution bandwidth, as a function of the temperature variations applied on the FBG in a thermo-climatic chamber. Linear fit is used to estimate the sensitivity as 10.22 pm/°C.

the grating length; as in [1, 5], strain variations within 1,000 $\mu\varepsilon$ do not alter the spectrum in visible way. Instead, strong strain values (over 1% elongation of the grating) can trigger nonlinear behaviors, while strain values higher than 1,600 to 2,000 $\mu\varepsilon$ can introduce distortions of the FBG spectrum from its original shape. Similarly, within –40°C and 120°C, which covers most of the biomedical measurements, the spectral distortions are negligible.

Thus, FBG sensors evaluate a combination of strain and temperature. The most popular sensors, including the majority of commercial sensors, operate by detecting the reflection spectrum of the FBG, and estimate the Bragg wavelength. During the calibration, it is possible to obtain the k_ε and/or k_T coefficients, and therefore the relationship between strain/temperature and the readout Bragg wavelength shift, solving (5.6). The sensitivity values, for uniform FBGs inscribed on standard single-mode fibers or photosensitive fibers in the third optical window are always close to the aforementioned values, which can also be used as an approximate estimate; FBG manufacturers also often report the average sensitivity values for each batch of FBGs. Chapter 10 shows the systems (research and commercial based) that allow measuring the FBG spectral shift, and Chapter 11 (together with Appendix B) provides insights on the demodulation techniques for precision sensing (i.e., estimation of the wavelength shift $\Delta\lambda_B$ from the FBG spectrum).

The dual strain/temperature linear sensitivity is both a weakness and a strength of FBG sensors. In temperature-sensing applications, it is recommended to mount the FBGs in a loose format, such that they are not subjected to strain. However, strain-sensing applications that operate in nearly constant temperature do not require a specific compensation; the same consideration applies to detection of vibration or pulsed strain, such as in blood pressure sensing, whereas temperature variations are much slower than the rate of the strain under investigation, and therefore thermal effects are negligible. Applications that involve a dual strain and temperature variation are required to compensate thermal changes in order to operate the correct estimate. Thermal compensation can be performed through the addition of an additional FBG, mounted in a strain-insensitive form factor (such as a loose positioning), used to detect the only temperature, and subtract it from the strain-induced wavelength shift [1, 10]. In this case, we can exploit the fact that the relationships are perfectly linear, and practically hysteresis-free, on the operative range for any FBG, and therefore compensation can be easily implemented by estimating the wavelength shifts of the strain and temperature sensor and temperature-only sensor. In some contexts, particularly when temperature sensors are positioned in tissues under breathing conditions [11] that induce strain, it is necessary to perform the opposite compensation and remove the strain variations from a temperature reading. In absence of any compensation the detuning error is approximately 10 $\mu\varepsilon$/°C.

5.4 FBG Arrays

The FBG is a classic example of intrinsic sensor, inscribed inside an optical fiber, as in Figure 3.1; the effect of an FBG is such that it produces a strong backreflection around a central wavelength, slightly shifting as a result from physical parameter variations, while the structure is substantially transparent to all the other wavelengths where the reflectivity is close to zero. Most FBG interrogation systems operate on a 40-nm to 100-nm window, while the FBG notch has a bandwidth usually <0.5 nm. This suggests that it is possible to increase the efficiency of sensing systems, and therefore the sensing capacity, by inscribing multiple FBGs inside the same optical fiber, each having a different Bragg wavelength. This arrangement implements the multiplexed sensing configuration described in Figure 3.1: several sensors are fabricated along the same optical fiber, in different active points, each sensor having a different wavelength bandwidth and therefore with the capability of discriminating each sensor in the spectrum domain. This feature is extremely important in sensing networks and is a unique capability of FBGs, unlike most FOS and all MEMS technologies, as it allows functionalizing a single miniature optical fiber to several points of detection. Given the reflectivity $R_i(\lambda)$ of each ith element of the array, computed as in (5.4) in reference condition and perturbed as in (5.6) when strain/temperature is changed, and assuming that the coherency length of the system is much shorter than the distance between each FBG element, the reflectivity of an array of N FBGs can be expressed as the cascade of each individual element:

$$R_{array}(\lambda) = 1 - \prod_{i=1}^{N} \left[1 - R_i(\lambda)\right] \tag{5.7}$$

FBG arrays operate under the wavelength-division multiplexing (WDM) principle: each FBG has a different central wavelength, and the spacing between each

Bragg wavelength is such that, even when each FBG is subjected to strain and/or temperature variations, FBG spectra are sufficiently separated in wavelength such that each Bragg wavelength can be successfully discriminated and estimated. Chapter 10 describes the WDM-based systems applied to FBG sensor networks.

Whereas (5.7) applies to a generic FBG array, in sensing systems (particularly in biomedical engineering), FBG arrays are often designed in accordance to specific conditions. In a typical design in use in most systems, all FBGs are designed with identical characteristics (same length, strength, and reflectivity), and their wavelengths are equally spaced by the quantity $\delta\lambda$, that is, $\lambda_{B,i} = \lambda_{B,1} + i \cdot \delta\lambda$; in addition, each FBG composing the array has the same geometrical distance from the adjacent FBGs. In this arrangement, there is univocal correspondence between the physical location of the *i*th element of the array in the fiber, and the *i*th wavelength in the array spectrum; this is an extremely useful tool for multipoint sensing, whereas each FBG acts as a local sensor.

An example of this array form factor is shown in Figure 5.6: the array, fabricated by FBGS International (Belgium), consists of 5 FBGs having wavelengths 1,545.8 nm, 1,547.6 nm, 1,549.4 nm, 1,551.2 nm, and 1,553.0 nm (1.8-nm spacing between each wavelength). Each FBG, inscribed with the draw-tower technique, has 5-mm length, and the distance between each FBG center is 10 mm, thus leaving a 5-mm void region between each FBG. The gratings have 10% maximum reflectivity. The FBG array behaves as a set of temperature sensors, narrowly spaced between each other; due to the loose package, FBGs are not strained. As described in [12] and Chapter 17, the FBG is inserted in a phantom tissue and exposed to radiofrequency

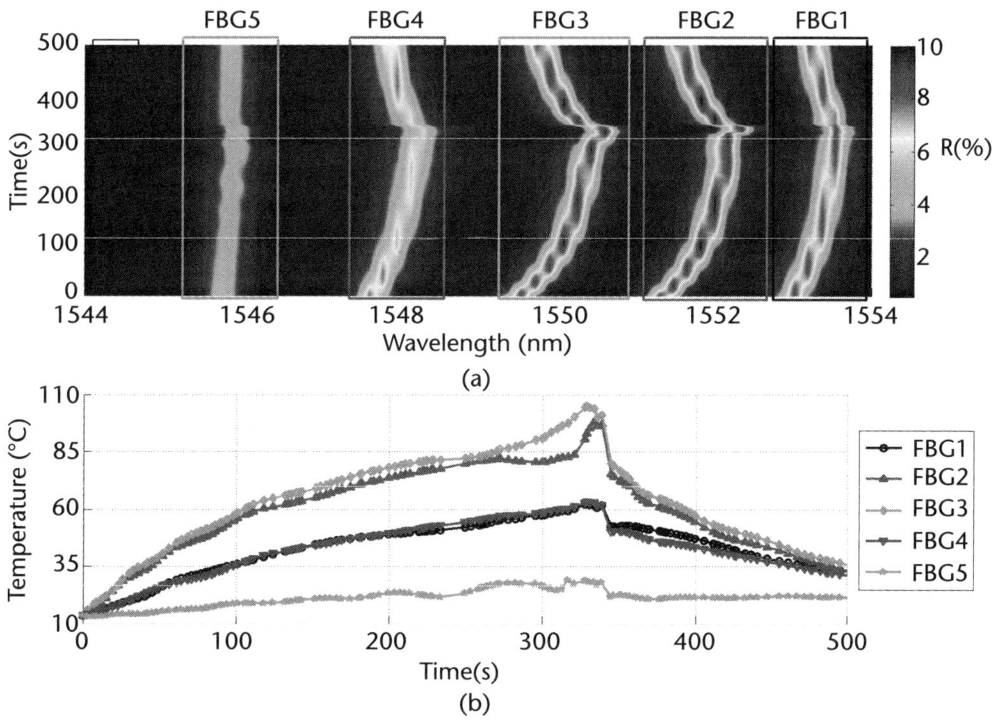

Figure 5.6 Behavior of an array of 5 FBGs (FBGS International, DTG-1550) exposed to an RF ablation-induced thermal pattern. (a) Temporal evolution of the spectrum of the FBG array, sampled with 1-second rate. (b) Temperature estimated for each FBG.

thermal ablation, which induces a temperature variation with strong spatial gradients; therefore, each FBG is subjected to a different temperature value. Figure 5.6(a) shows the evolution of the whole FBG array reflection spectrum, recorded with a spectrometer (Ibsen Photonics, I-MON-USB-512) over a 500-second time, consisting of an initial heating and subsequent cooling. All FBGs have sensitivity k_T = 10.2 pm/°C, and are labeled FBG1-5. The central FBG, positioned in correspondence of the temperature peak, is subjected to a strong wavelength shift, in excess of 1 nm, during the heating phase; the adjacent FBGs are also exposed to a temperature change, with a smaller shift, whereas the wavelength shift of FBG5 is the smallest. The chart shows the working principle of WDM: each individual FBG spectral component can be isolated from the overall FBG array spectrum, with no significant cross-interference. Figure 5.6(b) shows the temperature estimated by each FBG sensor, after separating the 5 reflection spectra and estimating each wavelength shift. Since the physical distance between each grating element in the fiber and the spectral distance between two Bragg wavelengths are both constant (10 mm and 1.8 nm, respectively), the configuration of Figure 5.6 can easily perform a thermal/strain mapping, which includes detection and localization of sensor data.

FBG arrays also are useful configurations for in-line temperature or strain compensation. An example of such arrangement is [11]. In this work, an array of FBGs serving as temperature sensors is perturbed by strain induced by breathing; to compensate, an additional sensor is inscribed in the array at a further location, such that it is close enough to be positioned in the same tissue and therefore be subject to a nearly similar strain pattern, but far enough to be unexposed to thermal variations. In general, FBG arrays with ad hoc design of grating location can perform a single compensation with two sensors, N-point sensing each with its compensation (with $2N$ gratings), or N-point sensing with a single compensation point ($N + 1$ gratings), the latter configuration being popular for biomedical sensors for in vivo use or smart textiles, whereas all FBGs are held at the same time-varying temperature [13]. Temperature or strain compensation is a specific case of generic multiparametric sensing techniques: in this case, Section 10.4 provides the general format for multiparametric detection.

5.5 Chirped and Apodized FBG

So far, we focused on uniform gratings, which have a modulation of the refractive index constant in amplitude and period. Special types of FBGs have nonconstant parameters throughout the FBG length, as described in depth in [4]. Apodized grating have nonconstant amplitude of the modulation pattern: rather than a perfect sine wave, the modulation can assume a Gaussian shape or a raised cosine profile. Apodized FBGs are often used to attenuate the sidelobes of the FBG or to flatten the central peak; however, their direct applications in biomedical engineering are marginal, also because low-cost spectrometers as in Figure 5.6 detect the FBG spectra on a coarsely sampled grid, and sidelobes are practically invisible. In addition, π-shifted FBGs exhibit a 180° phase shift in the modulation, usually at the center of the grating; the resulting spectrum is similar to Figure 5.2, but a narrow transmission hole is present at the center of the FBG. More interestingly, for medical applications, is the use of chirped FBGs, whereas the amplitude of the refractive index modulation is (ideally) constant, but the period is different in each portion of the grating [9, 14]. In other words, the grating period is not constant along z, but is a function of the position in the grating.

A formal description of chirped FBGs (CFBGs) that operate with broadband sources having short coherency length is in [9]. Since the modulation period is function of z, also the Bragg wavelength of each grating composing the cascade is different. Maintaining the previous notation for the Bragg wavelength, and adding the dependence on z, we define the chirp function $\phi(z)$ as:

$$\lambda_B(z) = \lambda_{B,0} + \phi(z)\big|_{0<z<L_g} \tag{5.8}$$

where $\lambda_{B,0}$ is the Bragg wavelength at $z = 0$. The most popular case is a linear chirp, which yield a linearly chirped FBG (LCFBG): in this case $\phi(z) = \zeta z$ and ζ, expressed in nm/mm, is the chirp rate coefficient which determines the spatial rate of change of the Bragg wavelength through the length of the grating. The LCFBG then behaves as a broadband reflector, with a nearly flat-top spectral response within $\lambda_{B,0}$ and $\lambda_{B,0} + \zeta L_g$. In order to analytically model the LCFBG, we can discretize the grating into a cascade of N uniform gratings, each having equal length $L = L_g/N$: typical N values are 100 to 500 [9]. Then we can apply the CMT to each discrete grating of length L, which will behave as a weak grating since the grating length is short and therefore the local kL coefficient has a small value. Thus, we can evaluate the nth grating reflectivity as in (5.4):

$$R_n(\lambda) = \frac{\sinh^2\left(L\sqrt{k^2 - {\sigma_n}^2}\right)}{\cosh^2\left(L\sqrt{k^2 - {\sigma_n}^2}\right) - \dfrac{{\sigma_n}^2}{k^2}} \tag{5.9}$$

assuming that k is constant in all sections of the grating (this condition is verified in flat-top LCFBGs). The wavelength-dependent term σ_n will have the following expression:

$$\sigma_n(\lambda) = \frac{\pi}{\lambda}\delta n_{eff} + 2\pi n_{eff}\left(\frac{1}{\lambda} - \frac{1}{\lambda_{B,n}}\right) \tag{5.10}$$

and the term in the right parenthesis will resonate at a different Bragg wavelength for each grating. As for an FBG array, we can evaluate the overall spectrum of the LCFBG as:

$$R_{CFBG}(\lambda) = \prod_{n=1}^{N}\left[1 - R_n(\lambda)\right] \tag{5.11}$$

Equations (5.9) through (5.11) can be applied to any CFBG, even when the chirp function is not linear, by simple discretization of the grating parameters (k, δn_{eff}, λ_B). Figure 5.7 shows some examples of LCFBG spectra simulated with 200

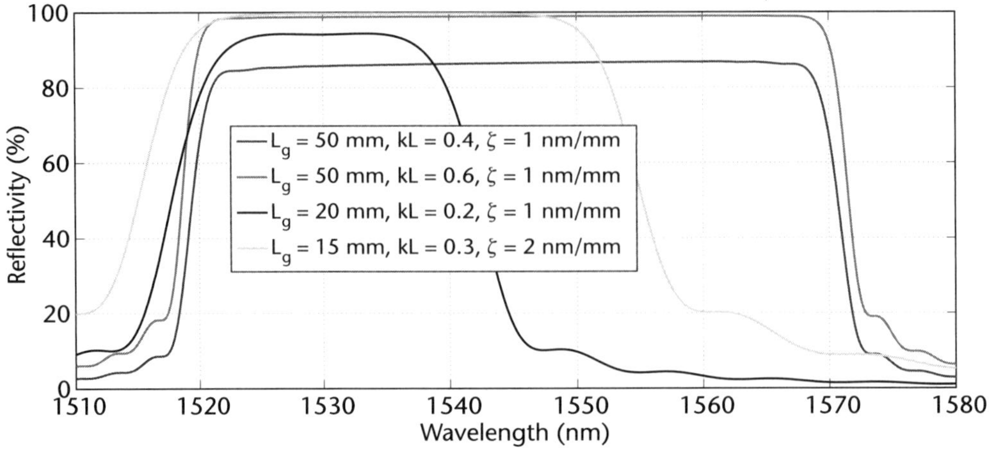

Figure 5.7 Examples of LCFBG spectra, simulated with different grating length Lg (15 to 50 nm), local grating strength kL (0.2 to 0.6), and chirp coefficient ζ (1 to 2 nm/mm). All simulations are accounted for λ_B,0 = 1,520 nm, N = 200, n_{eff} = 1.5, δneff = 10–5.

discretization steps, as well as (5.8) through (5.11). In this ideal condition, the LCFBG behaves as a flat-top reflector, with the bandwidth dominated by the chirp rate, and reflectivity depending on the grating strength. The value of kL determines the maximum reflectivity, achieved throughout the flat-top region. The bandwidth rapidly drops at the sides of the grating.

The sensing principle of an LCFBG is more advanced than standard FBGs; while the sensitivity coefficients k_T and k_ε are the same of standard FBGs, the LCFBG spectrum is sensitive to the temperature and strain values throughout the entire grating length. In case of a spatially uniform variation of strain and temperature, that is, $\Delta\varepsilon(z) = \Delta\varepsilon$ and $\Delta T(z) = \Delta T$, the LCFBG exhibit a spectral shift towards the higher wavelengths for positive variations, and to lower wavelengths for the opposite trend. When variations are not spatially uniform, but exhibit a gradient [9, 14], each nth element of the cascade is subjected to a different wavelength shift, and the LCFBG spectrum results from the perturbations introduced in each element of the cascade. If we assume that strain and temperature can be discretized on the same N-point grid, from $z = 0$ to $z = L_g$, we can express the nth wavelength shift as

$$\Delta\lambda_{B,n}\left(\Delta T_n, \varepsilon_n\right) = k_T \Delta T_n + k_\varepsilon \Delta\varepsilon_n \tag{5.12}$$

where $\Delta\varepsilon_n$ and ΔT_n refer to the nth value of temperature and strain. Usually, LCFBG has been used in biomedical engineering, since [14], mainly as temperature sensor, therefore it is assumed that the strain is negligible. It is then important to observe how an LCFBG spectrum changes, when subjected to different and nonuniform temperature distributions. Figure 5.8 shows the effect of temperature variations on an LCFBG spectrum due to different temperature profiles, applied to the first grating shown in Figure 5.7 (length 5 cm, bandwidth 50 nm). In case of a linear profile Figure 5.8(a–c), with temperature rising from the left to the right of the grating, we observe a progressive enlargement of the bandwidth [4]. The variation of the flat-top level is slight, and in a realistic system it is hard to detect due to noise.

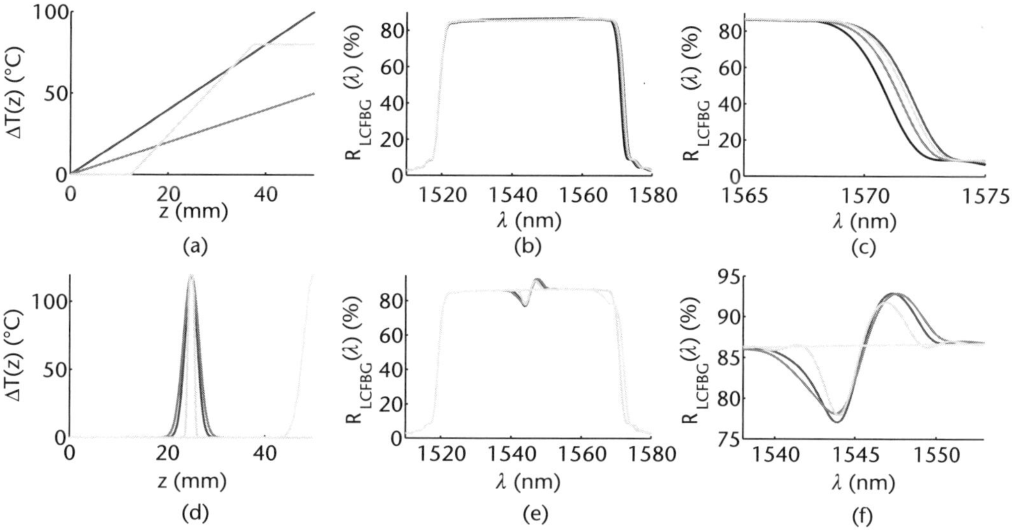

Figure 5.8 Behavior of an LCFBG when affected by nonuniform temperature gradients. (a) Three examples of linear and piecewise linear gradients are considered. (b) The resulting spectra computed with CMT are shown as well as the reference spectrum in black. (c) An inset on the edge bandwidth is shown. (d) Four examples of narrow Gaussian and super-Gaussian temperature pattern are shown, reproducing patterns similar to thermal ablation profiles. (e) The corresponding simulated spectra are shown. (f) An inset on the center of the spectrum is shown. Each pattern in (a–c) and (d–f) is reproduced in the same grayscale.

When a piecewise linear temperature is applied, mimicking, for example, thermal transitions during thermal treatments, the effect is observed both on the grating flat-top and on the bandwidth, depending on the edges of the transition. In case of a linear variation, a successful algorithm to estimate the temperature pattern takes into account the enlargement (or restriction, in case of a negative slope) of the grating bandwidth, precisely measured with an OSA [15]. A more modern application of LCFBG, introduced by [14], is for thermal pattern identification during thermal ablation (TA). The effect of different temperature profiles mimicking TA patterns is shown in Figure 5.8(d–f). In the pioneering work [14], the sensor is positioned such that the temperature profile is monotonic, growing from the left to the right edge of the LCFBG (rightmost curve); the resulting spectrum is nearly unperturbed on the left side of the LCFBG, while there is a shift on the right part of the grating with a visible smoothing of the spectrum. In the most recent works, in order to detect the temperature through the whole tissue exposed to hyperthermia, we aim at detecting temperature profiles similar to Gaussian or super-Gaussian patterns, in accordance to models [16] that show that similar patterns are observed in real time through the tissue. In case of a Gaussian or super-Gaussian pattern, of different widths, applied at the center of the LCFBG [9], we observe a variation of the spectrum in proximity of the center: since some wavelengths right-shift more than other ones, the resulting effect as in Figure 5.8(f) is a two-sided ripple of width and shape dependent on the width of the Gaussian function and super-Gaussian order. In this case, the demodulation, that is, the estimation of the temperature profile from the LCFBG spectrum is harder, and a preliminary iterative model for Gaussian and super-Gaussian function was presented in [9].

In Appendix A, we provide a MATLAB script (*LCFBG.m*) that simulates the behavior of an LCFBG in reference conditions, and when a temperature variation spatially nonuniform is applied. Users can design the grating, or adapt it from manufacturer parameters, by setting the grating length, optical parameters, strength, and bandwidth. The grating is discretized in a cascade of N gratings, each analyzed with the previous CMT function.

Figure 5.9 reports, as in [14], the behavior of an LCFBG when exposed to a thermal ablation pattern. The grating is a commercial device (Technica S.A., fabricated on SMF28 fiber), having $L_g = 15$ mm, $\zeta = 2.22$ nm/mm, $k_T = 10.2$ pm/°C, and maximum reflectivity 94.5%. The chart reports the grating spectrum, sampled every 3 seconds, with a spectrometer (Ibsen I-MON-512). At first, it is possible to observe that the grating does not have a flat-top profile, but a 2.0-dB ripple is present in the center of the spectrum; this is due to the tolerances in manufacturing and to a possible nonuniformity of the δn_{eff} term within the inscription of the FBG. This ripple can be artificially corrected by inserting a digital equalizer. The positioning of the grating ensures that the right side is heated more than the left side, thus with a positive-slope gradient. The resulting spectrum exhibits a progressive shift of the right edge of the bandwidth, whereas the left side results in a much smaller shift. When cooling down, the right edge of the grating reduces its bandwidth, reversing the process. The algorithm shown in [14] has been used to reconstruct the thermal map (time – space – temperature) from the spectra in Figure 5.9 and can be used for similar tasks.

Compared to standard FBG, CFBGs are harder to be used due to their broadband spectrum; indeed, they have been mainly used in mechanical applications [15] to measure and localize a strain differential concentrated in one point. The emergence of hyperthermia treatments and microsurgery in medicine requires sensors capable to measure a temperature (hyperthermia) or strain (biomechanical) distribution, with submillimeter resolution [12] and CFBG start to appear as a promising technology for this task.

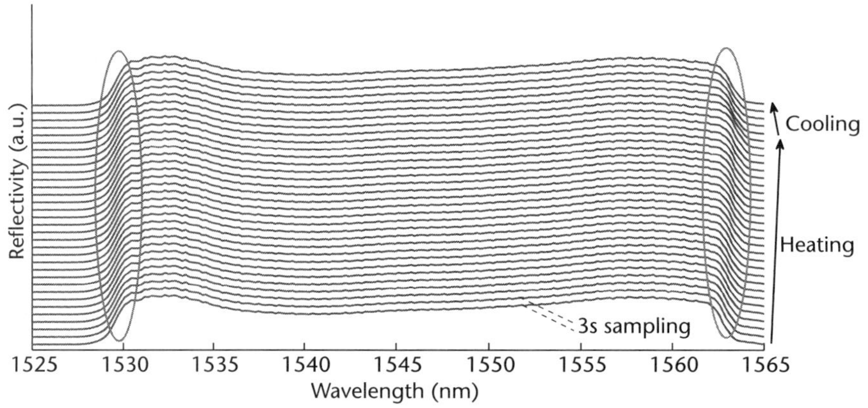

Figure 5.9 Evolution of the spectrum of a commercial LCFBG (Technica S.A., 1,529.3–1,562.7 nm), when exposed to thermal ablation, having strong heating in correspondence of the right edge of the grating. The chart reports the LCFBG spectrum, sampled every 3 seconds by a spectrometer, during heating and cooling, from the bottom to the top of the chart. Ellipses highlight the grating edges, having slight variations (left) and strong variation (right) of the spectrum.

5.6 Tilted FBG

Standard FBGs, in accordance with the CMT framework, form resonant structures that create an interference pattern between waves forward-propagating and back-propagating in the fiber core; the fact that waves travel in the inner core ensures that the corresponding mode is guided by the fiber, and this property applies to both transmitted and reflected fields. A tilted FBG (TFBG), instead, combines the standard characteristics of the uniform FBG (including thermal and strain sensitivity), but is also able to excite cladding modes that are not transmitted through the grating structure, but are rather evanescent in the fiber [17].

In a TFBG, like in a standard FBG, a modulation of the refractive index with uniform period and amplitude is inscribed into the fiber core, using the same manufacturing techniques described in the following section. However, rather than imposing a modulation parallel to the fiber propagation plane z, the refractive index modulation is purposely inscribed with an angle θ (tilt angle), measured between z and the refractive index modulation axis. Typically, θ varies from 2° to 20° and the properties of the grating vary as a function of the tilt angle. The spectral properties of TFBG are resumed in Figure 5.10, which is based on a CMT simulator described in [18, 19], and available also for MATLAB [20].

The TFBG spectrum results as a combination of the core mode, ghost mode, and cladding modes. The core mode is identified as the resonant mode with the longest wavelength and has the broadest bandwidth; it maintains the strain and temperature sensitivity of uniform FBGs, with the same sensitivity values. The ghost mode, instead, is inherently induced in a TFBG resulting in the overlap of low-order cladding modes. Ghost mode is usually observed at a wavelength about 2 nm shorter than the Bragg wavelength and can appear as either a single resonance or a comb of resonances, depending on the tilt angle. Ghost modes have been used mainly for mechanical sensing and recently demonstrated for rotation detection [21], exploiting their marginal dependence on the rotation imposed to the fiber. A growing importance for biosensors is posed on cladding modes; these are further

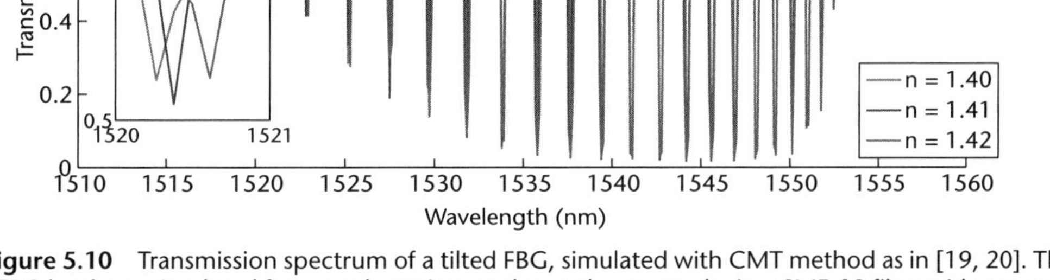

Figure 5.10 Transmission spectrum of a tilted FBG, simulated with CMT method as in [19, 20]. The TFBG has been simulated for a weak grating, and a weak core mode, in a SMF-28 fiber without outer jacket, at a tilt angle $\theta = 5°$ for different values of external refractive index n ranging from 1.40 to 1.42. The inset highlights the dependence of one cladding mode on refractive index.

modes, usually of a higher order, that resonate at decreasing wavelength and appear in the leftmost part of the transmission spectrum. Unlike the previous modes, cladding mode amplitude changes significantly when the boundary between the cladding and the outer environment changes. By removing the fiber jacket around the TFBG, and exposing the fiber cladding surrounding the active region to liquids with different values of refractive index, the amplitude (and in a minor form, the resonance condition) of several cladding modes changes, giving the access to refractive index sensing.

The transmission spectrum of a TFBG, simulated as in [19] for a standard single-mode fiber (diameters 10/125 μm for core/cladding), is shown in Figure 5.10 and results in a combination of the three types of resonance. In this case, both the core mode and the ghost mode appear as weak resonance, whereas a plurality of cladding modes is well visible. The cladding modes, due to their dependence on the outer boundary, vary with the external reflective index, yielding slight spectral shifts and/or variations of the peak amplitude. The reflection spectrum does not exhibit cladding modes, as they are forward but not backward propagated [17].

The tilt angle θ, which can be set during the manufacturing process, strongly controls the spectral profile of the grating: increasing the tilt angle progressively promotes the interactions occurring between core and cladding, as well as those at the outer boundaries. As a main reference, [17] showed that the cladding modes extend for about 20 nm for $\theta = 2°$, 40 nm for $\theta = 5°$, 60 nm for $\theta = 10°$, 100 nm for $\theta = 15°$, and 140 nm for $\theta = 20°$; effectively, the spectrum depends also on the design including size of core and cladding, core ellipticity (which may facilitate the recoupling of higher-order modes), external refractive index, and other variables. For refractive index sensing, it is a common choice to choose θ within 8° and 10°; this ensures a sensitivity of cladding modes to refractive index in the order of 1.33, which corresponds to water and therefore is suitable for measuring aqueous solutions [22].

An emerging application for TFBG that has been recently disclosed is for curvature sensing. Feng et al. [21] reported a compact sensor, formed by a pair of TFBGs inscribed on orthogonal planes, that is used for shape sensing [i.e., detection of curvature radius in the two orthogonal directions (x and y)]. The setup is shown in Figure 5.11. The two TFBGs operate at different wavelength ranges (1,550 to 1,569 nm for TFBG1 and 1,570 to 1,589 for TFBG2). The fiber is polished at the fiber end; hence, the system is capable to detect the Fresnel reflection. The architecture is described in Figure 5.11(a): TFBGs are weakly tilted (2°) making compact devices and are inscribed at 3-mm distance on the orthogonal directions. Each TFBG has a 10-mm active length. The system analyzes the ghost modes, as shown in Figure 5.11(b): the amplitude of the main ghost mode exhibits the highest variations to the curvature radius, with sensitivity ranging from 0.21 dB/m^{-1} to 0.33 dB/m^{-1}. Since the gratings are inscribed in orthogonal planes, it is possible to discriminate the bending in the directions orthogonal to the fiber plane, whereas in the axial direction z, the core mode of both TFBGs is sensitive to strain by 1.06 pm/$\mu\varepsilon$. The dual TFBG system acts as a compact three-dimensional shape sensor, suitable for biomechanics applications.

TFBGs are an excellent architecture for sensing, since they merge the properties of FBGs (strain, temperature sensitivity) to specific sensing features related to cladding modes' amplitude and distribution, enabling refractive index and bending

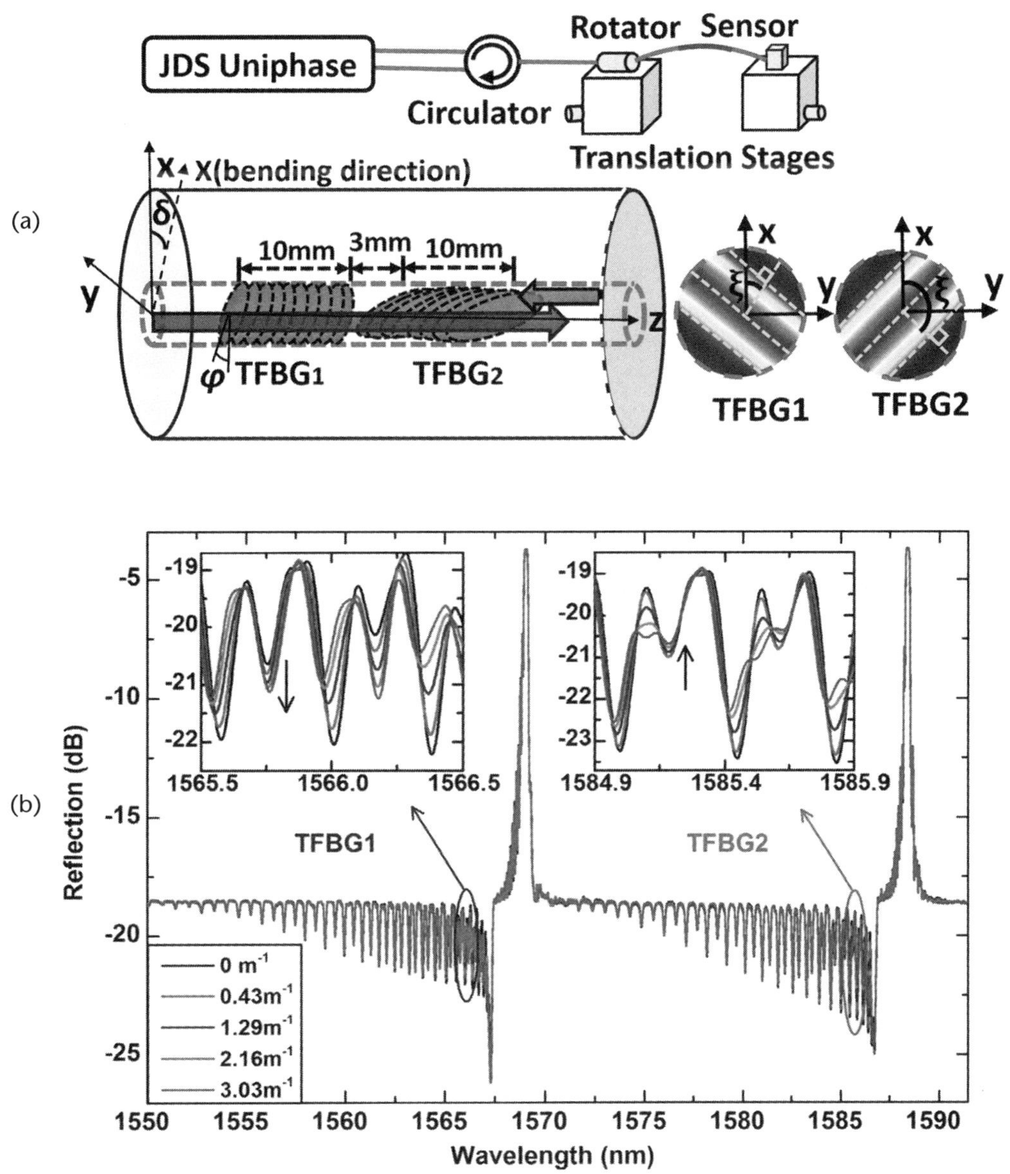

Figure 5.11 Dual-TFBG configuration reported by Feng et al. for shape sensing. (a) Setup and description of the orthogonal TFBG architecture. (b) TFBG spectra for different curvature radii. (Picture adapted from [21].)

detection [22]. Recent works have emphasized the latter aspects, and TFBG sensors [23] start to appear in biomedical engineering mainly for biodetection. There is a very recent research trend in combining TFBG sensors with surface plasmon resonance (SPR) sensors, forming a TFBG-SPR in which the plasmonic resonance modulates the amplitude of few cladding modes of the TFBG [24]; these systems and their biological application such as aptamer-based sensors are outlined in Chapter 18. It is likely that the potential of TFBG sensors is still far away. From the application standpoint, however, TFBGs have three major weaknesses. At first, most of the advantageous properties rely on the transmission spectrum, whereas the reflection spectrum is relatively similar to standard FBGs. This complicates the design of endoscopic or percutaneous sensors, which require microsized catheters that do now allow an optical fiber to travel in and out the catheter (the bending would be excessive even for bend-insensitive fibers). This drawback can be mitigated by inserting a highly reflective and perfectly polished mirror at the fiber end, such as a gold-coating, which is able to reflect also the transmission spectrum; however, this arrangement partially loses the low-cost characteristic of the grating and requires the reflector to have a non-negligible distance from the active region. Alternatively, we can exploit the baseline 4% at the fiber end, but only if the fiber is perfectly polished, a quality hard to maintain in a catheterized device. Guo et al. [17] suggested other alternatives to recouple the ghost modes in reflection. The second disadvantage is that cladding modes appear as tight resonances, narrower than the FBG spectrum: typical Q values are ~10^5, opposite of the ~10^4 of FBG peaks. This means that TFBGs are effectively invisible under low-cost spectrometers and FBG interrogators, which are the preferred unit for FBG and hybrid sensors thanks to their low cost and fast response, but require interrogators based on tunable lasers or OSA, which have a slower response (1 Hz or slower) and cost 6–20 times higher. Finally, as the TFBG has a broadband response, it cannot be easily combined with other WDM sensors, and unlike CFBG (which measure a distribution, therefore inherently a multipoint response), they are single-point sensors. It is envisioned that future research will address the realization of multipoint TFBG sensors based on a single device, making a more efficient use of the TFBG cladding-mode spectral resonances (both in amplitude and shift) and to develop signal processing techniques with a high efficiency, such as those described in Sections 11.3 and 11.4, which are able to operate even when cladding modes are not clearly visible on the spectrum, using different yet reliable metrics. To date, recent scientific literature on TFBG [22] has mainly been concerned with the sensing principle of TFBGs (and spectra reported on OSA instruments are acceptable only for scientific, yet not operative, purpose); thus, the algorithmic limits of detection have yet to be discovered.

5.7 Fabrication of FBGs

Fabrication of an FBG implies inscribing the desired modulation of the effective refractive index into an optical fiber [25]. To date, four different methods for FBG fabrications have been reported and are compared in Table 5.2.

5.7.1 Photosensitivity

Photosensitivity refers to the permanent change of the refractive index of an optical fiber, when exposed to light with specific wavelengths (usually in the UV, ultraviolet, range) and high intensity [5, 26]: it is the underlying process of FBG fabrication. For FBG fabrication, usually solid-state UV lasers or KrF excimer lasers are used, either continuous wave or pulsed. Exposure to strong and focused UV light yields a permanent change of the refractive index due to the damages induced in the fiber at atomic level. Standard single-mode fibers exhibit insufficient levels of photosensitivity to inscribe strong gratings. For this reason, Ge-doped photosensitive fibers are used. These fibers are designed to have high levels of germanium doping and

Table 5.2 Comparative review of FBG fabrication techniques

Method	*Key Approach*	*Industrial?*	*Strengths*	*Weaknesses*	*Maintains Jacket?*	*FBG Arrays*	*Maintains Fiber Jacket?*	*CFBG and TFBG*
Phase mask	Phase mask encodes diffraction	Yes	Repeatability, volume/cost, performance	Flexibility, fiber jacket, needs many masks	No	No or coarse	No	Yes (CFBG with ad-hoc phase mask)
Interfer-ometric	Focused beam split and combine	Yes	Flexibility, fiber selection, performance	Repeatability, hard to develop, fiber jacket	No	Yes	No	Yes (depending on software)
Draw-tower	Write FBG during fiber drawing	Yes	Fiber jacket, dense FBG arrays, cost	FBG perform., flexibility, hard to set up	Yes	Yes	Yes	No
Point by point	Directly focused laser	No	Fiber jacket, flexibility, multi-sensors	FBG spectrum, repeatability, efficiency	Possible	Yes	Yes (ideally)	Not yet

are index-matched to standard fibers; thus, they represent a good platform for FBG inscription. Alternatively, hydrogenation (or hydrogen loading) can be used to raise the photosensitivity as demonstrated by [27]: fibers are soaked in a hydrogen gas chamber, at 20°C to 75°C, at high pressure (150 atm typically), resulting in a high concentration of hydrogen that promotes photosensitivity. Hydrogenation allows fabricating gratings on virtually any fiber.

5.7.2 Phase Mask Inscription

The phase mask inscription is the most popular technique for FBG inscription and was successfully demonstrated in [28]. The phase mask is a thin silica diffractive element fabricated holographically or with electron-beam lithography; when exposed to incident UV light, it cancels the 0-order diffraction (perpendicular to the beam) and promotes the ±1-order diffraction components (around 70% of the total UV power is confined in the ±1-diffraction). The phase mask surface is corrugated with a periodic pattern (period Λ_{pm}) in the fiber direction z. This pattern allows, when exposed to the UV light, one to produce a near-field interference fringe, with maxima and minima repeated at a periodicity of $\Lambda = \Lambda_{pm}/2$, in the optical fiber, forming an FBG with Bragg wavelength $2n_{eff}\Lambda = n_{eff}\Lambda_{pm}$; for this reason, FBGs around 1,550-nm use phase mask with period slightly higher than 1 μm.

As the phase mask modulates the refractive index, inscription is performed by scanning with the UV laser the phase mask length (total or partial), transferring the modulation to the fiber core; a phase-mask setup, arranged at Politecnico di Torino, is sketched in Figure 5.12 and contains all element for control. The light source used for inscription is a UV laser, argon-ion frequency-doubled to 244 nm (Coherent, Innova 300 Fred) with 0–20-A driving current; UV light is coupled in an air path consisting of steering mirror, a beam expander, and a cylindrical astigmatic lens. A two-dimensional xy translation stage is used to perform the scanning: the photosensitive fiber (after stripping) is positioned within a V-groove located on the top of the translation stage, and the phase mask is laid on the V-groove in close contact. During inscription, the fluorescence power is monitored with a power

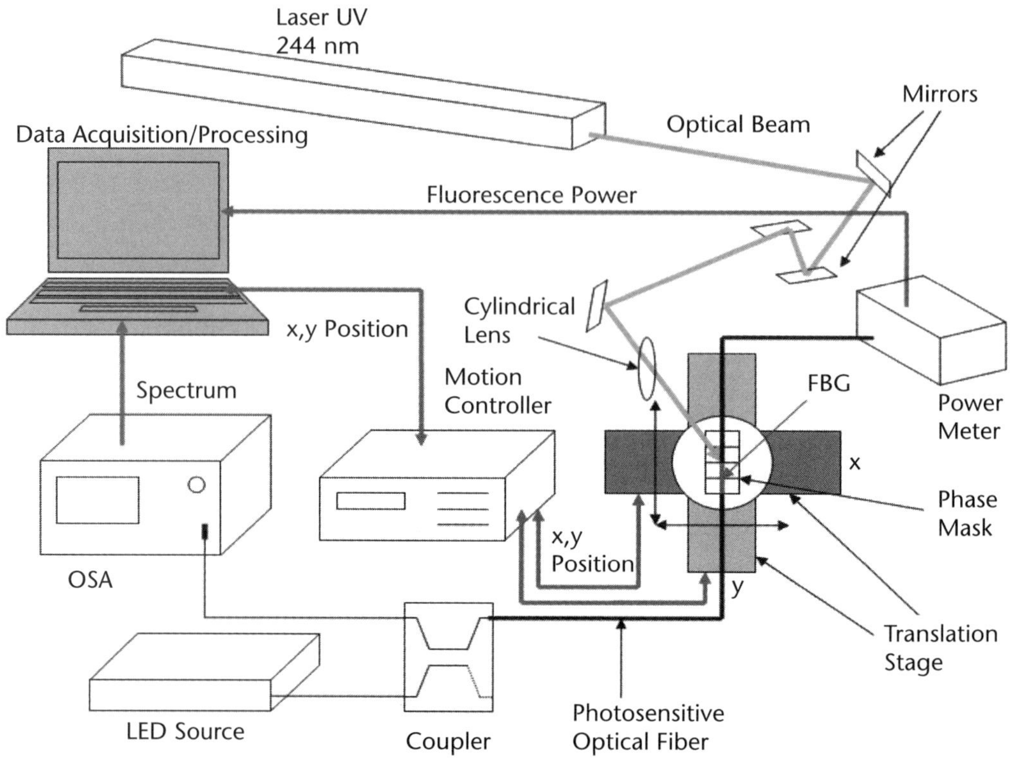

Figure 5.12 Phase mask setup for FBG inscription, developed at Politecnico di Torino.

meter having Si photodiode, in order to isolate the third window spectrum; this value is used as an efficiency marker. When the UV light is focused on the fiber core, the fluorescent power is maximum. The *xy* translation is controlled by a motion controller (Newport stepper motor) via GPIB in real time, with 0.1-μm resolution. Along the fiber axis (x), the motion speed is set to constant value, typically 10 μm/s; on the perpendicular axis (y), the stage is slightly adjusted in order to maintain the perfect UV focusing, maximizing the photosensitive power detected. During inscription, the spectrum of the FBG is monitored with an OSA, using a broadband LED source as input and a 50/50 coupler. The setup is completed by data acquisition software developed in LabVIEW; the software controls the motion controller in real time and reads the data from the OSA and fluorescence power meter.

The phase mask technique allows controlling the FBG strength and, within a short range, its Bragg wavelength. Figure 5.13 shows the evolution of an FBG spectrum during inscription, showing the Bragg wavelength (left) and the peak amplitude in decibels (right) measured each minute. During the first 50 minutes of the inscription process, the FBG changes significantly its properties: due to the growing modulation of the refractive index, the Bragg wavelength grows from the initial value of 1,559.86 nm to 1,560.43 nm, and then it stabilizes at 1,560.4 nm. Meanwhile, the FBG amplitude grows rapidly to 10 dB after 19 minutes and then smoothly to 15 dB after 40 minutes and stabilizes at 16 dB without any further change.

With the phase mask technique, the refractive index modulation is entirely encoded in the phase mask. It is possible to inscribe several gratings with the same

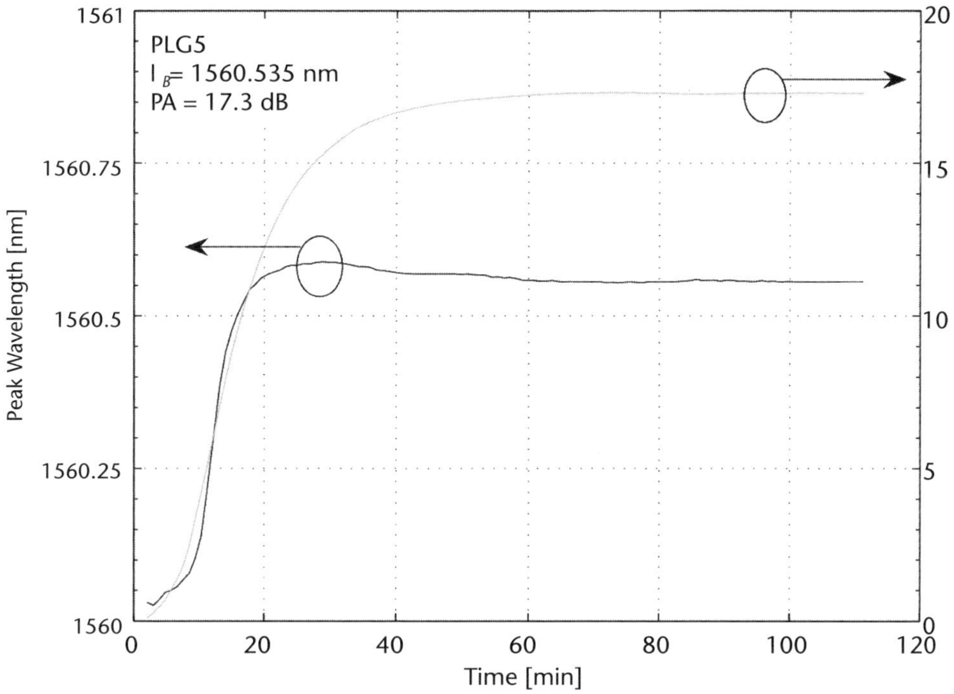

Figure 5.13 Evolution of the FBG parameters during inscription.

Bragg wavelength and same index pattern, using one phase mask; this feature makes this technique attractive to fabricate individual gratings, all having the same spectrum. Phase mask can also be chirped, for CFBG inscription, or it is possible to use a rotation stage [17] to tilt the phase mask at the angle θ and fabricate TFBGs. Commercially, this technique has been extensively used and leads to the lowest cost per single FBG, providing good repeatability even in large volume manufacturing. Its main weaknesses are: the lack of flexibility, as one phase mask can only inscribe one pattern; thus, one different phase mask is needed for each Bragg wavelength; the need to remove the fiber protective jacket, which weakens the fiber tensile strength (although packages in acrylate and polyimide are effective in recoating the FBG); and the difficulty to fabricate dense FBG arrays, since each FBG is fabricated as a single element (an array of FBG is often fabricated by splicing several FBGs, with critical limitations to the compactness of the array).

5.7.3 Interfering Beams

The setup based on interfering beams, described in [29], makes use of a free-space interferometer to encode the modulation pattern into the FBG. The UV laser source is split, by means of a 50/50 free-space beam splitter, into two arms. The two beams are separately propagated in air, steered with a mirror at the incidence angle θ_i, and focused by means of a pair of cylindrical lenses into the fiber at the FBG inscription point. Using a low-coherency source such as an excimer laser for the inscription, and assuming that both mirrors are aligned to the incidence angle θ_i, this structure produces an interference fringe in the fiber with period $\Lambda = \lambda_L/(2\ \sin\theta_i)$, where λ_L

is the excimer laser wavelength [5]. Moving the fiber with a translational stage ensures that the desired length of the FBG is inscribed.

FBG fabrication through interfering beams is harder to set up and maintain, compared to phase mask inscription, because it requires a fine control of the mirror orientation, and it strongly depends on the tolerances on each component (beam splitter, mirrors, translation stages), making the setup hard to be established and easily detuned. In addition, it is necessary to ensure that both beams, propagated in air, are perfectly focused in the fiber core at a known angle. Modern interferometric setups are developed with the support of ad hoc LabVIEW software, which controls the laser input or a shutter (to activate and deactivate the laser source), mirror orientation, fiber translation stage, and performance indicators (such as fluorescence power); also, they are able to set a target angle θ_i to control the FBG wavelength. Flexibility is the main asset of this method: virtually, any grating with different modulation pattern and Bragg wavelength can be fabricated, including FBG arrays and CFBG, provided that the software to control the interfering beams is effective in delivering the correct incidence angle to the fiber; it can also work with fibers with different thickness, adjusting the focus of the cylindrical lenses. In terms of performance, it is harder to achieve the same grating strength of phase mask method, because of focusing errors. FBG repeatability, in serial manufacturing, is much lower since slight variations of the beam control results in a significant change of Bragg wavelength. This method also requires removing the fiber protective jacket prior to fabricate the FBG. By controlling the inscription software or using a rotation stage, it is also possible to fabricate CFBG and TFBG, respectively.

5.7.4 Draw-Tower Method

The fabrication of FBG using the drawing tower has been industrialized by few companies, including FBGS International DTG products (Draw-Tower Grating); the methodology is sketched in Figure 5.12. The underlying principle is to fabricate the FBG arrays altogether with the fiber hosting the devices, starting from the glass preform. This approach can be implemented either using an interferometric setup or phase masks; interferometric setup is preferred for dense arrays of gratings. This method has two key advantages compared to previous methods. First, it is extremely suitable to fabricate FBG arrays and all-grating fibers (fibers with a long series of grating, all at the same wavelength). The complexity needed to fabricate a single FBG and an array of FBGs, with each sensor at a specific active location, is similar: thanks to the synchronization between the UV laser used to inscribe the FBGs, and the fiber speed through the drawing tower, it is possible to fabricate FBG arrays with equal spacing (such as Figure 5.6), series of arrays, or other combinations, with high versatility and throughput. The second key advantage is that the final FBG or array has the original protective coating of the optical fibers, and therefore its tensile strength is preserved: as seen in Figure 5.14, the FBG inscription takes place before the ultimate coating process. The process can also be implemented on several fiber compounds and thicknesses, by controlling the drawing tower; this feature allows, for example, using bend-insensitive fibers or thinner fibers (80-μm diameter). Performance-wise, to date, this process produces relatively weak gratings, and arrays have a minimum spacing (FBG-to-FBG distance) of 1 cm

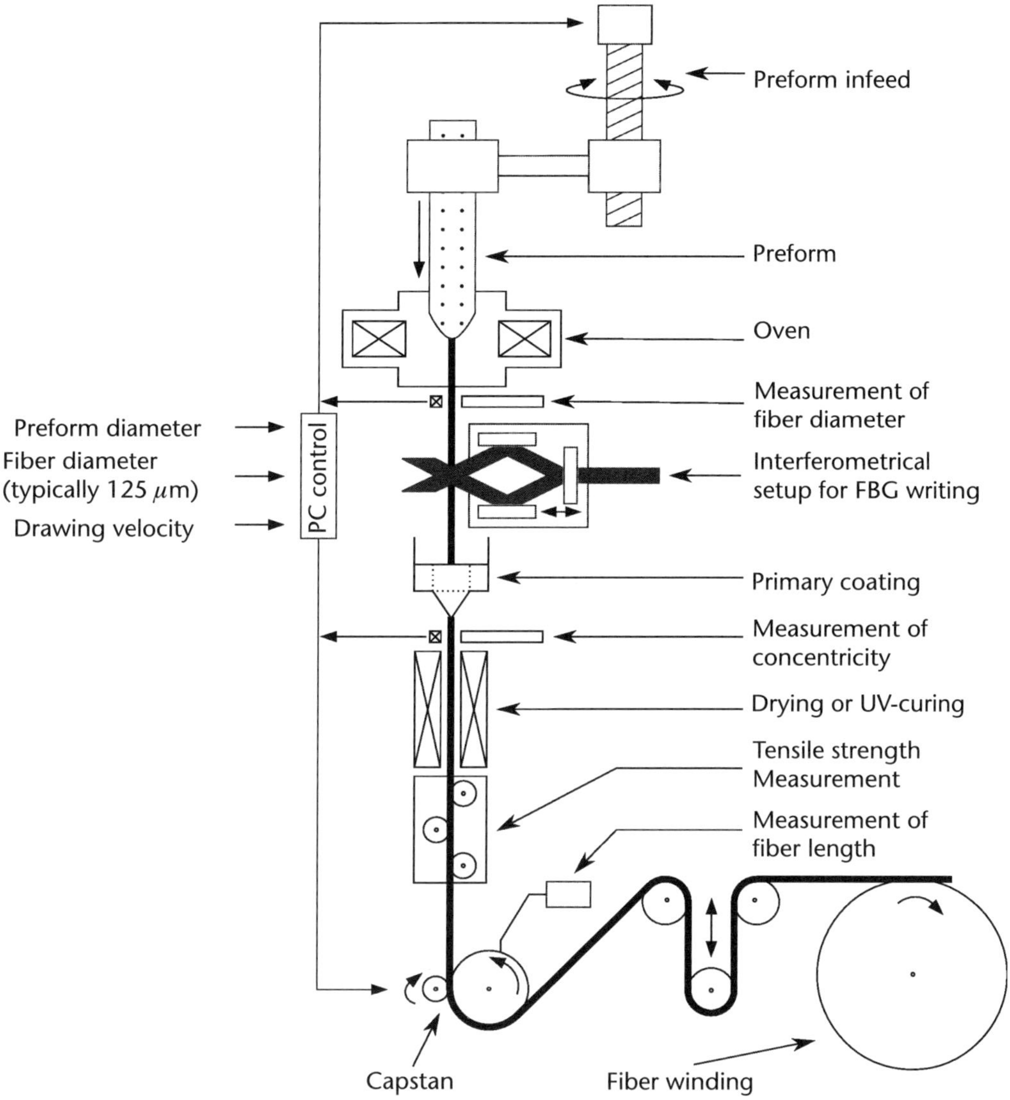

Figure 5.14 Schematic of a draw-tower fabrication technique for FBG arrays, industrialized by FBGS International (Belgium). (Courtesy of FBGS. Picture from www.fbgs.com.)

and are limited to uniform FBGs; it is likely that technological updates can improve current limitations in the near future.

5.7.5 Point-by-Point

The point-by-point technique was initially demonstrated in the 1990s and has strongly improved in the recent years, thanks to the advent of highly focused femtosecond laser (FSL) systems. An FSL source, capable of delivering pulses with strong energy, is focused into the fiber core with high precision; focusing is often achieved through index-matching oils. Then inscription is performed by rapidly scanning the laser across the fiber core. The method developed by Cyprus Institute of Technology [30] has proven to be the current state of the art and has been applied to different typologies of fibers, to FBG arrays [31], and to multisensors [32]. In this methodology, the FSL (HighQ femtoREGEN IC 355) is focused through an air-path composed of beam expander, collimator, and steering lenses with a high-resolution camera, on the fiber that is positioned on a three-stage translator. The laser pulse energy is 10 μJ with 300-fs width, at 1,064 nm, with a repetition rate of 1 kHz [32]. The three-stage controller performs a continuous scan, either following a linear motion along the fiber axis, or performing a saw-tooth pattern. The narrow duration of the pulse ensures that the permanent index change in the core remains confined, whereas the scanning speed determines the Bragg wavelength.

Inscription with an FSL is a recent achievement, and its performances are currently low compared to other techniques, both in terms of reflectivity capability of avoiding core-cladding interference. The results in [31, 32] showed that FBGs fabricated with this method have a spectrum that exhibits strong sidelobes, with negative implications for WDM and Bragg wavelength estimation. However, this technique has key assets, provided that the focusing of the FSL through the several layers is achieved. It is possible to fabricate an FBG in direct contact (or even overlapping) another sensor, as demonstrated in [32] when an FBG sensor has been fabricated over an EFPI sensor and later in [31] where an array of 7 FBGs was fabricated. Also, it is possible to inscribe FBG in innovative fiber shapes. Virtually, FSL laser can inscribe, through the fiber jacket, very short FBGs with sufficient strength and densely packed FBG arrays; these results will hopefully be soon achieved, pending the focusing improvements and the better control of the translation unit.

5.8 FBG Package

The FBG package plays a significant role in sensing, particularly for thermal detection, as the package acts as a measurement artifact (see Figure 5.15). FBGs inscribed with the draw-tower or the direct inscription methods maintain the original package of the optical fiber, which is the best solution in terms of tensile strength and

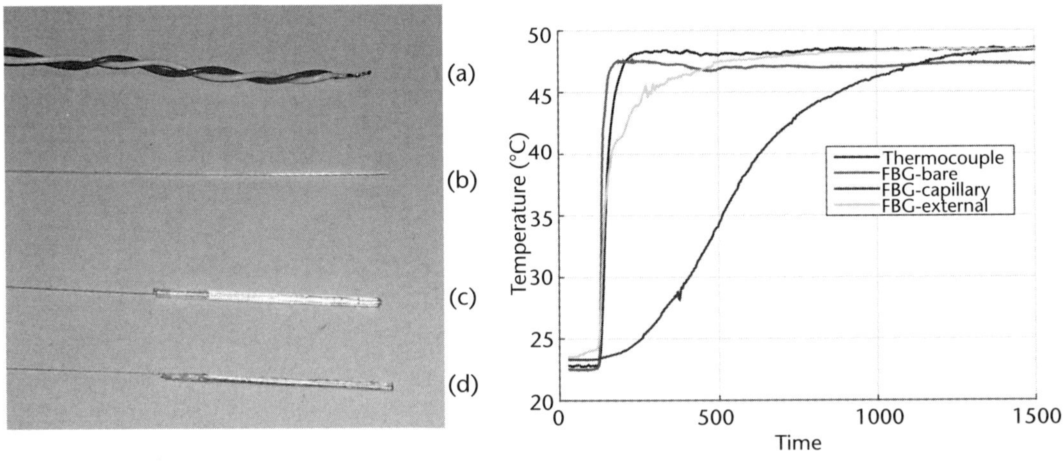

Figure 5.15 FBG packages for thermal detection. Left chart shows the picture of different packages: (a) micro thermocouple, (b) bare fiber, (c) capillary, and (d) fiber glue to capillary. Right chart shows the response time to a steep thermal variation.

thickness for biomedical sensors. When the fiber jacket is removed, a package is often applied to protect the fiber and preserve its sensitivity. Acrylate coating is the most popular solution for commercial gratings; layers of external thickness of 125 to 250 μm can be used to protect the fiber within the FBG length. The main drawback of acrylate is its poor resistance to high temperatures which can be an issue in thermotherapy monitoring. For strain sensors, a popular approach is a polyimide coating: the FBG is sandwiched between two layers of polyimide tape or paste, usually with ~1-cm thickness. This mount can be used for external sensors, such as tactile or smart textiles.

Micropackages suitable for catheters require a further investigation. In [33], several FBG packages are analyzed, compared with a microthermocouple. A bare FBG has the smallest transient time, but its fragile form factor requires a catheter that negates any axial strain; otherwise, the FBG can easily break due to strain exposure. The FBG package in a capillary is also investigated: in this case, a quartz capillary is used to enclose the FBG through its surface, protecting it from perturbations. This is an epoxy-free design that guarantees a good resistance and a minimum impact on the thermal sensitivity (–10%); its main effect is the slow response time when exposed to thermal variations (around 1.3 seconds), which may prevent fast detection. The third design is based on a capillary, by introducing the fiber inside the capillary tube, and then the gap between the fiber and the inner walls of the tube has been filled with epoxy resin. This experimental design maintains good mechanical properties and a fast response time (~0.4 second); future design will consider the use of a UV-curing biocompatible epoxy.

5.9 Recent Advances

Few notable recent achievements may soon impact FBG sensors in biomedical engineering:

- FBGs are sensitive to both strain and temperature, and this is an issue in many applications, which requires a compensation that is often unpractical. Van Roosbroeck et al. [34] demonstrated a technique to demodulate both strain and temperature on a single FBG, through the inscription of draw-tower gratings on a highly-birefringent polarization maintaining (PM) fiber developed at IPHT (Jena, Germany). The resulting FBG spectrum is composed by two, not one, main lobes, with a spectral separation of ~0.5 nm, due to the effect of the dual polarizations. By monitoring the spectral distance change between the two lobes, a sensitivity of 0.77 pm/°C has been obtained, whereas the FBG maintains the same k_ε coefficient.
- The University of Mons demonstrated in 2016 for the first time the sensitivity of FBGs to the magnetic field [35]. The analysis is based on the Poincare sphere, considering the FBG in transmission mode, and results in 0–1-T sensing range. Albeit pioneering, the work paves the road to optical sensors combined with MRI and other hybrid sensor-imaging applications.

References

[1] Kersey, A. D., et al., "Fiber Grating Sensors," *Journal of Lightwave Technology*, Vol. 15, No. 8, August 1, 1997, pp. 1442–1463.

[2] Liu, Z., and H. Y. Tam, "Industrial and Medical Applications of Fiber Bragg Gratings," *Chinese Optics Letters*, Vol. 14, No. 12, December 2016, p. 120007.

[3] Abushagur, A. A., et al., "Advances in Bio-Tactile Sensors for Minimally Invasive Surgery Using the Fibre Bragg Grating Force Sensor Technique: A Survey," *Sensors*, Vol. 14, No. 4, April 9, 2014, pp. 6633–6665.

[4] Erdogan, T., "Fiber Grating Spectra," *Journal of Lightwave Technology*, Vol. 15, No. 8, August 1997, pp. 1277–1294.

[5] Othonos, A., and K. Kalli, *Fiber Bragg Gratings: Fundamentals and Applications in Telecommunications and Sensing*, Norwood, MA: Artech House, 1999.

[6] Skaar, J., L. Wang, and T. Erdogan, "On the Synthesis of Fiber Bragg Gratings by Layer Peeling," *IEEE Journal of Quantum Electronics*, Vol. 37, No. 2, February 2001, pp. 165–173.

[7] Tosi, D., M. Olivero, and G. Perrone, "A Low-Cost Fiber Bragg Grating Vibro-Acoustic Sensor for Voice and Heartbeat Detection," *Applied Optics*, Vol. 47, 2008, pp. 5123–5129.

[8] Tosi, D., M. Olivero, and G. Perrone, "Dynamic Strain Measurement System with Fiber Bragg Gratings and Noise Mitigation Techniques," *Measurement Science and Technology*, Vol. 20, 2009.

[9] Tosi, D., et al., "Towards Inline Spatially Resolved Temperature Sensing in Thermal Ablation with Chirped Fiber Bragg Grating," *IEEE International Symposium on Medical Measurements and Applications (MeMeA)*, Benevento, Italy, May 15–18, 2016.

[10] Zhou, Z., and J. Ou, "Techniques of Temperature Compensation for FBG Strain Sensors Used in Long-Term Structural Monitoring," *International Society for Optics and Photonics Microtechnologies for the New Millennium 2005*, June 2005, pp. 167–172.

[11] Cavaiola, C., et al., "Error Caused by Respiratory Movements on a Probe for Distributed Temperature Monitoring During Cancer Ablation: Ex Vivo and In Vivo Analysis," *IEEE Sensors Journal*, Vol. 16, No. 15, 2016, pp. 5934–5941.

[12] Tosi, D., E. G. Macchi, and A. Cigada, "Fiber-Optic Temperature and Pressure Sensors Applied to Radiofrequency Thermal Ablation in Liver Phantom: Methodology and Experimental Measurements," *Journal of Sensors*, Vol. 2015, 2015, p. 909012.

[13] Ciocchetti, M., et al., "Smart Textile Based on Fiber Bragg Grating Sensors for Respiratory Monitoring: Design and Preliminary Trials," *Biosensors*, Vol. 5, No. 3, 2015, pp. 602–615.

[14] Tosi, D., et al., "Fiber-Optic Chirped FBG for Distributed Thermal Monitoring of Ex-Vivo Radiofrequency Ablation of Liver," *Biomedical Optics Express*, Vol. 5, No. 6, 2014, pp. 1799–1811.

[15] Okabe, Y., R. Tsuji, and N. Takeda, "Application of Chirped Fiber Bragg Grating Sensors for Identification of Crack Locations in Composites," *Composites Part A: Applied Science and Manufacturing*. Vol. 35, No. 1, January 31, 2004, pp. 59–65.

[16] Saccomandi, P., E. Schena, and S. Silvestri, "Techniques for Temperature Monitoring During Laser-Induced Thermotherapy: An Overview," *Int. J. Hyperthermia*, Vol. 29, No. 7, 2013, pp. 609–619.

[17] Guo, T., et al., "Tilted Fiber Grating Mechanical and Biochemical Sensors," *Optics & Laser Technology*, Vol. 78, 2016, pp. 19–33.

[18] Lu, Y. -C., et al., "Improved Full-Vector Finite-Difference Complex Mode Solver for Optical Waveguides of Circular Symmetry," *IEEE/OSA J. Lightwave Technol.*, Vol. 26, No. 13, 2008, pp. 1868–1876.

[19] Lu, Y. -C., W. -P. Huang, and S. -S. Jian, "Full Vector Complex Coupled Mode Theory for Tilted Fiber Gratings OSA," *Opt. Express*, Vol. 18, No. 2, 2010, pp. 713–726.

[20] Lu, Y. -C., "Fiber Grating Solver," MATLAB, Mathworks, 2014, https://cn.mathworks.com/matlabcentral/fileexchange/45809-fiber-grating-solver.

[21] Feng, D., et al., "Compact Optical Fiber 3D Shape Sensor Based on a Pair of Orthogonal Tilted Fiber Bragg Gratings," *Sci. Rep.*, Vol. 5, 2015, p. 17415.

[22] Albert, J., L. Y. Shao, and C. Caucheteur, "Tilted Fiber Bragg Grating Sensors," *Laser & Photonics Reviews*, Vol. 7, No. 1, January 1, 2013, pp. 83–108.

[23] Guo, T., et al., "Reflective Tilted Fiber Bragg Grating Refractometer Based on Strong Cladding to Core Recoupling," *Opt. Expr.*, Vol. 17, No. 7, 2009, pp. 5736–5742.

[24] Shevchenko, Y. Y., and J. Albert, "Plasmon Resonances in Gold-Coated Tilted Fiber Bragg Gratings," *Optics Letters*, Vol. 32, No. 3, February 1, 2007, pp. 211–213.

[25] Hill, K. O., and G. Meltz, "Fiber Bragg Grating Technology Fundamentals and Overview," *IEEE J. Lightw. Technol.*, Vol. 15, No. 8, 1997, pp. 1263–1276.

[26] Hill, K. O., et al., "Photosensitivity in Optical Fiber Waveguides: Application to Reflection Filter Fabrication," *Appl. Phys. Lett.*, Vol. 32, 1978, pp. 647–649.

[27] Lemaire, P. J., et al., "High-Pressure H_2 Loading as a Technique for Achieving Ultrahigh UV Photosensitivity and Thermal Sensitivity in GeO_2 Doped Optical Fibres," *Electron. Lett.*, Vol. 29, 1993, pp. 1191–1193.

[28] Hill, K. O., et al., "Bragg Gratings Fabricated in Monomode Photosensitive Optical Fiber by UV Exposure Through a Phase Mask," *Appl. Phys. Lett.*, Vol. 62, 1993, pp. 1035–1037.

[29] Meltz, G., W. W. Morey, and W. H. Glenn, "Formation of Bragg Gratings in Optical Fibers by a Transverse Holographic Method," *Opt. Lett.*, Vol. 14, 1989, pp. 823–825.

[30] Lacraz, A., et al., "Femtosecond Laser Inscribed Bragg Gratings in Low Loss CYTOP Polymer Optical Fiber," *IEEE Photon. Technol. Lett.*, Vol. 27, No. 7, April 1, 2015, pp. 693–696.

[31] Theodosiou, A., et al., "Modified fs-Laser Inscribed FBG Array for Rapid Mode Shape Capture of Free-Free Vibrating Beams," *IEEE Photon. Technol. Lett.*, Vol. 28, No. 14, 2016, pp. 1509–1512.

[32] Poeggel, S., et al., "Femtosecond-Laser Based Inscription Technique for Post Fibre-Bragg Grating Inscription in an Extrinsic Fabry-Perot Interferometer Pressure Sensor," *IEEE Sensors Journal*, Vol. 16, No. 10, 2016, pp. 3396–3402.

[33] Chen, W., et al., "Performance Assessment of FBG Temperature Sensors for Laser Ablation of Tumors," *2015 IEEE International Symposium on Medical Measurements and Applications (MeMeA)*, May 7, 2015, pp. 324–328.

[34] Van Roosbroeck, J., et al., "Stretching the Limits for the Decoupling of Strain and Temperature with FBG Based Sensors," *International Society for Optics and Photonics International Conference on Optical Fibre Sensors (OFS24)*, September 28, 2015, pp. 96343S–96343S.

[35] Descamps, F., et al., "Magnetic Field Sensing Using Standard Uniform FBGs," *Optics Express*, Vol. 24, No. 23, 2016, pp. 26152–26160.

CHAPTER 6

Distributed Sensors

6.1 Introduction

The sensing systems described so far are based on modulation of amplitude or frequency: it implies that a sensor, based on a fiber device or a simple optical fiber span, physically located in an active area, alters the propagating light waves either by attenuating their amplitude, or by filtering their spectra. FBGs, in particular, thanks to their small length and narrow spectrum, enable dense WDM and allow the realization of sensor networks based on a plurality of sensors arranged on the same fiber: multiplexing can become dense, both spectrally (stacking of several sensors in a narrow wavelength range) and physically (short physical distance between each sensing element). Chirped FBGs extend this concept: the behavior of the device is influenced by the sensing parameter changes across the whole device length, which implies that we can read a distribution of temperature or strain, rather than a set of localized values. The CFBG device still has a short active length (1 to 5 cm), because it is able to detect only in the region exposed its refractive index change, but within this length we could ideally detect the variation of temperature or strain with infinitely low spatial resolution, only limited by the technological impairments and hardware/software.

Distributed sensors extend this concept to the next level [1]. In this case, there is no physically confined active region, but the whole optical fiber used to propagate light waves behaves as the sensing element [2]. Whereas different technologies for distributed sensing are available, they all have in common the grounding aspects:

- The sensing element is the fiber itself, or a continuously long fiber device. Any fiber reflects a very weak fraction of the input light, due to Rayleigh backscattering [1]; the amount of backscattered light due to Rayleigh scattering is in the order of one millionth of the input power. Other scattering phenomena, such as Raman or Brillouin scattering, can also be used. Recently, in order to increase the amount of backreflected light, it has become a common practice to fabricated continuously reflective devices such as all-grating-fibers [3], a cascade of weak reflectors (0.001% to 0.1% of reflectivity) inscribed on the whole fiber length.

- The sensing principle is weak or moderate reflection, and is always based on a pulsed light emission. The propagating waves are reflected in each part of the fiber, and therefore we detect a continuous set of reflection components.
- Since each component travels different distance in the fiber, from the source forward to the reflecting element and backwards to the detector, the round-trip time (RTT) for each component of the pulse backreflected is different for each reflecting element; thus, the RTT allows discriminating the spatial dependence of the sensing parameters.
- Performances are strongly tied to a trade-off among spatial resolution, accuracy, and measurement time. Spatial resolution, in particular, is defined as the minimum distance between two locations that can be unambiguously resolved by the instrument.

Distributed sensors are a well-established technology for long-haul sensing, with tens of kilometers of sensing region and 10–100-cm resolution, when based on optical time-domain reflectometry (OTDR) [1]. Hardware limitations still prevent us from abating this limit: the RTT in a glass optical fiber is 10 ns for a length of 1m, which corresponds to a data acquisition and processing hardware operating at 100-MHz speed; to achieve 1-mm space resolution, we would need 10-GHz or faster hardware, which is an aggregate capacity available only for supercomputers (which are, however, expensive, cumbersome, and hardly usable as a sensing unit). However, systems operated with optical frequency-domain reflectometry (OFDR) nowadays can reach the centimeter resolution down to submillimeter and start to have significant impact on biomedical engineering: distributed sensors have been demonstrated in hyperdense thermal monitoring by Macchi et al. [4] and shape sensing [5], and most significantly a distributed shape-sensing system has been industrialized for robotic surgery, as described in Section 19.4. In this chapter, we will focus on the technologies suitable for biomedical engineering, and their implementation: coherent OFDR, microwave photonics (MWP), and their sensing elements.

6.2 OFDR Theory

The principle of operation of OFDR relies on Fourier transform of complex spectral envelopes and was initially developed in 1998 in a theoretical paper by Froggatt and Moore [6]; to date, it finds the best implementation in the Luna Technologies class of instruments (also described in Chapter 10, and compared in Table 12.2). The key principle of operation is that a change $\Delta\varepsilon$ of the permittivity ε of the optical fiber dielectric material, distributed along the fiber propagation axis z, is tied to the Fourier transform of the backscattered signal; permittivity is related to the refractive index n as $n = \varepsilon^{1/2}$.

6.2.1 Demonstration

In a 1996 technical note [7], Froggatt shows that the standard wave equation for the electric field, well known in electromagnetics theory:

$$\frac{\partial^2 E}{\partial z^2} + \beta^2 E = 0 \tag{6.1}$$

turns into

$$\frac{\partial^2 E}{\partial z^2} + \beta^2 \left[1 + \frac{\Delta\varepsilon(z)}{\varepsilon}\right] E = 0 \tag{6.2}$$

in presence of the perturbation of the permittivity $\Delta\varepsilon(z)$ distributed over the propagation axis; E is the electric field, and $\beta = 2\pi/\lambda$ is the wavenumber in the waveguide (λ being the wavelength). In an optical fiber, the electric field can be decomposed into the forward propagating component (E_F), and the backward propagating term (E_B). The forward propagating component, in distributed sensors, acts as the feeding signal, while the backward propagating waves correspond to the backscattered components, which are reflected back towards the generator by effect of scattering [1, 8]. In the formula:

$$E = E_F e^{j\beta z} + E_B(z, \beta) e^{-j\beta z} \tag{6.3}$$

Scattering phenomena are characterized by an extremely small reflectivity: therefore, if the laser source coherency length is longer than the measurement length and fiber attenuations are negligible (as always occurring in OFDR), the forward wave E_F is substantially not affected by the permittivity change, and is z-independent, and therefore its derivative is null. A further reduction can be applied: in case of slow-varying waves, we can approximate $\partial^2 E_B/\partial z^2 \cong 0$ and therefore limit our analysis to the first derivative. Thus, replacing (6.3) into (6.2) and solving for $\partial E_B/\partial z$, we obtain:

$$\frac{\partial E_B}{\partial z} \cong \frac{\beta}{2j} \frac{\Delta\varepsilon(z)}{\varepsilon} \left[E_F e^{j2\beta z} + E_B(z, \beta)\right] \simeq \frac{\beta}{2j} \frac{\Delta\varepsilon(z)}{\varepsilon} E_F e^{j2\beta z} \tag{6.4}$$

In (6.4), we apply the inequality $|E_B| << |E_F|$: in this case only we are able to remove the E_B term on the right side of the expression, and directly relate the rate of change of the scattering signal E_B to the forward propagating wave and the permittivity change. In fact, this is the key assumption of the OFDR, which consists of weak to very weak reflective phenomena. Now we can integrate (6.4) on the axis z:

$$E_B = E_F \frac{\beta}{2j} \int_{z=-\infty}^{+\infty} \frac{\Delta\varepsilon(z)}{\varepsilon} e^{j2\beta z} dz \tag{6.5}$$

This step shows that the (complex) amplitude of the backscattered wave is proportional to the inverse Fourier transform (IFT) of the function $\Delta\varepsilon(z)/\varepsilon$, evaluated at the wavenumber 2β; and we can rewrite it as

$$E_B = E_F \frac{\beta}{2j} \Im^{-1}\left[\frac{\Delta\varepsilon(z)}{\varepsilon}\right](2\beta) \tag{6.6}$$

The relationship (6.6) is the demonstration of the OFDR: we relate the (inverse) Fourier transform of the local permittivity change in the optical fiber medium, evaluated at twice the wavenumber (half the wavelength) of the feed signal E_F to the detected backscattered field E_B.

6.2.2 Detection

The implementation of the OFDR principle is complex and was formally described in [6]. While (6.6) shows the analog components of the field, in reality with a coherent detector we are able to detect a signal proportional to $E_B E_B{}^*$, with * being the complex conjugate operator. Coherent detection is required in order to evaluate both the power spectral density of the backscattered wave, and its phase; indeed, modern OFDR systems are substantially the extension of the optical vector analyzer (OVA) to the Fourier analysis. The hardware of Luna Technology optical backscatter meter (OBR 4600) has a very similar front end to the OVA manufactured by the same company (http://lunainc.com/ova), used to characterize components in terms of losses and group delay, and represents the unanimous state of the art. Most importantly, the detector does not integrate over the full wavenumber (wavelength) range, but only within a spectral range λ_1 to λ_2 and the Fourier transform condenses into a discrete Fourier transform, algorithmically implemented with the fast Fourier transform (FFT).

The detection problem cannot be solved with simple algebra, but [6] identified a closed-form solution in a special case, when a lumped reflector with reflection coefficient ρ and infinitely short length is positioned at the location $z = z_0$. In this case, the reflected wave is the overlap of the backscattering signal and the localized reflection, which has amplitude (ρ is known) and phase (z_0 is known) identifiable a priori. The measurement device is a swept wavelength interferometer that allows obtaining for each wavenumber (wavelength) the complex optical spectrum of the backscattered signal. In this condition, [6] showed, by a complex analytical solution, that the integral of $E_B E_B{}^*$, evaluated in a range of wavenumbers $\Delta\beta$ centered at β_0 is proportional to the fluctuation of dielectric permittivity $\Delta\varepsilon$ centered around the point z_0; in formulas:

$$\int_{-\Delta\beta/2}^{+\Delta\beta/2} P_d(\beta - \beta_0)e^{-j\beta x} d\beta \cong \frac{E_F^2 r c \pi \beta_0}{j n_{eff}} \Delta\varepsilon\left(z_0 + \frac{x}{2}\right) \tag{6.7}$$

where P_d is the detected power spectral density, $P_d = E_B E_B{}^*$; the 2β term in (6.6) is substantially converted in a $x/2$ term in this formula, as the IFT is applied to twice the wavenumber (half the wavelength) at which the permittivity change is applied.

From the hardware perspective, (6.7) shows the computational task that is performed. The fluctuations of permittivity are detected through the IFT of the signal at the photodetector. A convenient choice is also to select $z_0 = 0$ and use the small reflectance ($r \sim 0.001$) of a fiber-optic connector leading out of the instrument to obtain the right-side expression of (6.7) [9, 10].

6.3 Microwave Photonics

Microwave photonics (MWP) working principle is similar to OFDR: several backreflected components travelling with different RTT collected by a detector. The main differences are that MWP operates with incoherent detection, that is, does detect $|E_B|^2$ and not $E_B E_B^*$, and that backreflected components are not due to fluctuations of the permittivity (refractive index), but are generated by reflective elements such as FBGs. Modern MWP configurations decode the backreflected signals in the microwave domain, using a vector network analyzer (VNA). The work proposed by Ricchiuti et al. [11] described the current state of the art of MWP system used to interrogate a cascade of reflectors, which can be either weak FBGs arranged in WDM [12], all-grating fiber [13], or other configurations. The working principle is to convert the set of reflections being generated by each reflective element, each travelling a different length in the fiber, into a microwave filter.

Consider the case of a generic reflector: the output of a continuous-wave (CW) light source is electro-optically modulated with an electrical signal. The modulated light is input to the sensing array, which reflects a fraction of the travelling light at every distance. From the signal processing point of view, this is equivalent to have a continuous of replicas of the input signal, one for each value of fiber length, and each one multiplied by its own weight coefficient that corresponds to the reflectivity at each distance. However, since we work with a light source with a coherency length L_c correspondent with a coherency time t_c, and with an incoherent detection, we will not see a continuous of reflections, but rather a discrete set of N reflections each with a different propagation time T_n and a different amplitude a_n. Thus, as well explained in [11], the set of reflective elements in a fiber behaves, formally and experimentally, as a digital filter of N taps, each having amplitude a_n and progressive delay T_n. For a digital filter, we can write its transfer function $H(f)$, where f is the frequency:

$$H(f) = \sum_{n=0}^{N} a_n e^{-j2\pi f T_n} \tag{6.8}$$

where $T_n = [0, T, 2T, \ldots, NT]$. Being (6.8) the expression of a microwave filter, it can be measured with a VNA, connected to the photodetector to measure the transmission parameter (S_{21} for electromagnetics S-parameters). This measurement is similar to the measurement of the frequency response of a transmission channel, in telecommunication engineering, or to the observation of the echoes of a transmitted signal in a wireless communication system [12]. The VNA modulates the light source with a sine wave from 10 MHz to 10 GHz, allowing measuring $H(f)$ on the largest bandwidth. Then, in order to extract the coefficients a_n, which corresponds to the variation of reflectivity, we can apply the IFT, similarly to OFDR systems. Coherency time is ultimately the limiting factor to the spatial resolution: we can approximate the multiple fiber reflections to a digital filter with distinct taps only for a period $T >> t_c$, at least 10 times larger. The spatial resolution L_{res} is equal to $cT/(2n_{eff})$, and thus a short coherency time (i.e., strongly incoherent source) is preferred. Using a 40-nm LED, the coherency time is 0.2 ps and the minimum spatial resolution is 20 μm.

The MWP technique is engineered to detect the position and the amplitude (i.e., peak temperature or strain) of a hot spot, within a chain of WDM FBGs, or a long weak reflector.

6.4 Sensing Elements

OFDR and microwave photonics are both principles of operation used to extract the distributed profile of weakly reflective element in the optical fiber. The sensing element is therefore the fiber itself, in the case of Rayleigh scattering, or a cascade of weak reflectors. Similarly to FBGs, distributed sensing architectures are used to detect temperature and strain distributions.

For Rayleigh backscattering, the fiber itself acts as a sensing element. It has been observed [10] that Rayleigh scattering is substantially a chaotic phenomenon, with a spectrum substantially equal to a random function and behaving similarly to a random walk. Figure 6.1 shows an example of the spectrum of a Rayleigh backscattering evaluated in the third optical window, detected via OFDR; the spectrum itself does not encode any particular temperature or strain dependence, but is just the result of a random fluctuation of the glass permittivity $\varepsilon = n^2$, with n being the refractive index. The sensing information is, however, encoded in the fluctuation of the Rayleigh backscattering pattern when a temperature and/or strain variation is applied. In Chapter 5, (5.5) and (5.6) described the shift of the FBG spectrum when a temperature variation or strain is applied; the reference spectrum is produced by an FBG, and its shift is due to the linear response to the changes of refractive index and physical length of the fiber due to temperature and strain changes. For

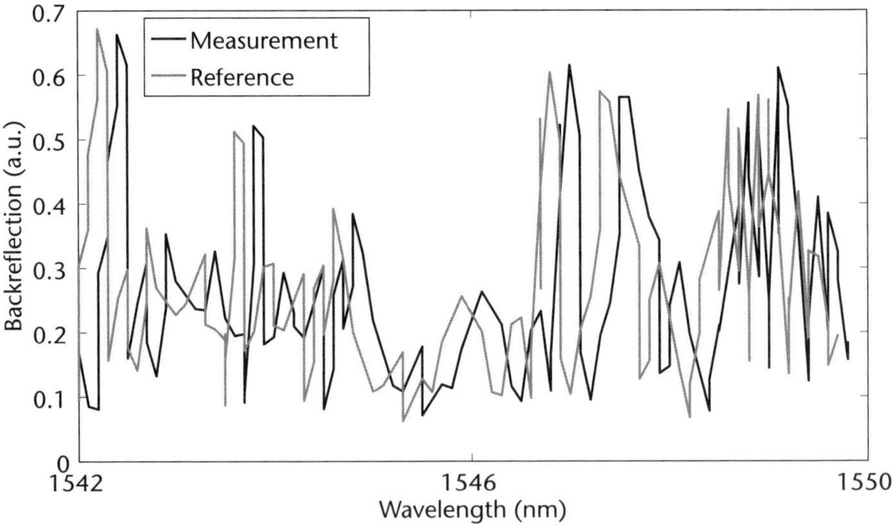

Figure 6.1 Chaotic spectrum of Rayleigh backscattering pattern, acquired on Luna OBR 4600 instrument [10]. The spectra are acquired in reference condition and with a temperature variation of ~26°C; the wavelength shift is estimated by means of a cross-correlation.

a Rayleigh-based OFDR, we have the same working principle, but applied to a chaotic spectrum: as shown in Figure 6.1, when a temperature or strain variation is applied, the chaotic spectrum shifts from its reference waveform. The sensitivity is, as in (5.6) condensed in two parameters, which are the same for FBGs because they make use of the same material: sensitivity values are ≈10 pm/°C and ≈ 1 pm/$\mu\varepsilon$ respectively for temperature and strain variations. In order to detect the wavelength shift, between the measured and reference spectra, we can simply compute the cross-correlation between the two waveforms over the whole measurement range of the detector [10].

MWP-based sensing systems work at best with an array of reflective elements. Although virtually the system can work in presence of Rayleigh scattering, the signal is usually too weak to be processed with a VNA. FBGs are therefore used as a sensing element: in [14], MWP is used to detect the wavelength shift of an FBG array, using the decoding technique described in Section 11.3; in another work [13], a cascade of FBGs at the same wavelength is used as a sensing element.

From the application perspective, MWP becomes relevant when using a continuous distribution of weakly reflective elements. In this case, the limiting factor to the spatial resolution becomes the coherency time of the source, and performances are comparable with OFDR. In this case, it is desirable to have a long reflector (several cm of length) with low reflectivity, in order to have an equal amplitude of the a_n coefficients in absence of strain or temperature variations. In [11] the MWP is applied to weak FBGs having 10-cm length, and 6% reflectivity; the reflectivity per unit length is in the order of 0.6%/cm, making it a long and weak reflective element, well suited for MWP. The system is functionalized for one or two hot-spot detection, whereas the hot spot is defined as a single point within the grating characterized by a sharp peak of temperature or strain. The IFT applied on the VNA output allows returning an $H(f)$ estimate with a periodic pattern; the measurement of its free spectral range, which can be directly performed on the VNA, is proportional to the position of the hot-spot within the grating, and it can be achieved with sub-0.1-mm accuracy. It is also possible to estimate the position of a second hot-spot through further processing. Recently, progress in draw-tower gratings allowed development of all-grating fibers, which extend the weak FBG concept: an optical fiber with FBGs inscribed in its portion, with weak reflectivity, and suitable for MWP (or OFDR) interrogation. Notably, FBGS International (Belgium) has recently industrialized an all-grating-fiber with each FBG having 1–10-mm length and dense FBG spacing, fabricated on a biocompatible low-loss fiber and suitable for distributed interrogation (http://www.fbgs.com/productsadv/be-en/12/detail/item/28/page/1/). All-grating fibers are fabricated with standard FBGs; an efficient alternative is the use of Gaussian-apodized gratings, described in [14], which allows reducing the range of variability of the MWP tap amplitude.

Spatial division multiplexing (SDM) represents the next frontier for MWP, and the pioneering work from iTEAM Research Institute (Universidad Politecnica de Valencia, Spain) [15] shows that the use of a multicore fiber can increase the number of sensing points per single optical fiber, detectable through MWP. The principle is shown in Figure 6.2. In this work, SDM combines with WDM: multiple transmitters operating in WDM mode are coupled in a single optical fiber, with a single modulator (always controlled by the VNA and sweeping from 10 MHz to 10 GHz). The sensing fiber is a multicore fiber (MCF), which contains N cores. FBGs

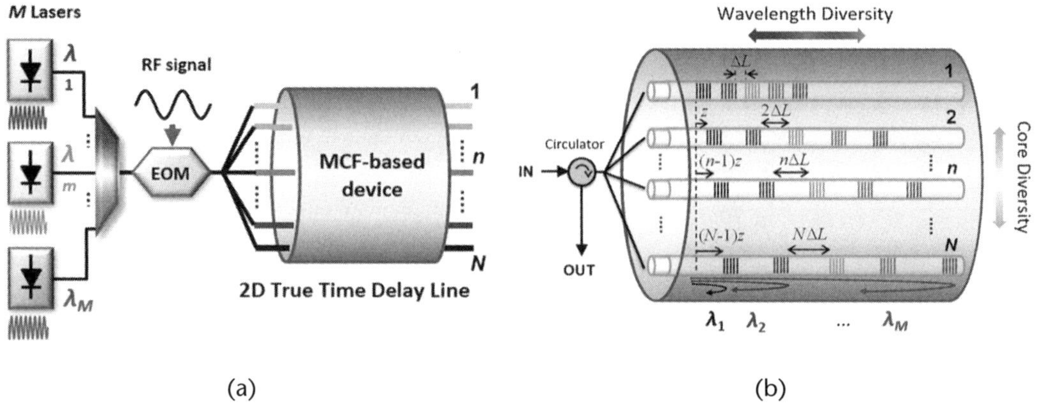

Figure 6.2 Spatial division multiplexing applied to MWP. (a) Layout of the front end: an array of *M* continuous-wave (CW) lasers, having different emission wavelength, is input to a single fiber, which is externally modulated via RF signal. The optical power is coupled to a multicore fiber (MCF), based on *N* cores. (b) Layout of the wavelength and spatial multiplexing: each core contains a plurality of reflectors, matching each laser wavelength and distributed at different lengths; the total (theoretical) number of sensing points is $M \times N$. Image adapted from [15].

are inscribed, with WDM technique, into each core of the fiber, using a different distance (ΔL in the figure) between each sensing element. This configuration allows using each core of the fiber as a distributed sensor with different spatial resolution. The demonstration of Gasulla et al. is performed with a 7-core fiber and 10 FBGs per core, and paves the road to unprecedented sensing networks considering the possibility of combining strain and temperature sensing on the same fiber with multiple compensations.

6.5 Practical Considerations

Rayleigh-based OFDR, industrialized by Luna Technologies, is the present state of the art of short-range (20 to 70m), high spatial resolution (<1 mm) distributed sensors, and it can measure both strain and temperature, like FBGs. The sensing element, as previously described, is a simple standard single-mode fiber: from the biomedical perspective, where a sensor is often a disposable device, it is a great advantage the possibility of using a standard fiber with cost <$1/meter, as opposite to sensing elements that are far more expensive. The interrogator, however, presents a much larger complexity, and commercial interrogators are based on the setup described in Section 10.4, dedicated to the hardware. In particular, in order to obtain a calibration function, the OFDR needs to have a fiber (lead span) with calibrated length internal to the interrogator, with no strain applied and held at ambient temperature. A trigger optical circuit is also needed in order to align the wavelength axis of the interrogator; this is particularly important considering that the wavelength uncertainty on swept-wavelength lasers is not negligible. The Rayleigh pattern is chaotic but somehow deterministic on a short fiber length, as described in the previous section; however, the fiber polarization states are subject to change during the measurement time. This issue, technologically, is solved by detecting both polarizations separately (through a polarization beam splitter) and reconstructing the detected signal P_d as a vector sum of both polarization components. This additional optical circuitry complicates the interrogator design and requires significant hardware, making the system not portable and expensive.

The calibration of an OFDR system requires starting the measurement in reference conditions (i.e., no temperature and strain variations) in order to obtain the reference Rayleigh pattern, and thus correlate it to the patterns acquired in each measurement. Prior to start the measurement it is necessary to calibrate the OFDR with a reference reflector, with known reflectivity and distance from the lead fiber and with a good polishing quality (to avoid diffusive reflections). Metallic mirrors are a good solution, due to high broadband reflectivity; a gold-tipped fiber pigtail (such as https://www.fiberinstrumentsales.com/specs/Gold-Tipped-Pigtail.pdf) is a good instrument for OFDR calibration.

In OFDR, it is very important to design the fiber sensing length, particularly its reflective elements. Since the Rayleigh backscattering is an extremely small signal, in the order of 10^{-6} times the feed power, any reflective element overlapping to the Rayleigh scattering results in a strong variation of the output waveform, which is also integrated by the photodetector. For this reason, OFDR systems work better if the sensing fiber is terminated on a nonreflective element, or at least polished to avoid strong backreflections. In some applications, however, this is not feasible, particularly if the fiber is follows a percutaneous or endoscopic insertion through a catheter, which may cause to scratch the fiber surface.

Figure 6.2 shows the use of an OFDR as temperature sensor, to measure the temperature profile in a thermal ablation procedure, following the setup described by Macchi et al. [4]. The figure shows the application of an OFDR, from the user's perspective. The left chart shows the Rayleigh backscattering pattern, as a function of distance. The length of 0m corresponds to the OFDR lead-out connector, which is visualized on the interrogator as a peak with reflectivity 18 dB higher than the Rayleigh signal intensity. Negative lengths correspond to the lead fiber, which is maintained at constant ambient temperature. To the interrogator output, a 3.75-m fiber length is connected. Since the fiber is not cleaved or polished at the end, we observe a rough and strong backreflection, with intensity 50 dB higher than the Rayleigh scattering; when integrated by the photodiode bandwidth, this results in a 20-cm region where the Rayleigh signal is covered by the reflection effect. The highlighted region contains the length that we analyze in our measurement, and only on this portion of signal, the autocorrelation, and therefore temperature estimation, is applied; this is shown on the right chart, as 4 distinct temperature peaks corresponding to 4 fiber sections inserted in the tissue in proximity of the heating source. The abrupt temperature variation, despite its abrupt spatial gradient, is still detectable by the OFDR, still satisfying the assumption $\partial^2 E_B/\partial z^2 \cong 0$.

In this OFDR, the spatial resolution is different in online and offline processing. Raw data are acquired and stored for postprocessing and correspond to data extracted by the autocorrelation with ~1-Hz sampling time (corresponding to the frequency scan of the laser source). The spatial resolution depends on the length of fiber highlighted as sensing region, and it is 20 μm for the measurement of Figure 6.3. Real-time data instead are averaged and displayed with 1-mm spatial resolution, in order to improve temperature accuracy.

Coherent OFDR can be applied to very weak signals, using a suitable detection method, and even Rayleigh scattering is enough to perform accurate sensing. MWP

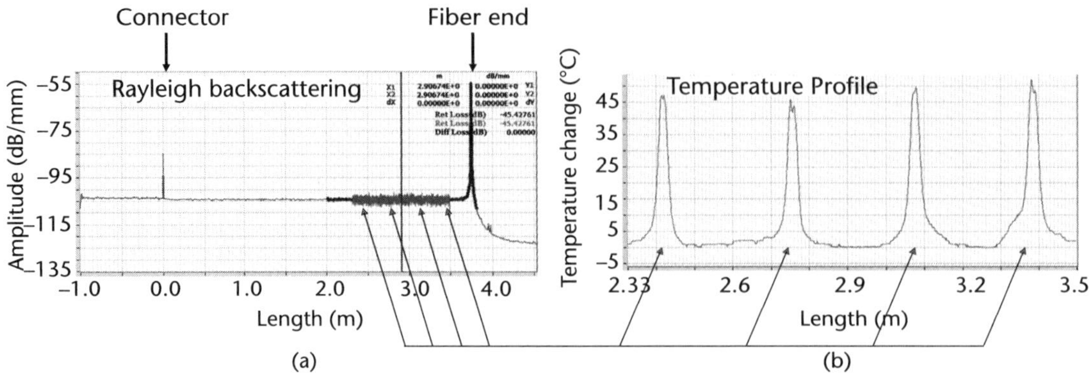

Figure 6.3 Typical backscatter pattern for an OFDR-based distributed sensor, exposed to thermal ablation pattern; experiments are described in [4]. The left chart shows the Rayleigh backscattering pattern, while the right chart shows the temperature profile reconstructed through autocorrelation.

(as well as incoherent OFDR) instead is applied to reflective elements, due to its incoherent nature; thus, it cannot detect Rayleigh scattering, but it needs a set or continuous of reflective devices. Its sensing range can be adjusted even to remote sensing; thus, it is possible to use a lead fiber in front of the sensing element, and there is an active region where sensing is confined. There is no need to calibrate the MWP with a reflective element. However, we cannot simply use an inexpensive fiber as a sensor, but the cost of a sensing array is similar to that of FBGs.

6.6 Perspectives

Distributed sensing is a unique feature of FOS technology, and represents the present state of the art for physical sensing in terms of number of sensing points and distribution. OFDR and MWP techniques achieve submillimeter/millimeter (offline/online) spatial resolution and can detect temperature and strain, similarly to an FBG but with thousands of sensing points. We can also foresee that with the advent of supercomputing, also OTDR techniques (that have sensing range up to several kilometers and can better discriminate temperature and strain using Raman and Brillouin scattering) might reach the centimeter or millimeter resolution in the future.

The gap between the potential of the technology and its application is presently huge, due to the high hardware cost, the difficulty in calibrating the instruments for users, the lack of design of sensing probes that do not have an active region, and the relatively long measurement time. In terms of biomedical applications, only two recent works pioneer the use of OFDR in biomedical sensing. In the first, proposed by Macchi et al. [4], an investigation has been performed in order to evaluate temperature distribution in human tissues exposed to thermal ablation; the sensor is designed to operate in vivo, embodied in a catheter. The second work, proposed by Polyechnique Montreal (Canada) [5] makes use of an OFDR system to measure the position and the bending of radiofrequency ablation surgical instrument, through strain measurement. Both works use a Luna OBR 4600 and relate to thermal ablation. Despite the early stage, it is foreseen that OFDR sensors will emerge as a concrete alternative to the single point of multiplexed sensors, thanks to their capability to measure distributions rather than single quantities. MWP sensors instead are rather unknown to the biomedical sensors community, but the key innovation proposed by iTEAM [15] is a game-changer because not only does it allow extending MWP to multiple hot spots, but also it allows arranging an array of reflective sensors all with a different spatial resolution, making it possible to have different spatial resolutions in the same sensing device. For these reasons, we expect a plurality of applications will emerge, once OFDR and MWP have settled into established systems, with the possibility of designing both the software for detection and the sensing element(s).

References

[1] Bao, X., and L. Chen, "Recent Progress in Distributed Fiber Optic Sensors," *Sensors*, Vol. 12, No. 7, June 26, 2012, pp. 8601–8639.

[2] Posey, R., G. A. Johnson, and S. T. Vohra, "Strain Sensing Based on Coherent Rayleigh Scattering in an Optical Fibre," *Electronics Letters*, Vol. 36, No. 20, September 28, 2000, pp. 1688–1689.

[3] Van Hoe, B., et al., "Distributed Strain and Curvature Measurements Based on Tailored Draw Tower Gratings," *2016 IEEE Avionics and Vehicle Fiber-Optics and Photonics Conference (AVFOP)*, October 31, 2016, pp. 285–286.

[4] Macchi, E. G., et al., "Optical Fiber Sensors-Based Temperature Distribution Measurement In Ex Vivo Radiofrequency Ablation with Submillimeter Resolution," *Journal of Biomedical Optics*, Vol. 19, No. 11, 2004, p. 117004.

[5] Parent, F., et al., "3D Shape Tracking of Minimally Invasive Medical Instruments Using Optical Frequency Domain Reflectometry," *International Society for Optics and Photonics SPIE Medical Imaging*, March 18, 2016, pp. 97862J–97862J.

[6] Froggatt, M., and J. Moore, "High-Spatial-Resolution Distributed Strain Measurement in Optical Fiber with Rayleigh Scatter," *Applied Optics*, Vol. 37, No. 10, 1998, pp. 1735–1740.

[7] Froggatt, M., "Distributed Measurement of the Complex Modulation of a Photoinduced Bragg Grating in an Optical Fiber," *Appl. Opt.*, Vol. 35, 1996, pp. 5162–5164.

[8] Barnoski, M. K., et al., "Optical Time Domain Reflectometer," *Appl. Opt.*, Vol. 16, 1977, pp. 2375–2379.

[9] Soller, B. J., et al., "High Resolution Optical Frequency Domain Reflectometry for Characterization of Components and Assemblies," *Optics Express*, Vol. 13, No. 2, January 24, 2005, pp. 666–674.

[10] Gifford, D. K., et al., "Swept-Wavelength Interferometric Interrogation of Fiber Rayleigh Scatter for Distributed Sensing Applications," *International Society for Optics and Photonics Optics East 2007*, September 26, 2007, pp. 67700F–67700F.

[11] Ricchiuti, A. L., J. Hervas, and S. Sales, "Cascade FBGs Distributed Sensors Interrogation Using Microwave Photonics Filtering Techniques," *Optics & Laser Technology*, Vol. 77, 2016, pp. 144–150.

[12] Ricchiuti, A. L., et al., "Long Fiber Bragg Grating Sensor Interrogation Using Discrete-Time Microwave Photonic Filtering Techniques," *Opt. Express*, Vol. 21, No. 23, 2013, pp. 28175–28181.

[13] Ricchiuti, A. L., et al., "Long Weak FBG Sensor Interrogation Using Microwave Photonics Filtering Technique," *IEEE Photonics Technol. Lett.*, Vol. 26, No. 20, 2014, pp. 2039–2042.

[14] Cheng, R., et al., "Ultra-Short FBG Based Distributed Sensing Using Shifted Optical Gaussian Filters and Microwave-Network Analysis," *Optics Express*, Vol. 24, No. 3, 2016, pp. 2466–2484.

[15] Gasulla, I., et al., "Spatial Division Multiplexed Microwave Signal Processing by Selective Grating Inscription in Homogeneous Multicore Fibers," *Scientific Reports*, Vol. 7, 2017.

CHAPTER 7

Fabry-Perot Interferometers

Interferometers are based on the interference of light waves traveling in resonant structures on different paths. Among the several structures, the Fabry-Perot interferometer (FPI) is the most important for biomedical FOS [1], because it allows fabricating pressure sensors, which are a key element of medical measurements. In this chapter, we will mainly discuss FPI sensors and a few other fiber-optic interferometers with applications in biomedical engineering.

7.1 Fabry-Perot Interferometer Principle

The principle of the Fabry-Perot interferometer (FPI) is shown in Figure 7.1(a). The FPI is formed by a cavity or etalon, that is, a device consisting of two reflective mirrors spaced by the short-distance L_{FP} [2, 3]. The several interferences obtained between waves traveling within the device, through the infinite reflections occurred at each mirror interface, form a wavelength-selective device. The implementation of the FPI concept into fiber-optic sensors is introduced in Figure 7.1(b, c). The first one shows an inline configuration: the cavity is formed between the facets of two identical optical fibers leaving a short air gap of length L_{FP}; in alternative, it is possible to fabricate a microbubble inside an optical fiber serving as cavity. The air-glass interface operates as a mirror forming a low-finesse interferometer. The capillary is a hollow cylindrical element that allows sealing the structure, enclosing the air gap. This implementation is mainly used in temperature sensors; its inline denomination refers to the possibility of continuing the fiber link on both sides, eventually adding further sensors. The last schematic shows the most popular FPI design, mostly known as extrinsic FPI (EFPI) due to the support of an external element. This interferometer is fabricated on the tip of a standard optical fiber, and makes use of a diaphragm to close the cavity. The diaphragm has a length $d << L_{FP}$ and has marginal influence on the spectral properties of the device. In the EFPI, the diaphragm compresses or enlarges the inner air gap, modulating the FPI cavity length.

The transmission-matrix (or scattering-matrix) method can be used to analyze FPI systems, whether in single- or multicavity sensors [4–6]. Each element of the FPI (i.e., the reflectors at the cavity interfaces and the pure propagation through each cavity) is modeled with its correspondent transmission matrix $\underline{T}$. The transmission matrix corresponding to the propagation of a distance l is given by [5]:

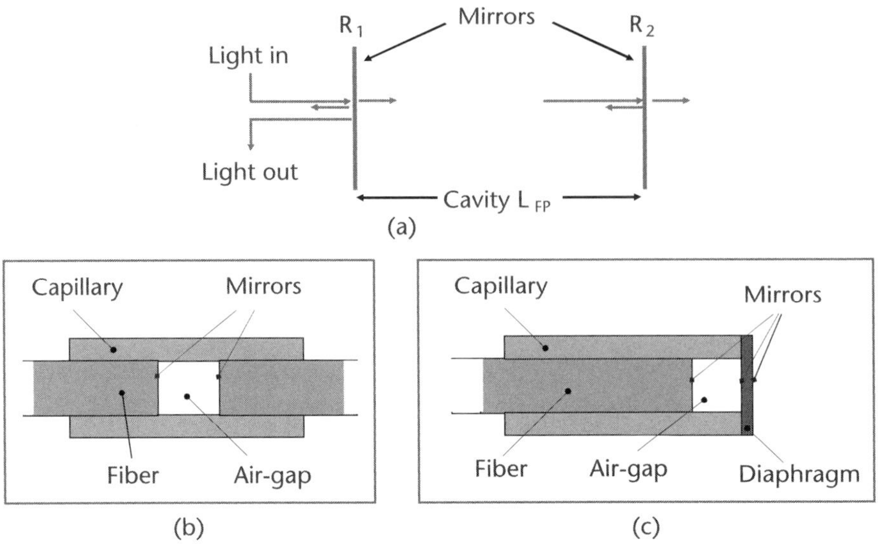

Figure 7.1 FPI and its implementation. (a) Schematic of a generic FPI; (b) implementation of a fiber-optic inline FPI; and (c) implementation of a diaphragm-based FPI tip sensor.

$$\underline{\underline{T}}_{prop}(l) = \begin{bmatrix} \exp\left(j2\pi n_{eff} l / \lambda\right) & 0 \\ 0 & \exp\left(-j2\pi n_{eff} l / \lambda\right) \end{bmatrix} \tag{7.1}$$

and this term includes the dependence on the wavelength λ; n_{eff} is the effective refractive index within the propagation medium. Since the cavity is short and often made with air or low-loss dielectric, we assume that the medium is lossless. At the reflection between two boundaries, at the sides of each cavity of the FPI, we observe a wave reflection and transmission; hence, each boundary behaves as a mirror, and its corresponding transmission matrix is:

$$\underline{\underline{T}}_{mirror}(\rho) = \begin{bmatrix} j/t & -j\rho/t \\ -j\rho/t & -j/t \end{bmatrix} \tag{7.2}$$

where ρ is the reflection coefficient between the two boundaries, and t is the transmittance; in the case of a lossless medium, which is the case considered in this approach, $t = (1 - \rho^2)^{1/2}$. The reflection coefficient, at the boundary between two media with refractive indices n_1 and n_2, is given by Fresnel's equations. In an FPI system fabricated on a single-mode fiber, like the vast majority of the sensors described in this chapter, light is incident on the boundaries of media with purely real refractive index (such as glass, air, dielectric layers) at a small angle θ_i, inferior to the critical angle. In this case we can use the normal incidence approximation, since $\cos\theta_i \approx 1$: for example, for a fiber with numerical aperture 0.08, $\cos\theta_i = 0.997$. Under normal incidence, Fresnel's reflection coefficient can be expressed as

$$\rho = \frac{n_1 - n_2}{n_1 + n_2} \tag{7.3}$$

In the transmission matrix approach, thus, we can estimate the transmission matrix of the entire structure by multiplying the transmission matrices of each element (mirrors and propagations) of the cascade, in the order they appear. For a single-cavity FPI with a cavity length L_{FP}, the FPI transmission matrix is:

$$\underline{\underline{T}}_{FPI} = \underline{\underline{T}}_{mirror}(\rho_1) \times \underline{\underline{T}}_{prop}(L_{FP}) \times \underline{\underline{T}}_{mirror}(\rho_2) \tag{7.4}$$

where ρ_1 is the reflection coefficient at the fiber-cavity interface, and ρ_2 is the reflection coefficient at the outer boundary, that is, the cavity fiber in Figure 7.1(b) and the cavity outer medium in Figure 7.1(c). For a double-cavity EFPI as in Figure 7.1(c),

$$\underline{\underline{T}}_{FPI} = \underline{\underline{T}}_{mirror}(\rho_1) \times \underline{\underline{T}}_{prop}(L_{FP}) \times \underline{\underline{T}}_{mirror}(\rho_2) \times \underline{\underline{T}}_{prop}(d) \times \underline{\underline{T}}_{mirror}(\rho_3) \tag{7.5}$$

taking into account also the propagation through the diaphragm length d, and the reflection coefficient at the diaphragm-outer medium ρ_3. Here we can see the advantage of the $\underline{\underline{T}}$ matrix approach, since we can isolate each element of the cavity and model it separately, whereas the matrix is arranged such to compute the multiple reflections and transmissions occurring within the cavities. In a sensing system, the FPI is fed with a broadband light source, and we collect the backreflection introduced by the FPI, measuring the reflectivity. From (7.4) and (7.5), we can compute the reflectivity as [5]:

$$R_{FPI} = \left| \frac{\underline{\underline{T}}_{FPI}(2,1)}{\underline{\underline{T}}_{FPI}(1,1)} \right|^2 \tag{7.6}$$

In Appendix A, we provide the MATLAB code for a function (*EFPIgeneration.m*) that estimates the reflection spectrum of an EFPI two-cavity of length L_{FP} with diaphragm length d, with effective refractive indices n_i (input fiber), n_{FP} (Fabry-Perot medium), n_d (diaphragm medium), and n_o (outer medium). The function can be used to reconstruct FPI spectra under different conditions: for example, all-glass cavity measuring pressure in water solutions has $n_i = n_d = 1.5$ (glass), $n_{FP} = 1$ (air), and $n_o = 1.33$ (water), d ranges from 2 μm to 3 μm, and L_{FP} ranges from 20 μm to 100 μm. These parameters are also used for the following simulations.

Figure 7.2 shows the spectrum of a single-cavity FPI structure, based on a launch fiber (refractive index 1.5), and air cavity, and an outer medium. The reflection spectrum has an almost periodical pattern, alternating reflection peaks and dips. The wavelength difference $\Delta\lambda_{FSR}$ between two reflection dips corresponds to the free spectral range (FSR) of the cavity; the FSR depends on the cavity length as:

$$\Delta\lambda_{FSR} = \frac{\lambda^2}{L_{FP} n_{FP}} \tag{7.7}$$

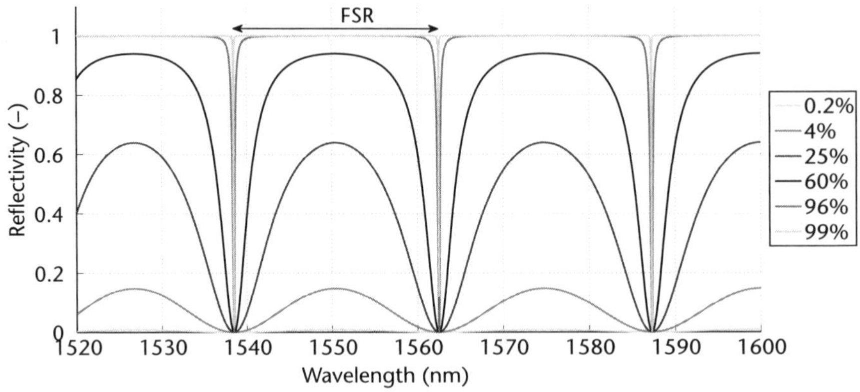

Figure 7.2 Simulated spectrum of a single-cavity FPI structure, having cavity length LFP = 50 μm, air cavity (nFP = 1), and optical fiber as input (ni = 1.5). Spectra are simulated for different values of reflectivity between the air cavity and the outer edge, listed in the legend.

and constitutes the periodicity of the FPI pattern. The FPI spectrum amplitude and shape depend on the reflectivity between the air cavity and the edge. For a small reflectivity value, such as 0.2%, the FPI cavity effect is practically invisible; as the reflectivity increases, the spectrum gradually takes a sine-wave shape around 4%, progressively distorted for 25% to 60% reflectivity. Highly reflective FPIs have a strong reflectivity, with the exception of a narrow transmission peak in correspondence of the FSR; the notch bandwidth decreases as the reflectivity increases.

FPI sensors usually mount a diaphragm, and are based on a glass-air-glass structure [7], or with other dielectric materials with low refractive index (~1.4 to 1.8); the majority of fiber-optic FPI units operate with low-reflective layers. Figure 7.3(a) shows FPI spectra, for an all-glass sensor, with different cavity lengths. Low values of cavity length (20 μm, 45 μm) imply a spectral envelope characterized by a small (1 to 2) number of periods, within the observation range of the detector, usually 40 to 100 nm. Increasing the cavity length to 95 to 120 μm results in a larger number of periods in the detector range. It is necessary to design the length of the FPI cavity for the specific type of detector. The chart, compared to Figure 7.2, shows also the effect of the diaphragm, which acts as a further spectral modulation: the spectral peaks do not have all the same amplitude but a further cavity with a small length acts as a modulator. The sensing principle of the FPI sensors is tied to the variations of the cavity length; variations are usually small, in the order of nanometers, particularly for pressure sensors. A change of L_{FP} of 0 to 100 nm is shown in Figure 7.3(b): small-cavity variations result in a shift of the spectrum towards the longer wavelengths.

7.2 FPI-Based Sensors

The extrinsic FPI (EFPI) sensor is the most popular fiber-optic pressure sensing unit, in modern systems [8–10] because of its microscopic size that allows fitting inside narrow catheters and the inherent sensitivity to pressure variations; EFPI pressure sensors have been demonstrated in urology [9, 11], cardiovascular diagnostic [12], neurology [13], and laser and RF thermotherapy [14]. EFPI pressure sensors are

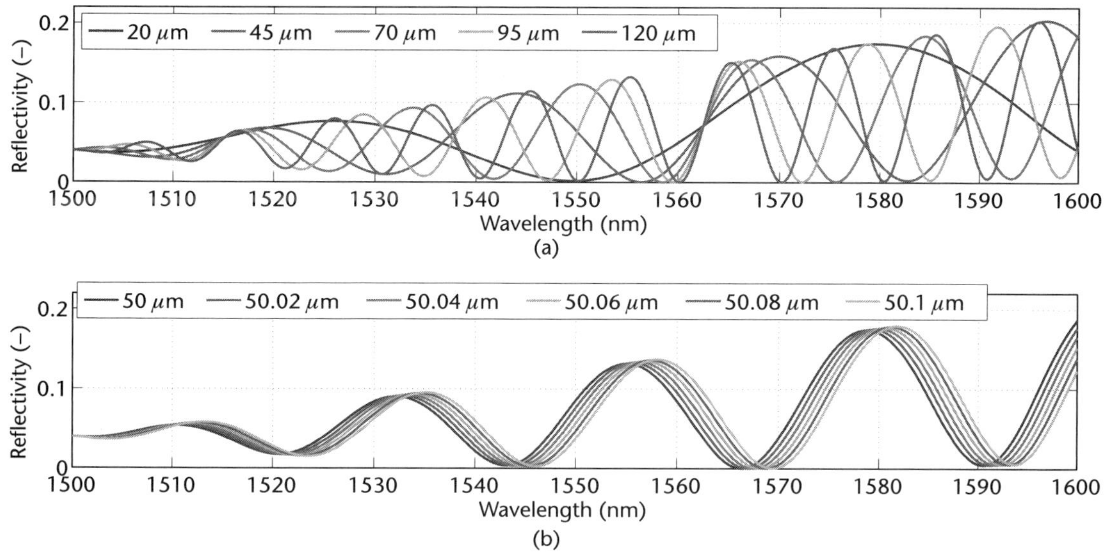

Figure 7.3 Simulated spectrum of an all-glass EFPI sensor [5, 8] having ni = nd = 1.5, nFP = 1, no = 1, d = 2 μm, and different values of cavity length LFP, listed in the legend. (a) Spectra for LFP ranging from 20 μm to 120 μm; and (b) small cavity variations, from 50 μm to 50.1 μm.

based on a diaphragm structure, as in Figure 7.1(c) and [11, 14]. The diaphragm is a thin cylindrical layer, made of dielectric, metal, or organic material, which exerts the sensitivity to the pressure applied outside of the sensor. When the pressure is changed by the quantity ΔP from the reference value, the thin diaphragm bends, resulting in a compression of the inner Fabry-Perot cavity. Since most cavities are made in air, the resulting effect of a pressure increase is the compression of the Fabry-Perot cavity by the quantity ΔL, whereas the diaphragm thickness variation is unperceivable. The sensitivity term k_P depends on the geometry of the pressure probe, and the diaphragm length d, and is independent from the cavity length; for a cylindrical diaphragm compressing an air gap, the sensitivity can be expressed as [7]:

$$k_P = \frac{\Delta L}{\Delta P} = \frac{3}{16} \frac{(1-\mu)^2 r^4}{E d^3} \tag{7.8}$$

where E is the Young modulus, μ is the Poisson ratio, and r is the radius of the diaphragm. It is important to notice that the sensitivity is proportional to d^{-3}; thus, it is essential to minimize thc diaphragm thickness in order to obtain a large sensitivity; however, a short diaphragm results in a more fragile probe, and it poses significant challenges in manufacturing. For example, a silica diaphragm (E = 73.7 GPa, μ = 0.17), having a thickness r = 100 μm and a length d = 2 μm, the sensitivity is 21.9 nm/kPa.

Another important factor to consider is the temperature sensitivity. Temperature accounts for a linear sensitivity, which affects both the thermal expansion of the diaphragm and the launch fiber, and the thermo-optic coefficient, which results in a variation of the refractive index due to thermal changes. The overall effect depends on the geometry and composition of the entire probe, including the capillary used to enclose the cavity. The work from Bremer et al. [7] exhibits a thermal sensitivity of 1.13 nm/°C, while [14] shows a sensitivity of 52 pm/°C, using a different capillary. As shown in [15], it is possible to use an intrinsic FPI structure to exploit the thermal sensitivity, developing a compact temperature sensor: compared to an FBG, the FPI has a smaller active length, but its broadband spectrum prevents from easily arranging an array of sensors and it is hard to fabricate dense arrays. In addition to pressure and temperature, EFPI structures have also been fabricated using an external moving target as reflective element modulating the cavity [9]; this arrangement was popular for the first generation of pressure sensors [16], in which the Fabry-Perot cavity is formed by the fiber end and the moving target. EFPI sensors have also been demonstrated for refractive index sensing [17]: in this case, the cavity maintains the same length but the reflectivity of the outer layer changes as a function of the outer refractive index, varying the finesse of the interferometer. However, SPR and TFBG techniques appear to be better performing than EFPI for refractive index detection.

Figure 7.4 shows the results of an all-glass sensor fabricated at University of Limerick by Poeggel et al. [10]. The left chart shows the variation of the cavity length when small pressure variations in the range 0 to 5.8 kPa are applied. It is possible to show that, with a very high linearity coefficient (>0.999), the cavity length estimated as in [10] linearly varies, reducing its length when pressure increases due to the diaphragm bending on the cavity. The cavity has a length L_{FP} = 22.154 μm and progressively shortens with the sensitivity coefficient k_P = 1.08 nm/kPa. The right chart shows a typical calibration, performed with a water burette: this technique is effective in calibrating biomedical pressure sensors, because it is effective for small pressure variation. The sensor is positioned at the bottom of a burette, filled with water to its top; in Figure 7.4(b), a burette with a 60-cm height is used. Then the burette is emptied and the water voids at a constant rate (17.5 mm/s), reducing the pressure level acting on the sensor. This way, it is possible to easily obtain a reference curve, both measuring the response to water levels

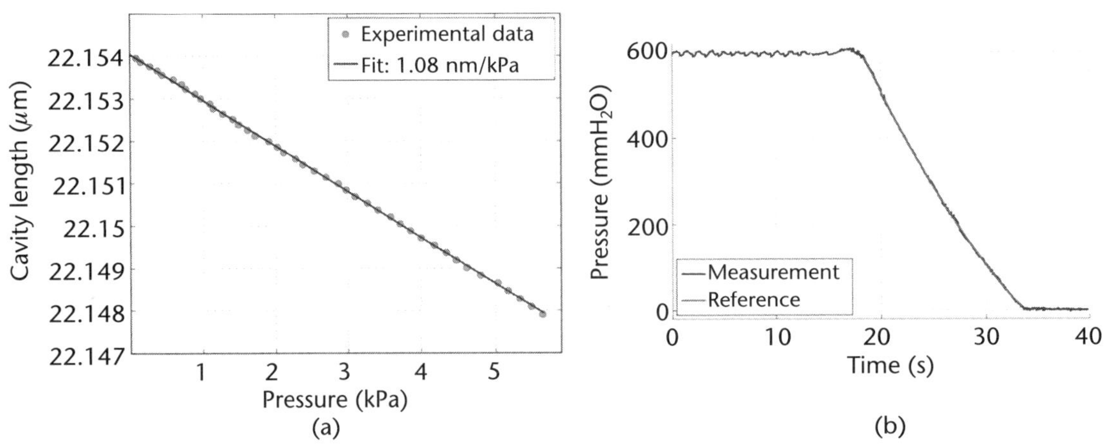

Figure 7.4 Response of an all-glass EFPI sensor [10, 12] to pressure variations. (a) Sensitivity chart, showing the Fabry-Perot cavity length as a function of pressure variations. (b) Calibration curve, in a water burette, dynamically changing the pressure level by voiding.

(considering the equivalence 1 mmH_2O = 9.807 Pa), and the accuracy during the static pressure condition.

In general, FPI probes are sensitive to the elongation or compression of the inner cavity, which is due to the pressure variation (in air, gas, or fluid) in an EFPI configuration due to the diaphragm effect and to temperature due to materials expansion. It is possible, for some applications, to functionalize the inline FPI structure to the detection of axial strain, with a suitable transducing effect that converts strain/force into a compression of the FP cavity. An example is shown in [18], whereas a set of FP cavities are fabricated within a robotic surgery device, packaged in expansile material. In this setup, axial forces acting on the device tip result in a compression of the FP cavity due to the relative movement of the fiber proximal to the tip with respect to the fiber on the opposite side; the sensitivity observed in this configuration is ~6 nm/mN, referring to forces exerted on the FPI tip parallel to the fiber direction.

7.3 FPI/FBG Dual Sensors

A modern trend in FOS technologies is to combine EFPI sensors, fabricated on the tip of an optical fiber, with proximal FBG sensors, positioned in vicinity of the EFPI [7, 14, 19]. The EFPI/FBG combination allows multiparameter sensing: the EFPI is sensitive to pressure and, in most structures, it also has a temperature sensitivity, while the FBG has a temperature or strain sensitivity and a nearly zero pressure sensitivity (in the order of 0.003 pm/kPa). This probe design achieves the task of measuring the pressure and temperature, while compensating for the cross-sensitivity to temperature of the EFPI sensor [7, 14]. In a dual EFPI/FBG probe, the dual task is to estimate the cavity length variation of the EFPI sensor ΔL, and the FBG wavelength shift $\Delta\lambda$. In this framework, we exploit the different spectral profile of the two sensors, particularly when the EFPI probe is fabricated as a low-finesse interferometer. The EFPI spectrum is broadband and exhibits a periodicity in the wavelength domain that allows estimating the FSR in different portions of the spectrum; the FBG spectrum on the other side has a very steep resonance, thus appearing as a narrowband element that is practically unaffected by the EFPI. Using the simple software methods described in Chapter 11, it is straightforward to separate the two spectral contributions due to the two sensing elements, and estimate ΔL and $\Delta\lambda$ as two separate tasks. In a recent work [21], it was demonstrated how the FBG and EFPI spectral contributions can be isolated from each other, using signal processing.

Figure 7.5 shows a dual EFPI/FBG pressure/temperature sensor fabricated by University of Limerick, and applied to thermal ablation, in a medical procedure where temperature and pressure both vary abruptly [14]. The first chart shows the spectrum of the probe, which results in a combination of a broadband envelope due to the EFPI and a narrow reflection peak around 1,540 nm due to the FBG. The two spectra can be almost perfectly separated, as shown in the insets: the FBG is completely transparent outside of its inner bandwidth, while the low reflectivity of the EFPI does not affect the FBG peak. This allows estimating the two spectral variations independently from each other, and using the calibration coefficients (1.60 nm/kPa and –0.0519 nm/°C for the EFPI; 10.70 pm/°C for the FBG) it is possible to obtain the pressure and temperature estimations.

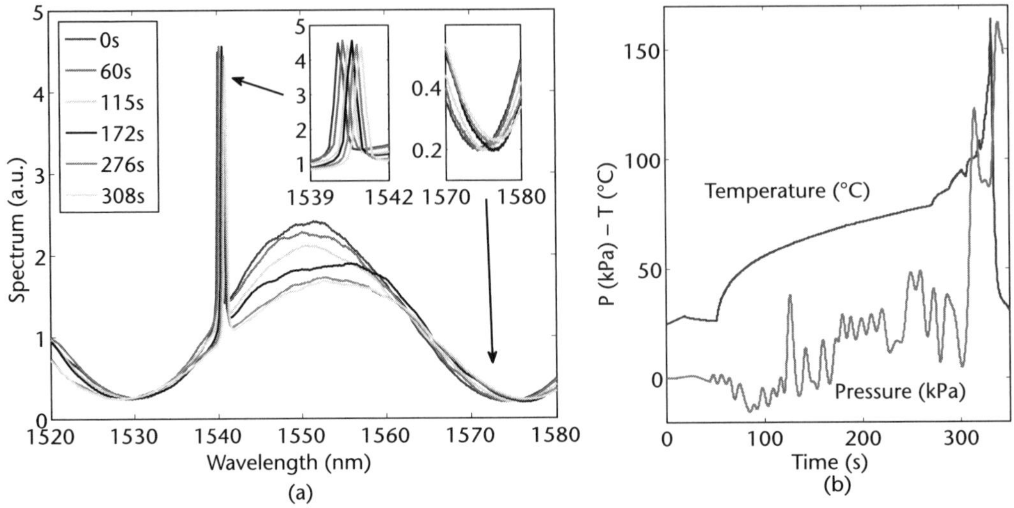

Figure 7.5 Application of a dual EFPI/FBG all-glass probe to RF thermal ablation for dual pressure-temperature measurement [14]. (a) Spectrum of the probe, sampled at different times during the ablation process; the insets show the FBG wavelength shift and the variations of the edge spectrum of the EFPI. (b) Estimated pressure and temperature patterns.

Recently, the EFPI sensor was combined with an FBG array [21] obtained an increased sensing capacity. In Appendix A, we include a MATLAB function (*FBGandEFPI.m*) that, based on FBG and EFPI generators, is able to provide the spectrum of a dual probe, allowing users to design an EFPI/FBG probe and its interrogation system.

7.4 Fabrication of FPI Sensors

Diaphragm-based EFPI sensors are substantially based on 4 elements: the input/output fiber, used to deliver light to the cavity and collect its backreflection; the Fabry-Perot cavity, having a specific length and built as a void air-gap region; the thin diaphragm that closes the Fabry-Perot cavity and provides the pressure sensitivity; and a sealing structure that encapsulates the sensor and avoids any leakage into the cavity. For biomedical applications, the all-glass EFPI configurations have particular importance; in this design, all components are made of glass compounds, which guarantees a miniature footprint (the maximum probe diameter is typically 125 to 200m), achieves biocompatibility thanks to an epoxy-free procedure that makes use of arc-discharge fusion to assemble all the silica components, and guarantees satisfactory performance and sensitivity. The design reported by University of Limerick [11] makes use of standard single mode (SMF-28) or bend-insensitive ormoceramic fiber as input (10/125-m core/cladding diameter), a multimode glass fiber (62.5/200-μm core/cladding diameter) as diaphragm, and quartz capillary with cylindrical hollow shape (Fiberware pure fused silica tube, 130/200-μm inner/outer diameter). This design operates in the third optical window (1,520 to 1,600 nm). The manufacturing equipment includes: a fiber splicer operating in manual mode (Siecor X77), or alternatively a modern fully customizable splicer (such as

Fujikura LazerMaster LZM-100) and fiber cleaver/stripper; a white light interrogation setup that allows monitoring in real time the spectrum of the EFPI during its building steps (in the described process, an Exalos EXS SLED and Ibsen Photonics I-MON-512-USB are used); fiber polishing sheets, with finesse ranging from 1 μm to 1/20 μm, and an SMA connector; hydrofluoric acid (HF), with 40% to 48% concentration or diluted; and a chemical fume hood to operate with HF.

Figure 7.6 shows the manufacturing steps for an EFPI/FBG probe, as described in [22]. During the whole fabrication, the EFPI spectrum is continuously detected with the white light setup. Two techniques are exposed; in the first one, the FBG is pre-inscribed using the phase-mask technique on a standard single-mode fiber, and the EFPI is fabricated in proximity of the FBG, around a 1-cm distance to avoid excessive heating on the grating region during splicing. As a first step, the fiber is cleaved at ~1-cm distance from the FBG. Then the capillary is polished at the external side (on the right, in the figure); in order to polish the capillary, this is inserted in a SMA connector and manually lapped on polishing paper, progressing from 1-μm to 1/20-μm grit to achieve a <λ/10 polished surface. After polishing, the capillary is spliced to the single-mode fiber, at a certain distance from the FBG (usually 1 to 3 cm); splicing is performed in manual mode, using a medium-power arc discharge in order to completely seal the structure, which ensures that no liquid can leak inside the cavity when sensing pressure in fluids even at microscopic level. Then the multimode fiber serving as diaphragm is fused to the capillary: the fiber is polished with the same method of the capillary and spliced manually to the capillary, leaving the air-gap within the structure. This second splice between the capillary and the multimode fiber requires a strong and extended arc discharge, because the both elements have a larger diameter. Eventually, it is possible to swap the order of the splices, performing first the multimode/capillary splice and later the capillary/single-mode fusion; this arrangement allows a better control of the

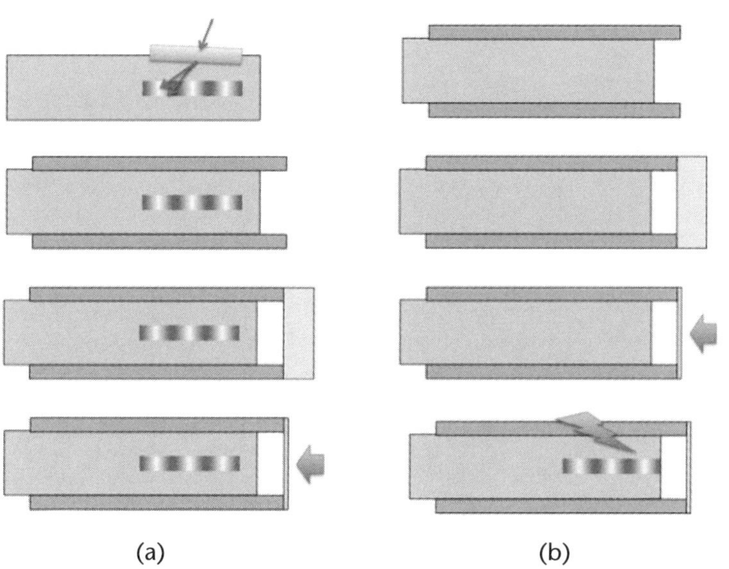

Figure 7.6 Process for manufacturing a dual EFPI/FBG with an all-glass design. (From: [22].) (a) FBG pre-inscribed with the phase mask; and (b) FBG post-inscribed with femtosecond (fs) laser.

cavity gap length but is more unpractical for operating with the splicer motors when the launch fiber slides inside the capillary. At this stage, a Fabry-Perot cavity is obtained; to functionalize it for pressure sensing, the diaphragm d needs to be shrunk to few micrometers in length according to (7.8). First, the diaphragm is cleaved and polished, using the same polishing paper with 1/20-μm grit, until the diaphragm approaches $d \sim 10$ μm; this can be verified by detecting the reflected EFPI spectrum. During polishing, it is essential that the tilt angle is <0.5°. Then, in order to obtain a fine reduction to $d \sim 2$ to 3 μm, wet etching in HF is performed. The fiber end is immersed in HF, which performs an isotropic etching of all glass surfaces (including the capillary, and for this reason etching can only be performed with a diaphragm already prepolished); in the meanwhile, the EFPI spectrum varies periodically, allowing detecting the etching rate. When the diaphragm reaches the desired thickness, in correspondence to the target sensitivity, the fiber is removed from HF and cleaned with deionized water. The etching speed depends on HF concentration. In the second method, proposed in Figure 7.6(b), the EFPI is fabricated firstly with the same technique, and at the end the FBG in inscribed using the direct inscription with an fs laser [21].

Diaphragm etching is the most critical part, as it determines the sensitivity of the probe, but also its tensile strength and resistance, particularly when the probe is embodied in a catheter and inserted in vivo. The HF etching rate is highly variable: it is necessary to control the diaphragm length in other ways. Figure 7.7 shows the evolution of the EFPI spectrum during etching in 48% HF a diaphragm with an initial length of 16 μm. Since the cavity length does not significantly vary, by (7.5) as the diaphragm reduces its length, we obtain a periodic variation of the optical spectrum, which is confirmed in Figure 7.7. The period is $/(2n_d) \approx 0.5$ μm, which means that every time the spectrum completes one period, the diaphragm length is 0.5 μm shorter. This provides an estimation of the diaphragm thickness independent on the etching rate. As the diaphragm is shorter and comparable with the wavelength, the spectrum abruptly diverges from its original shape: this is the

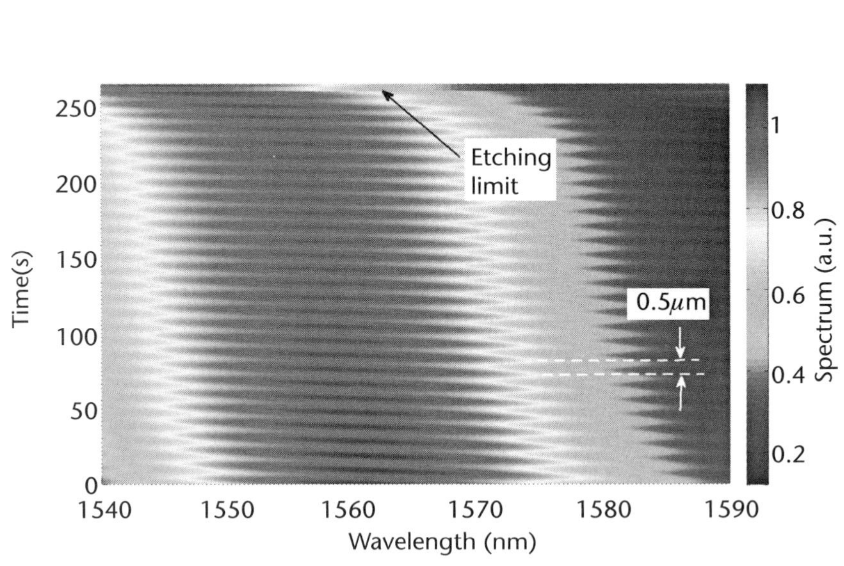

Figure 7.7 Spectrum of an all-glass EFPI probe during diaphragm etching.

empirical limit of etching, which is the minimum diaphragm length that preserves the integrity of the glass compound and the perfect sealing of the probe.

The final result of the manufacturing is shown in Figure 7.8, which shows a microscopic view of the side and front of the EFPI probe; the capillary enclosing the cavity is visible, while the thin diaphragm covers the Fabry-Perot cavity on the external side. The front view shows the polishing and etching effect on the diaphragm tip.

Table 7.1 compares the all-glass design hereby described with different alternatives, proposed from 2010 to 2016. Glass diaphragm provides an epoxy-free design, with medium difficulty of manufacturing, a stable structure (particularly in a catheter), and with the capillary choice as in [14], the thermal sensitivity is very low. The design of [19] provides an easy manufacturing, but it has a large thermal sensitivity and it is not biocompatible. Chen et al. [23] demonstrated a very thin and sensitive diaphragm made of chitosan, but its biocompatibility may not be up to medical standards and the fabrication is intricate. In [24], the diaphragm is a subwavelength silver layer, good for sensitivity and to increase the finesse, decoupling the reflection from the outer medium; the thermal sensitivity is high, and the diaphragm may be too delicate for in vivo placement. A simple and splicer-based manufacturing process is demonstrated in [25], in which the Fabry-Perot cavity is a microbubble fabricated by fiber splicing; however, the pressure sensitivity of a spheroid is much inferior to a disc-shaped diaphragm. Li et al. [26] demonstrated the operation of a grapheme diaphragm, achieving good results, but the design depends on a ferrule with a 1-mm diameter, making the sensor less miniaturized than all-fiber design. Very recent works show diaphragm design with silk fibroin [27] and index-matching liquids [28].

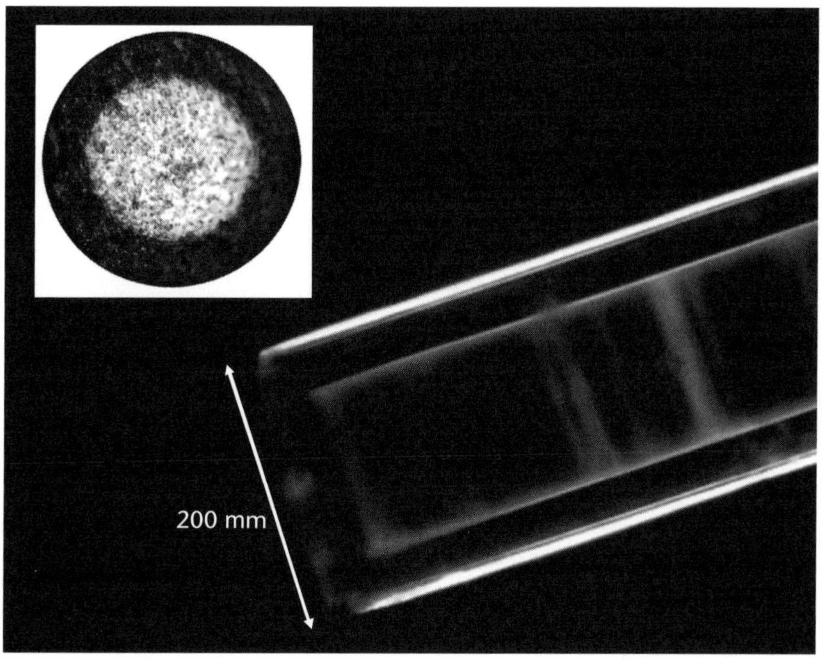

Figure 7.8 Side and front (inset) view of an all-glass EFPI probe with microscope (Zeiss AX).

Table 7.1 Comparison Chart of the Main EFPI Pressure-Sensing Architectures Proposed in the Latest Research Works

Reference	Sensitivity (nm/kPa)	Thermal Sensitivity (nm/°C)	Diaphragm Material	Diameter (m)	Epoxy-Free	Ease of Fabrication
[14]	1.60	0.052	Glass fiber	200	Yes	Medium
[7]	1.68	1.7	Glass fiber	200	Yes	Medium
[19]	1.53	15.8	UV-mold	1,000	No	Easy
[25]	1.03×10^{-3}	1.0×10^{-3}	Bubble	125	Yes	Easy
[23]	223	NA	Chitosan	125	Yes	Hard
[24]	70.5	39.3	Silver	≈1,000	Partially	Medium
[26]	1,096	NA	Graphene	1,000	Partially	Hard
[27]	12.3	NA	Silk fibroin	~4,000	No	Hard
[28]	150	6.95	Fluid	250	Yes	Hard

The table lists the pressure sensitivity, the temperature cross-sensitivity, the type and thickness of diaphragm, ease of fabrication, and whether the design is epoxy-free. NA = not available.

7.5 Self-Mixing Interferometry

The self-mixing interferometry (SMI), developed and established by Donati and Norgia [29], is based on the feedback generated by a laser illuminating a remote target (either reflective or diffusive). The SMI principle, in its simplest implementation, is described in Figure 7.9. The principle is based on the feedback generated by a coherent laser light illuminating the target and returning back to the source. The light reflected alters the laser equations, creating both an amplitude and a frequency modulation that depends on the linear displacement between the target and the laser; the SMI output can be directly measured with a photodiode mounted on the side of the laser, and in most commercial lasers already included within the laser package, hence without the need for any external device [30]. SMI is used in biomedical engineering mainly for noncontact measurements, but its fiber-optic implementation [31] (which belongs to the FOS category) is used to perform displacement measurement in cyclical measurements. The key asset of the SMI system is that it does not need any external element: only the laser source with its monitor photodiode and the standard single-mode fiber constitute the minimum viable structure to perform a displacement measurement. Alternatively, signal decoding is not straightforward and it depends on the amount of feedback.

Figure 7.9, as drawn in [32], shows the principle of operation of a SMI. The system behaves as the combination of a cavity laser and an external mirror; as in [29], we can express the output power of the laser as:

$$P = P_0\left[1 + m_A F(\varphi)\right] \tag{7.9}$$

where P_0 is the unperturbed power, that would be emitted in no-feedback condition. φ is the optical phase shift experienced through the propagation from the laser to the target and backwards; in a fiber-optic SMI running on a fiber with length L_{ext} and effective refractive index n_{eff}, $\varphi = 4\pi\, L_{ext}\, n_{eff}/\lambda$. The term m_A constitutes the amplitude modulation:

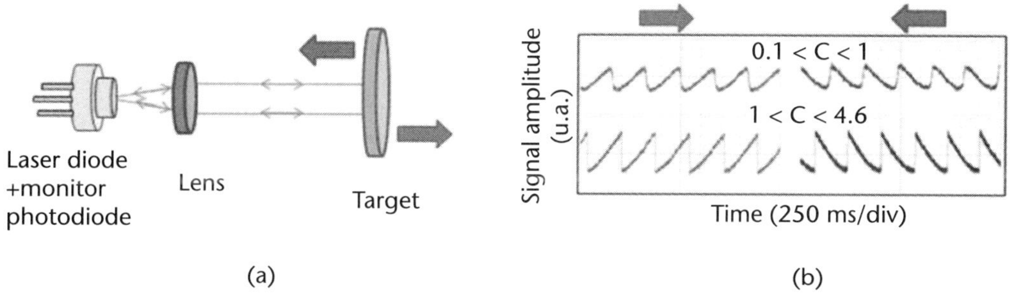

Figure 7.9 Principle of operation of a SMI-based sensor. (From: [32].) (a) Schematic of the working principle; and (b) amplitude of the self-mixing signal measured on the laser photodiode.

$$m_A = \frac{1}{\sqrt{A}}\left[\frac{c}{2L_{ext}}\left(\gamma - \frac{1}{\tau}\right)\right] \tag{7.10}$$

where γ is the gain per unit time of the medium, τ is the cavity decay time [29], and $1/A$ is the fraction of the power that is returned from the external target to the laser. In an SMI system, this value constitutes the amount of feedback, and is also depending on the laser isolation and focusing. The function F in Figure 7.9 is a periodic function with period 4π and argument φ, but its shape depends on the amount of feedback. The strength of the feedback is given by the injection parameter (C), which describes the coupling strength; its expression is given by:

$$C = \sqrt{\frac{1+\alpha^2}{A}} \cdot \frac{\kappa_{ext} L_{ext} n_{eff}}{n_{las} L_{las}} \tag{7.11}$$

where α is the laser linewidth enhancement factor, n_{las} and L_{las} are, respectively, the effective refractive index and the cavity length of the laser medium, and κ_{ext} as defined in [33] is the fraction of the output field interacting with the laser mode and is mainly dependent on the target reflectivity and fiber link attenuation. The parameter C, as shown in Figure 7.9(b), determines the shape of the function $F(\varphi)$. For a very low feedback ($C < 0.01$), the SMI behaves as a standard interferometer and the shape of F is a sine wave. Increasing the feedback up to $C = 1$, the waveform gets progressively distorted, while maintaining a sine-like envelope; this case is shown in Figure 7.9(b) (upper waveform) and is the typical working conditions for SMI sensing systems. When $1 < C < 4.6$, the distortions are increasing while the periodicity is partially maintained; then, for $C > 4.6$, the SMI operates in chaotic mode.

The SMI signal can be detected directly on the laser monitor photodiode, where it can be amplified and converted to digital waveform for estimation of the displacement. An example of detection is shown in Figure 7.9(b), which as reported in [32] shows the experimental values of F amplitude obtained for a moving mirror, with position varying at the rate of 1 mm/s on a linear motion stepper. As the external cavity length moves backward (right traces) or forward (left traces), the output signal has a periodic pattern and exhibits an opposite trend, which depends on the motion direction. The maximum periodicity occurs for $\varphi = 4\pi\, L_{ext}\, n_{eff}/\lambda =$

2π; hence, the system can effective discriminate displacement values ranging within the period $\lambda/2\ n_{eff} \approx 500$ nm for systems operated in the nearly infrared, making this technique very efficient in precision sensing. However, digital signal processing techniques have been proposed [29] to extend the sensing range up to millimeters when monitoring quasi-periodic signals, discriminating between forward and backward motions of the target thanks to their different effect on the F waveform. For this reason, extended-range SMI systems operate with some amount of feedback ($C \approx 1$) rather than with very low feedback. In biomedical signals, fiber-optic SMI has been demonstrated for the measurement of blood pulsation [33], and blood velocity [34]. Also, in an attempt to operate SMI in chaotic conditions, [35] showed how the SMI induced by a FBG reflector can be used to separate static and dynamic signals interrogation, hence separating strain and temperature sensing in FBGs.

7.6 Other Interferometers

The Mach-Zender interferometer (MZI), Michaelson interferometer, and Sagnac loop are high-performance interferometers, which can be applied in precision sensing of linear displacement, vibrations, or temperature according to their design. Fiber-optic implementations of these 3 interferometers are described in [36]. In the MZI, light is split by means of a 50/50 coupler into two different branches, and recombined using another 50/50 coupler. The optical phase difference between one branch, used as a reference, and the second interferometer branch, which works as a sensor, is the key parameter; the vibrations or displacement variations can modify the optical path, modulating the optical phase difference and therefore the interferometer output. The Michaelson interferometer works similarly to the MZI, but rather than having a transmission through two different optical paths, the light is reflected by two targets, one reference and one sensing mirror. The Sagnac loop, implemented as a gyroscope in fiber-optic sensors, is a ring interferometer in which two counterpropagating waves interfere. Despite their popularity for precision measurement, particularly during the 2000s, these interferometers are rarely applied to biomedical engineering due to the difficulty to maintain a stable optical path within the interferometer period.

References

[1] Islam, M. R., et al. "Chronology of Fabry-Perot Interferometer Fiber-Optic Sensors and Their Applications: A Review," *Sensors*, Vol. 14, No. 4, April 24, 2014, pp. 7451–7488.

[2] Yoshino, T., et al., "Fiber-Optic Fabry-Perot Interferometer and Its Sensor Applications," *IEEE Transactions on Microwave Theory and Techniques*, Vol. 30, No. 10, October 1982, pp. 1612–1621.

[3] Rao, Y. J., "Recent Progress in Fiber-Optic Extrinsic Fabry-Perot Interferometric Sensors," *Optical Fiber Technology*, Vol. 12, No. 3, July 1, 2006, pp. 227–237.

[4] Kriegsmann, G. A., "Scattering Matrix Analysis of a Photonic Fabry-Perot Resonator," *Wave Motion,* Vol. 37, 2003, pp. 43–61.

[5] Tosi, D., "Simultaneous Detection of Multiple Fiber-Optic Fabry-Perot Interferometry Sensors with Cepstrum-Division-Multiplexing (CDM)," *IEEE/OSA Journal of Lightwave Technology*, Vol. 34, No. 15, 2016, pp. 3622–3627.

[6] Fricke, T., "Solving an Optical Cavity Using Scattering Transfer Matrices," http://www.phys.lsu.edu/~tfricke/notebook/2010/08/29/Tmatrix.pdf.

[7] Bremer, K., et al. "Feedback Stabilized Interrogation Technique for EFPI/FBG Hybrid Fiber-Optic Pressure and Temperature Sensors," *IEEE Sensor Journal*, Vol. 12, 2012, pp. 133–138.

[8] Tosi, D., et al., "Adaptive Filter-Based Interrogation of High-Sensitivity Fiber Optic Fabry-Perot Interferometry Sensors," *Sensors and Actuators A: Physical*, Vol. 206, 2014, pp. 144–150.

[9] Roriz, P., et al., "Review of Fiber-Optic Pressure Sensors for Biomedical and Biomechanical Applications," *Journal of Biomedical Optics*, Vol. 18, No. 5, 2013, p. 050903.

[10] Poeggel, S., et al., "Fiber-Optic EFPI Pressure Sensors for In-Vivo Urodynamic Analysis," *IEEE Sensor Journal*, Vol. 14, No. 7, 2014, pp. 2335–2340.

[11] Poeggel, S., et al., "Differential In Vivo Urodynamic Measurement in a Single Thin Catheter Based on Two Optical Fiber Pressure Sensors," *Journal of Biomedical Optics*, Vol. 20, No. 3, 2015, p. 037005.

[12] Poeggel, S., et al., "Low Drift and High Resolution Miniature Optical Fiber Combined Pressure- and Temperature Sensor for Cardio-Vascular and Other Medical Applications," *IEEE Sensors Conference*, Baltimore, MD, November 2013.

[13] Chavko, M., et al., "Measurement of Blast Wave by a Miniature Fiber Optic Pressure Transducer in the Rat Brain," *Journal of Neuroscience Methods*, Vol. 159, No. 2, January 30, 2007, pp. 277–281.

[14] Tosi, D., et al., "Fiber Optic Combined FPI/FBG Sensors for Monitoring of Radiofrequency Thermal Ablation of Liver Tumors: Ex-Vivo Experiments," *Applied Optics*, Vol. 53, No. 10, 2014, pp. 2136–2144.

[15] Liu, G., M. Han, and W. Hou, "High-Resolution and Fast-Response Fiber-Optic Temperature Sensor Using Silicon Fabry-Pérot Cavity," *Optics Express*, Vol. 23, No. 6, March 23, 2015, pp. 7237–7247.

[16] Lindström, L. H., "Miniaturized Pressure Transducer Intended for Intravascular Use," *IEEE Transactions on Biomedical Engineering*, Vol. BME-17, No. 3, 1970, pp. 207–219.

[17] Gao, R., et al., "Filmed Extrinsic Fabry-Perot Inter-Ferrometric Sensors for the Measurement of Arbitrary Refractive Index of Liquid," *Sens. Actuat. B: Chem.*, Vol. 177, 2013, pp. 924–928.

[18] Liu, X., et al., "Miniature Fiber-Optic Force Sensor Based on Low-Coherence Fabry-Pérot Interferometry for Vitreoretinal Microsurgery," *Biomed. Opt. Express*, Vol. 3, No. 5, May 2012, pp. 1062–1076.

[19] Bae, H., and M. Yu, "Miniature Fabry-Perot Pressure Sensor Created by Using UV-Molding Process with an Optical Fiber Based Mold," *Opt. Expr.*, Vol. 20, 2012, pp. 14573–14583.

[20] Tosi, D., "Advanced Interrogation of Fiber-Optic Bragg Grating and Fabry-Perot Sensors with KLT Analysis," *Sensors*, Vol. 15, 2015, pp. 27470–27492.

[21] Poeggel, S., et al., "Multi FBG Femtosecond Laser Inscription in FPI Based Pressure Sensors for Temperature Distribution," *International Society for Optics and Photonics International Conference on Optical Fibre Sensors (OFS24)*, September 28, 2015, pp. 96342N–96342N.

[22] Tosi, D., E. G. Macchi, and A. Cigada, "Fiber-Optic Temperature and Pressure Sensors Applied to Radiofrequency Thermal Ablation in Liver Phantom: Methodology and Experimental Measurements," *Journal of Sensors*, Vol. 2015, 2015, p. 909012.

[23] Chen, L. H., et al., "High Performance Chitosan Diaphragm-Based Fiber-Optic Acoustic Sensor," *Sens. Actuat. A: Phys.*, Vol. 163, 2010, pp. 42–47.

[24] Xu, F., et al., "High-Sensitivity Fabry-Perot Interferometric Pressure Sensor Based on a Nanothick Silver Diaphragm," Opt. Lett., Vol. 37, 2012, pp. 133–135.

[25] Liao, C., et al., "Sub-Micron Silica Diaphragm-Based Fiber-Tip Fabry-Perot Interferometer for Pressure Measurement," *Opt. Lett.*, Vol. 39, 2014, pp. 2827–2830.

[26] Li, C., et al., "Interference Characteristics in a Fabry-Perot Cavity with Graphene Membrane for Optical Fiber Pressure Sensors," *Microsyst. Technol.*, Vol. 21, 2014, pp. 2297–2306.

[27] Cheng, L., et al., "Silk Fibroin Diaphragm-Based Fiber-Tip Fabry-Perot Pressure Sensor," *Opt. Expr.*, Vol. 24, No. 17, 2016, pp. 19600–19606.

[28] Bai, Y., et al., "Highly Sensitive Temperature and Pressure Sensor Based on Fabry-Perot Interference," *IEEE Photon. Technol. Lett.*, Vol. 28, No. 21, 2016, pp. 2471–2474.

[29] Donati, S., and M. Norgia, "Self-Mixing Interferometry for Biomedical Signals Sensing," *IEEE J. Select. Topics Quant. Electron.*, Vol. 20, No. 2, 2014.

[30] Giuliani, G., et al., "Laser Diode Self-Mixing Technique for Sensing Applications," *Journal of Optics A*, Vol. 4, 2002, pp. S283–S294.

[31] Lu, H., et al., "All-Fiber Self-Mixing Interferometer Based on DFB Laser and Phase Modulating Technique," *IEEE Photonics Technology Letters*, Vol. 23, No. 4, 2011, pp. 221–223.

[32] Ottonelli, S., et al., "Laser Self-Mixing Interferometry for Mechatronics Applications," *Sensors*, Vol. 9, 2009, pp. 3527–3548.

[33] Hast, J., et al., "Arterial Pulse Shape Measurement Using Self-Mixing Effect in a Diode Laser," *Quantum Electronics*, Vol. 32, No. 11, 2002, pp. 975–980.

[34] Rovati, L., A. Cattini, and N. Palanisamy, "Measurement of the Fluid-Velocity Profile Using a Self Mixing Superluminescent Diode," *Meas. Sci. Techn.*, Vol. 22, 2011.

[35] Tosi, D., "FBG/DFB-Induced Chaotic Self-Mixing Analysis and Sensing Applications," *Optics and Laser Technology*, Vol. 50, 2013, pp. 176–181.

[36] Udd, E., and W. B. Spillman Jr., (eds.), *Fiber Optic Sensors: An Introduction for Engineers and Scientists*. New York: John Wiley & Sons, 2011.

CHAPTER 8

Fiber-Optic Biosensors Principles

8.1 Introduction

In Chapter 4, two types of intensity-based sensors, those exploiting the modulation of propagation loss of guided light and those exploiting the coupling loss between two fibers, have been described. These implementations are best suited for the measurement of physical quantities, such as displacements or pressures.

In this chapter, two other types of sensors that rely on the variation of the received signal intensity are introduced: sensors based on the evanescent field absorption loss and sensor exploiting the surface plasmon resonance. In both cases, the typical application is in the detection of chemicals or biochemical materials.

8.2 Sensors Exploiting the Evanescent Field Absorption

In Chapter 1, we saw that the electromagnetic field of a guided mode is not going abruptly to zero at the core-cladding interface of a fiber but has an exponential decay, so it actually extends into the cladding for some distance that is related to the refractive index contrast between core and cladding materials and the wavelength.

In other words, we can also say that light propagating in the fiber interacts with the cladding material through the evanescent tails of the electromagnetic field. In particular, the guided light can experience additional propagation loss if in a portion of the fiber the low-loss material forming the cladding (e.g., ultrapure undoped silica for glass fibers) is substituted with a more lossy material. The substituting material can also have a wavelength-dependent behavior for which it absorbs a certain wavelength (hence, the increase in propagation loss at that wavelength), but it partially reemits the absorbed power at another wavelength, the fluorescence phenomenon.

This general approach has been exploited to fabricate intensiometric sensors in which the propagation loss is modulated with the change in the concentration of the quantity under measurement. This represents the basis of the evanescent-wave spectroscopy, which exploits the attenuated total internal reflection at the core-cladding interface to sense changes in the refractive index surrounding the transducer. It is therefore evident that this principle is particularly suited to sense chemicals [1].

The fabrication of such sensors requires totally or partially removing the cladding in a portion of the fiber and then either exposing the remaining fiber structure to the analyte or replacing the removed cladding with a suitable material that changes its transmission properties with the quantity under measurement. Some researchers call direct spectroscopic evanescent wave sensor the first approach and the reagent-mediated spectroscopic evanescent wave sensor the second approach.

Removing of the cladding over a given length is a critical fabrication phase because this has to be done without damaging the core or excessively weakening the structure. Large core fibers are favored from the mechanical stability point of view. Then, unless one has access to specific tools (e.g., milling systems or chemical etching systems), the easiest cladding removal is for fibers having a glass core and a polymeric cladding as the polymer can be dissolved with solvents that do not affect the glass core. Therefore, if one targets sensing probes that can be easily handled, a good compromise between size and mechanical strength could be to use multimode fibers having a large glass core (400 μm or more) and a thin polymeric cladding. However, single mode, fiber-based sensors have been also investigated for their high sensitivity and the possibility to control the evanescent field penetration depth by operating at small normalized frequency V values[1].

As the sensing mechanism is a field absorption (albeit of the evanescent tails), it is not dependent on how the variation in the propagation constant is generated; therefore, two substances inducing the same power loss will be indistinguishable. However, by suitable surface functionalization, it is possible to maximize the sensitivity to specific chemicals, thus to have a selective detection.

Surface functionalization means to modify the surface by introducing chemical functional groups: by properly choosing these functional groups it is possible to have selective binding with specific molecules only [2–4]. Figure 8.1 shows a schematic representation of the selective detection through surface functionalization. As an example of application, a recent paper dealing with the glucose detection using this approach is [5].

The theoretical evaluation of the working principle of these sensors has been the subject of many research works; for instance, [6–8] cite some of them.

As mentioned, variants of this sensing approach rely on the evanescent wave excitation of fluorescence instead of the increase in the absorption; examples can be found in [9, 10].

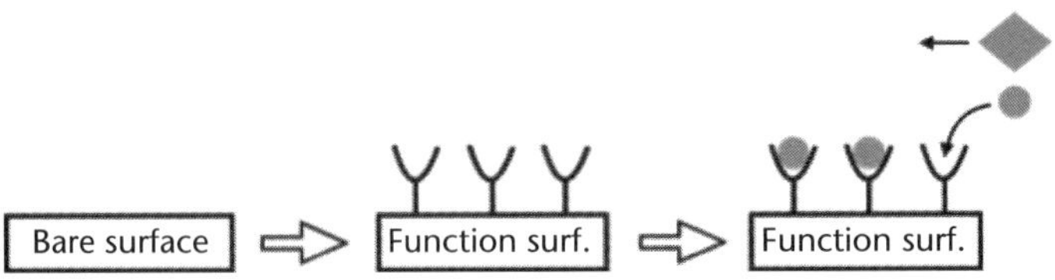

Figure 8.1 Schematic representation of the functionalization of a surface to induce a molecule selective behavior.

1. Recall that the confinement of a mode, the power fraction carried in core with respect to the overall power carried by the mode, increases as the V number moves away from the mode cutoff value.

The generic layout of the interrogation system is reported in Figure 8.2 for a sensor working in transmission.

As the sensing principle can be implemented by substituting the removed cladding with a broad variety of materials having different transmission behavior, the practical realization of the layout in Figure 8.2 could require different types of components. For example, the source can be broadband, like an LED, or narrowband, like an LD and a specific wavelength, and the receiver can be just a photodiode or a spectrometer. When using laser light sources and multimode fibers, the launch must be optimized (e.g., by launching at an angle or in an offset position) to maximize the sensitivity. This is because, as we have seen in Chapter 1, the penetration in the cladding is larger for higher-order modes, but a well-collimated laser light launched parallel to the axis of a multimode fiber will tend to excite more the lower-order modes. LEDs with their broad emission will typically excite many modes without further adjustment of the launch conditions.

The evanescent field extension, and thus the interaction with the surrounding environment, can be enhanced also by other techniques, as:

- *Fiber tapering*: Tapers are progressive change in fiber diameter produced by stretching it while it is heated to soften the glass. The propagation conditions change as consequence of the geometrical changes and this can be exploited to make the evanescent field extend for longer distances out of the core [11, 12].
- *Long period gratings (LPGs):* These are periodic structures inscribed in the fiber core and aimed at coupling forward-propagating modes in the core with cladding modes, therefore transferring power from the guided modes to cladding modes; as such, they can be used to enhance the evanescent field extension without having to remove the cladding (or at least reducing the thickness that has to be removed) [13, 14].
- *Bends*: We have already seen that bends enhance the losses by increasing the coupling between the guided modes and irradiated mode, which are field extending in the cladding; therefore, an optimum bending radius in dependence of the fiber numerical aperture and diameter can be used to couple more optical power in the evanescent field and thus enhance the sensitivity [15, 16].

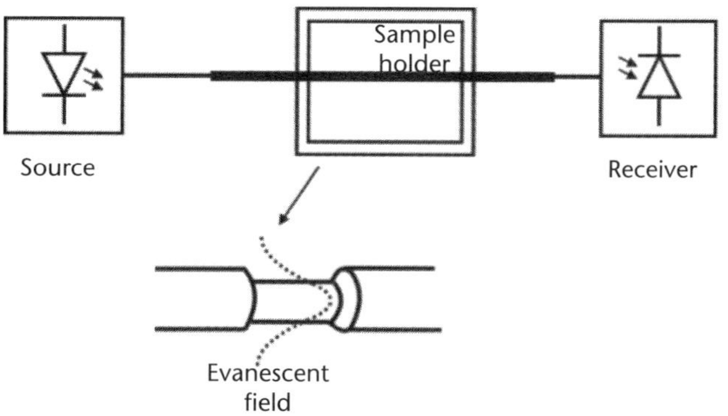

Figure 8.2 Schematic representation of a sensor working in transmission that exploits evanescent wave absorption.

In the simplest case, the setup in Figure 8.2 can be implemented with an LED (or a low cost LD) source and a photodiode-based receiver: as the attenuation in the cladding modified region changes, the received power changes.

In the literature different materials have been used as coating, targeting different quantities to be sensed. For example, substituting the cladding with a thin film of porous sol-gel containing analyte-sensitive dyes a range of sensors for pH, ammonia and oxygen have been demonstrated [17]. Others have coated the sensing region with materials that progressively but permanently change their absorption in presence of a given substance to fabricated cumulative sensors. For example, by coating with a thin silver layer, the total exposure to hydrogen sulfide can be evaluated [18, 19]; similarly, by coating with copper, fluid ingress in epoxy vinyl ester-based glass reinforced polymer (GRP) composite was measured [20].

In biosensing, this approach has been mainly used to detect specific bacteria by immobilizing suitable antigens on the fiber surface [21] or oligonucleotides in DNA hybridization assays [22]. Reviews are reported, among other types of fiber sensors, in [23, 24].

This sensing approach is very fast (the response is only limited by the reaction time of the chemical of interest with the material substituting the cladding) and nondestructive and allows in situ measurements. The latest aspect is more evident when using fiber sensors working in reflection as they can be configured in the form of probes. A word of advice, however, is important when aiming at the realization of sensors to use in clinical application: these are intensiometric sensors and, as such, they suffer from the cross-sensitivity issues due to parasitic additional losses already pointed out for the other intensiometric sensors; therefore, the performance achievable in a controlled environment, such as a laboratory, can be very different from what one can obtain in practical cases.

8.3 Sensors Exploiting Surface Plasmon Resonance

Surface plasmon resonance (SPR) can be considered a special case of the evanescent wave absorption, but, given its importance and slightly different working principle, it is discussed in a section of its own.

SPR occurs at the interface between a metal and a dielectric, when the light of a suitable polarization, wavelength, and angle of incidence hits this interface and manifests itself as a strong reduction of the light reflected from the interface. A detailed discussion of SPR-based sensors can be found in [25].

SPR sensors offer many advantages similar to the other evanescent wave sensors, such as high sensitivity, label-free, and the possibility of rapid, real-time detection.

8.3.1 SPR Sensors in Bulk Optics

Before discussing the implementation with fibers, it worthy to start with the discrete optical component version. Indeed, the easiest way to understand this phenomenon is using a configuration based on a prism, known as the Kretschmann configuration, in which a thin layer of metal is deposited on the prism as in Figure 8.3.

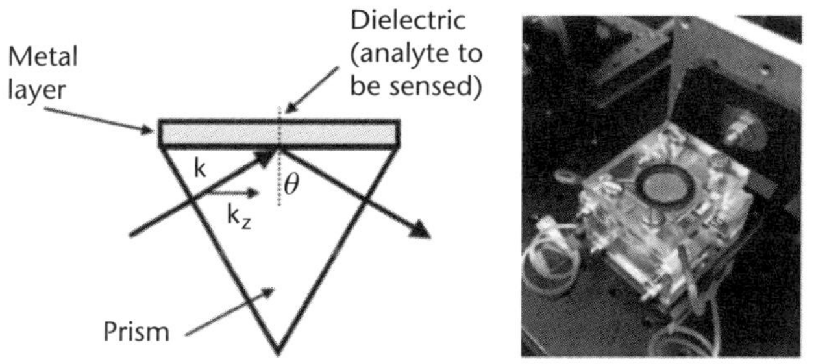

Figure 8.3 Schematic representation of the Kretschmann configuration used to excite surface plasma waves (left) and picture of a bench-top setup equipped with pipes to inject/drain a liquid analyte in contact with the metal (right).

Clearly, we expect that any incident beam from the prism side is totally reflected; however, a p-polarized (and only this polarization) incident beam at the right angle and wavelength can excite the resonant oscillations of free electrons in the metal (known as surface plasmons), resulting in the absorption of that light beam and a drop in the reflected power. Therefore, if the prism setup is illuminated by a nonmonochromatic or by a noncollimated beam (so it can be considered as composed of many incident angles), there will be a dip in the reflected light wavelength or angular spectrum corresponding to the beam components that have been absorbed. Another way to see this is that, for specific angles of incidence or wavelengths, the incident light becomes resonant with a surface plasma wave (SPW), a wave that propagates parallel to the metal-dielectric interface being evanescent in both media and characterized by high losses. In this picture, the excitation of the SPW corresponds to a matching between the component parallel to the interface of the incident wave propagation constant (k_z in Figure 8.3) and the SPW propagation constant.

The resonance conditions for which the SPW is excited, hence the angle or the wavelength for which light is absorbed and not reflected, are strongly dependent on the metal film (material and thickness) and, most important for the applications, on the refractive index of the media on either side of the metal film. Since the refractive index of the prism is defined and constant, it turns out that shifts in the dip position of the reflected spectrum can be related to changes in the refractive index (hence, in the composition) of the material in contact with the metal layer. We have therefore realized a sensor of chemical composition, actually a very sensitive sensor as it is possible to detect variations in the order of few units in 10^{-7} RIU (refractive index units, the usual way to express the refractive index) [26].

Figure 8.4 shows an example of the shift in the spectral response dip position versus the beam incidence angle (left) or wavelength (right) for two refractive index values.

Although the elective interrogation systems for these kinds of sensors are those known as the angular interrogation and the wavelength interrogation, other techniques such as the phase interrogation and the amplitude interrogation have been investigated as well in the literature [27]. The angular interrogation uses a laser source (therefore at a fixed wavelength) with the beam that spans a certain angular range of incidence and photodiode at the receiver; the obtained measurement result is similar to that in Figure 8.4 (left). The wavelength interrogation uses a

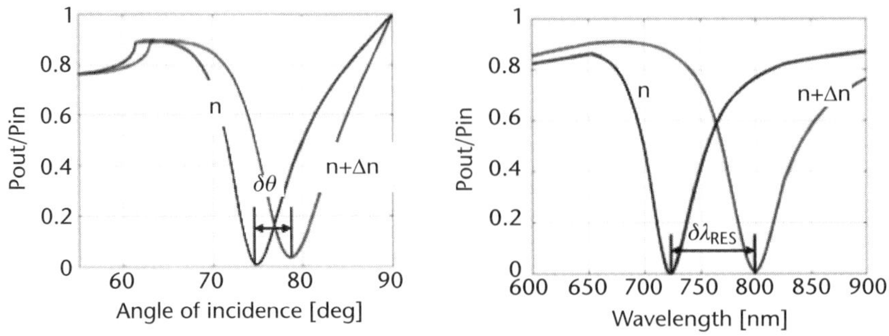

Figure 8.4 Example of the shift in the SPR spectral response dip position versus the beam incidence angle (left) or wavelength (right) for two refractive index values.

broadband light source launched at a fixed angle and spectrometer (or an OSA) at the receiver; the obtained measurement result is similar to that in Figure 8.4 (right).

Regardless of the interrogation method, as refractive index values are sensitive to the temperature, unless the temperature is actually the quantity of interest, it is important to perform the measurements at constant temperatures or to use compensation techniques [28]. The exact impact of the temperature on the measurement result depends on the sensor implementation: as an example, for the configuration considered in the cited paper, a temperature variation of ±20°C produces a shift in the SPR spectrum obtained measuring a water-based solution that is undistinguishable from changes in the analyte refractive index of ~$\pm 10^{-3}$ RIU, a value that is well above the resolution of SPR sensors.

As for the metal used for the realization of the sensor, it must have conduction band electrons capable of resonating with the incoming light at a suitable wavelength, it must be free from oxides and sulfides, and it must not react with other molecules in the atmosphere or in the liquid in contact. Very few metals satisfy all these requirements, with gold being the most practical. Another important parameter is the metal thickness: a too thick layer will cause the dip in the reflected spectrum become shallow, while a too thin layer will broaden the dip. In both cases, the detection of dip shifts is largely compromised. A typical value for gold layers is from 40 nm to 70 nm. The best detection results require a purely p-polarized light to obtain a sharp dip because all non-p-polarized light does not contribute to the SPR and thus increases the reflected light, causing a sort of background noise.

Similar to the case of evanescent wave spectroscopy, SPR sensors are sensitive to refractive index variations regardless of the specific molecules that caused them. However, also in this case, it is possible to functionalize the surface to allow a selective detection [29]. The prism surface can also be subdivided into different portions to allow parallel sensing of different molecules. Such approach has been implemented in commercial machines for the chemical analysis and DNA sequencing.

Besides the use of compact metal layers, another form of SPR can be obtained by using nanoparticles, the localized SPR (LSPR). In this case, the induced plasmons oscillate locally to the nanostructure, with a much shorter decay length (order of few nanometers) that produces a strong resonance absorbance peak less sensitive to bulk refractive index changes, including the temperature effect [30, 31]. Nanoparticles too can be properly functionalized to allow the selective detection of specific molecules only [32, 33].

8.3.2 SPR Sensors in Optical Fibers

The SPR principle described in the previous section has also been transferred in fiber structures [34], although the results are still more confined to the lab demonstrator because in fibers the performance is intrinsically worse due to the practical impossibility to control the wave polarization and the effects of unavoidable cross-sensitivities (for example, coming from strain due to deformations in the fiber, opposite from the prism structure that is fixed). However, the fiber approach is very attractive because it allows the realization of probes for in situ measurements.

Figure 8.5 shows the schematic representation of some possible implementations of SPR sensors in fibers.

There are many kinds of available fibers, which differ in size, materials (for example, glass or plastic) and behavior (for example, multimode or single mode), and there many proposed structures for SPR optical fiber sensors. A first rough classification can distinguish between transmission reflection types: the difference is that in reflection type the light comes out from the same side of the incident light; this makes the configuration more suitable for a probe-like sensor and light interacts twice with the sensing region, but requires couplers to separate incident and reflected light beams (and these may spoil the sensor performance if they have poor directivity).

Then tapers can be used to enhance the sensitivity of fiber SPR, similarly to the case of evanescent wave sensor discussed in the preceding section [35, 36].

The first step for the fabrication of a SPR fiber sensor is the same of that of the evanescent wave sensor: locally removing the cladding over a given length without damaging the core or excessively weakening the structure. Although SPR sensors in single-mode fibers have been reported [37], aiming at the development of probes or applicators, again, the best compromise is obtained with multimode fibers, such as those having a large glass core for the strength and a thin polymeric cladding for easy removal [38]. Alternatively, all plastic fibers have been successfully used [39, 40], especially to realize low-cost disposable sensors by exploiting the simplified connection with sources thanks to their extremely large multimode behavior.

The most common interrogation setup for these sensors makes use of a broadband source, typically an SLED, and a spectrometer. An example is shown in Figure 8.6: on the left there is a picture of a homemade setup to test fiber sensor, and

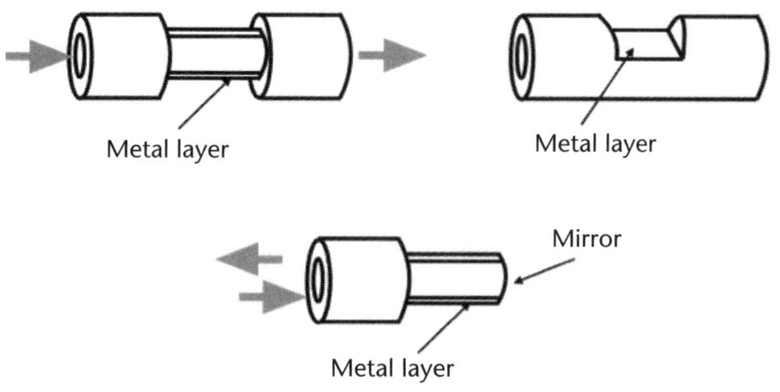

Figure 8.5 Schematic representation of three possible implementation of SPR sensors in fibers: the two above work in transmission, while that in the bottom in reflection.

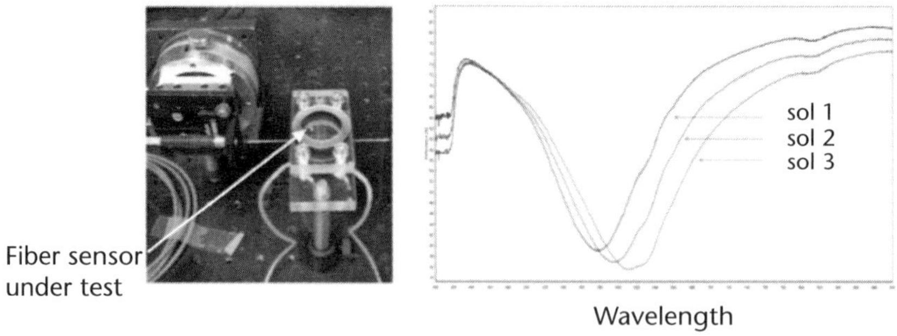

Figure 8.6 Picture of a lab setup to test fiber sensor (left) and examples of the readings obtained for spectral interrogation of a sensor made with a 400-μm core fiber coated with 50 nm of gold for three different solutions (pure water, and water and glycerol in two different concentrations, with refractive index values of 1.333, 1.346, and 1.357, respectively).

on the right the results obtained for a fiber sensor made with a 400-μm core fiber coated with 50 nm of gold for three different water-based solutions.

The spectrometer resolution, dynamic range, and integration time strongly influence the achievable sensor resolution: the usually quite poor performance of portable instruments can be improved by signal processing routines, such as filtering and fitting of the measurement with the theoretical models. However, in practice, it is very difficult to achieve the detection limits of prism-based setups.

In lower-cost implementation, the spectrometer can be avoided by using a cheap fiber coupled laser and a photodiode for receiver and relying for the measurement on the variation of the received power at a single wavelength. The principle, similar to one of the possible approaches used also to interrogate fiber Bragg grating sensors, is sketched in Figure 8.7. In practice, using the setup in Figure 8.7(a) a shift in the SPR spectral peak position $\Delta\lambda$ is converted into a change in the received power ΔP, Figure 8.7(b). The drawback of such an approach is that, as usual, any fluctuation in the received power in undistinguishable from actual wavelength shifts of the SPR sensor response. The laser diode power drifts can be compensated by

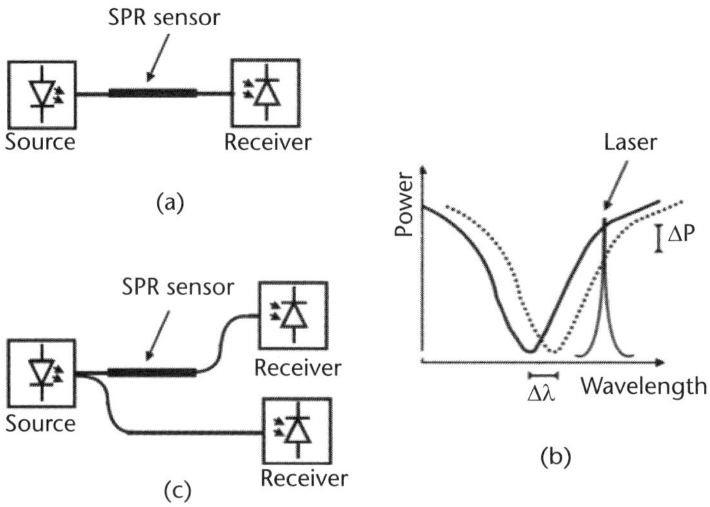

Figure 8.7 Sketch of a low cost interrogation setup using a fixed wavelength laser and a photodiode in which a variation in the peak position $\Delta\lambda$ is converted into a change in the received power ΔP.

tapping part of the power and directing it at a second receiver as in Figure 8.7(c), but wavelength fluctuations or noise in the SPR sensor curve severely reduce the sensor resolution and repeatability.

Aiming at provide multisensing capabilities, different sensing areas, with different spectral responses, can be cascaded along the fiber [28, 41].

The high dissipation occurring at the interface between dielectric and metal layers, besides for sensing, can also be exploited to realize a sort of laser-based hot knife, made by depositing a metal layer on the core of a fiber as for SPR sensors, but using a high-power laser diode to feed the applicator instead of a SLED for the interrogation. This way, the metal-coated tip of the probe by absorbing the laser light can reach a high temperature, finding possible application as alternatives to more classical laser approaches, for example, in the solid tumor ablation or in the varicose vein endovenous treatment [42].

8.3.3 SPR Sensors in Photonic Crystal Fibers

Highly sensitive fiber SPR sensors can also be implemented using photonic crystal fibers (PCFs), also known as holey fibers or microstructured fibers. PCFs are a relatively new type of fibers first reported in the mid-1990s [43], although early investigations date back of almost 20 years earlier [44], that exploit not only the refractive index difference to confine light as in ordinary fibers, but also structural modifications. A review of various kinds of PCFs with details on their working mechanism can be found in [45].

The simplest case is that depicted in Figure 8.8, which shows a schematic representation and a picture of the cross-section of a solid-core PCF. This is a fiber made of all of pure silica (hence, no difference in doping to define the core and cladding regions), but with some holes positioned in a periodic arrangement around the center of fiber. Some holes may be missing (the introduction of defects in the periodic arrangement) to define the equivalent of the core region in conventional fibers. In other words, we can say that in these kinds of fibers, the confinement is not obtained from the spatially varying glass composition, but from the arrangement of very tiny and closely spaced air holes. In particular, the fiber shown in Figure 8.7 is made with one hole missing.

In general, the guiding property of PCFs can be tailored by changing the hole size, shape, and position.

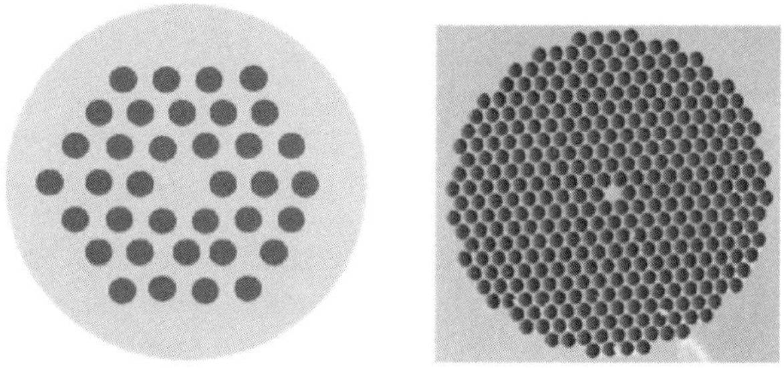

Figure 8.8 Schematic representation of the cross-section of a typical photonic crystal fiber (left) and picture of an actual fiber (right).

As seen in Chapter 1, conventional fibers guide light exploiting the total internal reflection (TIR); in PCFs, the light confinement properties can be derived from the theory of periodic structures, which explains how in these cases there can exist forbidden bands in the dispersion relation, similar to what happens in crystals (hence, the name of these fibers) for electrons. It turns out that three guidance mechanisms exist for PCF, the most common being a modified form of TIR and the presence of a photonic bandgap (PBG) [45].

The simple case of Figure 8.8 can be explained considering that the region with the missing hole has a higher effective refractive index than the surrounding region in which the index is a sort of average between that of the solid material and that of the air (or of the material filling the holes).

Some PCFs have a hollow-core surrounded by a microstructured cladding [also called photonic bandgap (PBG), fibers for the equivalence with the bandgap, the forbidden region in semiconductors]. In this case, the period structure allows light guidance in a low-index region, opposite of what happens in ordinary fibers. A periodic structure determining the photonic bandgap acts as a virtually loss-free mirror capable of confining light to the core region. As in conventional single-mode fibers, the fundamental mode in hollow-core PFCs has a quasi-Gaussian intensity distribution; however, they are more difficult to use as they require special equipment to be properly terminated (e.g., seal the ends to prevent moisture and dust from entering the hollow capillary structure) and coupled to standard fibers.

The combination of a holey structure that can be easily impregnated with gases or liquids with the possibility to tailor the fiber properties by changing the parameters that define the holes gives many degrees of freedom for the realization of sensors, mainly for chemicals, using PCFs.

Indeed, both the large sensitivity of the mode propagation constant and/or the absorption rate with the hole filling material (e.g., [46, 47]) and the surface plasmon resonance (e.g., [48–51], just to cite most recent works) have been exploited.

SPR sensors using PCFs can have quite a complex structure, but can lead to record sensitivities, exceeding 7,000 nm/RIU (compared to the typically few thousands of nanometers per RIU).

Other uses of PCF in biomedical field included the detection of interaction forces generated in surgical devices, although not exploiting intensity-based sensors [52].

References

[1] Bosch, M. E., et al., "Recent Development in Optical Fiber Biosensors," *Sensors*, Vol. 7, 2007, pp. 797–859.

[2] Marques, M. E., A. A. P. Mansur, and H. S. Mansur, "Chemical Functionalization of Surfaces for Building Three-Dimensional Engineered Biosensors," *Applied Surface Science*, Vol. 275, 2013, pp. 347–360.

[3] Hosseini, S., et al., "Recent Advances in Surface Functionalization Techniques on Polymethacrylate Materials for Optical Biosensor Applications," *Analyst*, Vol. 139, 2014, pp. 2933–2943.

[4] Reimhult, E., and F. Höök, "Design of Surface Modifications for Nanoscale Sensor Applications," *Sensors*, Vol. 15, 2015, pp. 1635–1675.

[5] Yetisen, A. K., et al., "Glucose-Sensitive Hydrogel Optical Fibers Functionalized with Phenylboronic Acid," *Advanced Materials*, Vol. 29, 2017.

[6] Gupta, B. D., and C. D. Singh, "Fiber-Optic Evanescent Field Absorption Sensor: A Theoretical Evaluation," *Fiber and Integrated Optics*, Vol. 13, 1994, pp. 433–443.

[7] Messica, A., A. Greenstein, and A. Katzir, "Theory of Fiber-Optic, Evanescent-Wave Spectroscopy and Sensors," *Applied Optics*, Vol. 35, 1996, pp. 2274–2284.

[8] Xu, Y., et al., "Theoretical Analysis of the Evanescent Wave Absorption Coefficient for Multimode Fibre-Optic Evanescent Wave Absorption Sensors," *J. Modern Optics*, Vol. 46, 2009, pp. 2007–2014.

[9] Yu, S. -H., C. -H. Chen, and H. K. Chiang, "Fiber Optic Evanescent Wave Biosensor for Biomolecular Detection," *2nd Asian and Pacific Rim Symposium on Biophotonics*, 2004.

[10] Taitt, C. R., G. P. Anderson, and F. S. Ligler, "Evanescent Wave Fluorescence Biosensors," *Biosensors and Bioelectronics*, Vol. 20, 2005, pp. 2470–2487.

[11] Shriver-Lake, L. C., et al., "The Effect of Tapering the Optical Fiber on Evanescent Wave Measurements," *Anal. Lett.*, Vol. 25, 1992, pp. 1183–1199.

[12] Golden, J. P., et al., "An Evanescent Wave Biosensor. II. Fluorescent Signal Acquisition from Tapered Fiber Optic Probes," *IEEE Transactions on Biomedical Engineering*, Vol. 41, 1994, pp. 585–591.

[13] Guo, S., and S. Albin, "Transmission Property and Evanescent Wave Absorption of Cladded Multimode Fiber Tapers," *Opt. Express*, Vol. 11, 2003, pp. 215–223.

[14] Waechter, H., et al., "Chemical Sensing Using Fiber Cavity Ring-Down Spectroscopy," *Sensors*, Vol. 10, 2010, pp. 1716–1742.

[15] Khijwania, S. K., and B. D. Gupta, "Maximum Achievable Sensitivity of the Fiber Optic Evanescent Field Absorption Sensor Based on the U-Shaped Probe," *Opt. Comm.*, Vol. 175, 2000, pp. 135–137.

[16] Elosua, C., et al., "Volatile Organic Compound Optical Fiber Sensors: A Review," *Sensors*, Vol. 6, 2006, pp. 1440–1465.

[17] Mac Craith, B. D., et al., "Fibre Optic Chemical Sensors Based on Evanescent Wave Interactions in Sol-Gel-Derived Porous Coatings," *J. Sol-Gel Sci. Technol.*, Vol. 2, 1994, pp. 661–665.

[18] Angelini, E., et al., "Plasma Modified POF Sensors for In Situ Environmental Monitoring of Museum Indoor Environments," *Appl. Phys. A*, Vol. 100, 2010, pp. 975–980.

[19] Grassini, S., et al., "Design and Deployment of Low-Cost Plastic Optical Fiber Sensors for Gas Monitoring," *Sensors*, Vol. 15, 2015, pp. 485–498.

[20] Milsom, B., et al., "Glass Optical Fibre Sensors for the Detection of Through Thickness Moisture Diffusion in Glass Reinforced Composites Under Hostile Environments," *Advances in Applied Ceramics*, Vol. 114, 2015, pp. S76–S83.

[21] Valadez, A. M., et al., "Evanescent Wave Fiber Optic Biosensor for Salmonella Detection in Food," *Sensors*, Vol. 9, 2009, pp. 5810–5824.

[22] Abel, A. P., et al., "Fiber-Optic Evanescent Wave Biosensor for the Detection of Oligonucleotides," *Anal. Chem.*, Vol. 68, 1996, pp. 2905–2912.

[23] Leung, A., P. Mohana Shankar, and R. Mutharasan, "A Review of Fiber-Optic Biosensors," *Sensors and Actuators B*, Vol. 125, 2007, pp. 688–703.

[24] Anderson, G. P., and C. R. Taitt, "Evanescent Wave Fiber Optic Biosensors," in *Optical Biosensors: Today and Tomorrow*, 2nd ed., F. S. Ligler and C. R. Taitt, (eds.), New York: Elsevier, 2008, pp. 83–138.

[25] Homola, J., (ed.), *Surface Plasmon Resonance Based Sensors*, New York: Springer, 2006.

[26] Fan, X., (ed.), *Advanced Photonic Structures for Biological and Chemical Detection*, New York: Springer, 2014.

[27] Zeng, Y., et al., "Recent Advances in Surface Plasmon Resonance Imaging: Detection Speed, Sensitivity, and Portability," *Nanophotonics*, Vol. 6, 2017, pp. 1017–1030.

[28] Fallauto, C., et al., "Compensated Surface Plasmon Resonance Sensor for Long-Term Monitoring Applications," *IEEE Transactions on Instruments and Measurements*, Vol. 63, 2014, pp. 1287–1292.

[29] Oliverio, M., et al., "Chemical Functionalization of Plasmonic Surface Biosensors: A Tutorial Review on Issues, Strategies and Costs," *ACS Appl. Mater Interfaces*, 2017.

[30] Sagle, L. B., et al., "Advances in Localized Surface Plasmon Resonance Spectroscopy Biosensing," *Nanomedicine*, Vol. 6, 2011, pp. 1447–1462.

[31] Hammond, J. L., et al., "Localized Surface Plasmon Resonance as a Biosensing Platform for Developing Countries," *Biosensors*, Vol. 4, 2014, pp. 172–188.

[32] Shynkarenko, O. V., and S. A. Kravchenko, "Surface Plasmon Resonance Sensors: Methods of Surface Functionalization and Sensitivity Enhancement," *Theoretical and Experimental Chemistry*, Vol. 51, 2015, pp. 273–292.

[33] Ma, T., et al., "Surface Functionalization of SPR Chip for Specific Molecular Interaction Analysis Under Flow Condition," *Journal of Innovative Optical Health Sciences*, Vol. 10, 2017, p. 1650040.

[34] Sharma, A. K., and B. D. Gupta, "Fiber-Optic Sensors Based on Surface Plasmon Resonance: A Comprehensive Review," *IEEE Sensors Journal*, Vol. 7, 2007, pp. 1118–1129.

[35] Srivastava, S. K., and B. D. Gupta, "A Multitapered Fiber-Optic SPR Sensor with Enhanced Sensitivity," *IEEE Phot. Technol. Lett.*, Vol. 23, 2011, pp. 923–925.

[36] Lin, H. -Y., et al., "Tapered Optical Fiber Sensor Based on Localized Surface Plasmon Resonance," *Optics Express*, Vol. 20, 2012, pp. 21693–21701.

[37] Slavık, R., J. Homola, and J. Ctyroký, "Single-Mode Optical Fiber Surface Plasmon Resonance Sensor," *Sensors and Actuators B: Chemical*, Vol. 54, 1999, pp. 74–79.

[38] Dhara, P., et al., "Reflectance-Based Low-Cost Disposable Optical Fiber Surface Plasmon Resonance Probe with Enhanced Biochemical Sensitivity," *Opt. Eng.*, Vol. 55, 2016, pp. 1–7.

[39] Cennamo, N., et al., "Low Cost Sensors Based on SPR in a Plastic Optical Fiber for Biosensor Implementation," *Sensors*, Vol. 11, 2011, pp. 11752–11760.

[40] Cennamo, N., et al., "An Easy Way to Realize SPR Aptasensor: A Multimode Plastic Optical Fiber Platform for Cancer Biomarkers Detection," *Talanta*, Vol. 140, 2015, pp. 88–95.

[41] Verma, R., S. K. Srivastava, and B. D. Gupta, "Surface Plasmon Resonance Based Multi-Channel and Multi-Analyte Fiber Optic Sensor," *Proc. SPIE*, Vol. 8351, 2012.

[42] Gassino, R., et al., "Comparison of Surface Micro-Structured and Plasmonic All-Fiber Delivery Probes for Laser-Induced Thermotherapy of Tumor Cells," *Proc. SPIE*, Vol. 9702, 2016, pp. 1–6.

[43] Knight, J. C., et al., "All-Silica Single-Mode Fiber with Photonic Crystal Cladding," *Opt. Lett.*, Vol. 21, 1996, pp. 1547–1549; Erratum, Vol. 22, 1997, pp. 484–485.

[44] Yeh, P., A. Yariv, and E. Marom, "Theory of Bragg Fiber," *J. Opt. Soc. Am.*, Vol. 68, 1978, pp. 1196–1201.

[45] St. J. Russell, P., "Photonic Crystal Fibers," *J. Lightwave. Technol.*, Vol. 24, 2006, pp. 4729–4749.

[46] Wu, B., et al., "A Photonic Crystal Fiber Sensor Based on Differential Optical Absorption Spectroscopy for Mixed Gases Detection," *Optik—International Journal for Light and Electron Optics*, Vol. 125, 2014, pp. 2909–2911.

[47] Morshed, M., et al., "Microstructure Core Photonic Crystal Fiber for Gas Sensing Applications," *Applied Optics*, Vol. 54, 2015, pp. 8637–8643.

[48] Fan, Z., et al., "High Sensitivity of Refractive Index Sensor Based on Analyte-Filled Photonic Crystal Fiber with Surface Plasmon Resonance," *IEEE Photonics Journal*, Vol. 7, 2015.

[49] Popescu, V. A., N. N. Puscas, and G. Perrone, "Simulation of the Sensing Performance of a Plasmonic Biosensor Based on Birefringent Solid-Core Microstructured Optical Fiber," *Plasmonics*, Vol. 12, 2016, pp. 905–911.

[50] Luan, N., and J. Yao, "A Hollow-Core Photonic Crystal Fiber-Based SPR Sensor with Large Detection Range," *IEEE Photonics Journal*, Vol. 9, 2017.

[51] Li, D., et al., "High Sensitivity Refractive Index Sensor Based on Multicoating Photonic Crystal Fiber with Surface Plasmon Resonance at Near-Infrared Wavelength," *IEEE Photonics Journal*, Vol. 9, 2017.

[52] Rajan, G., et al., "Photonic Crystal Fiber Sensors for Minimally Invasive Surgical Devices," *IEEE Transactions on Biomedical Engineering*, Vol. 59, 2-12, pp. 332–338.

CHAPTER 9

Optical Fiber Spectroscopy

The sensing systems described so far identify active regions, where the sensing mechanism is confined, and operate by modulating light spectrum, amplitude, or polarization within this region. Next to sensors, other techniques can be employed particularly in biological detection, measuring the spectral properties of tissues, biological fluids, or biomarkers [1]; these types of measurements are only improperly referred as sensors, because there is no transduction, and belong to the broad area of optical fiber spectroscopy [2]. Spectroscopy is not a main content of this book, but is important to understand through a brief discussion its key principles and applications because it paves the road for optical biosensors, able to detect changes in biological constituents.

The schematic of a typical spectroscopy layout is in Figure 9.1, as illustrated in [3]. A light source, emitting through the whole spectrum under consideration, is used to illuminate a fiber-optic probe; in alternative, a tunable laser diode or a supercontinuum light source can be used. Light is delivered to a probe, composed by one or multiple fibers; the output signal can be collected with a spectrometer, or a Fourier transform infrared reflectometer (FTIR). Depending on the tissue and its constituents, spectroscopy can operate in UV, visible (VIS), near-infrared (NIR), and infrared (IR) optical windows; instruments as [1, 2] can cover the entire optical range from UV to IR.

The probe can be arranged in different configuration, depending on how the unknown spectrum is measured. Reflection spectroscopy is measured without contact, exploiting Snell's reflection coefficients at the surface boundary; this configuration is often used in medicine for the analysis of hard tissues such as muscles content, as in Figure 9.1(b). A contact measurement can be performed with attenuated total reflectance (ATR), using two fibers as in bifurcated fiber bundle. The reflected component is analyzed both in the wavelength domain, and in amplitude; this technique is used for contact measurements of soft tissues, and to detect biological fluids. Another configuration, similar to ATR, involves the detection of diffused reflectance spectroscopy; this measurement involves the processing of the spectra diffusively reflected at the boundary of diffusive boundaries, and is used to discriminate biological constituents of solid tumors, as will be explained in Chapter 21. In absorption spectroscopy, the measurement is performed in transmission, detecting the transfer function through the measurand medium, and collecting the transmission with a second fiber aligned to the launch fiber. Most spectroscopy configurations resemble fiber bundle configurations, using Snell's law to collect

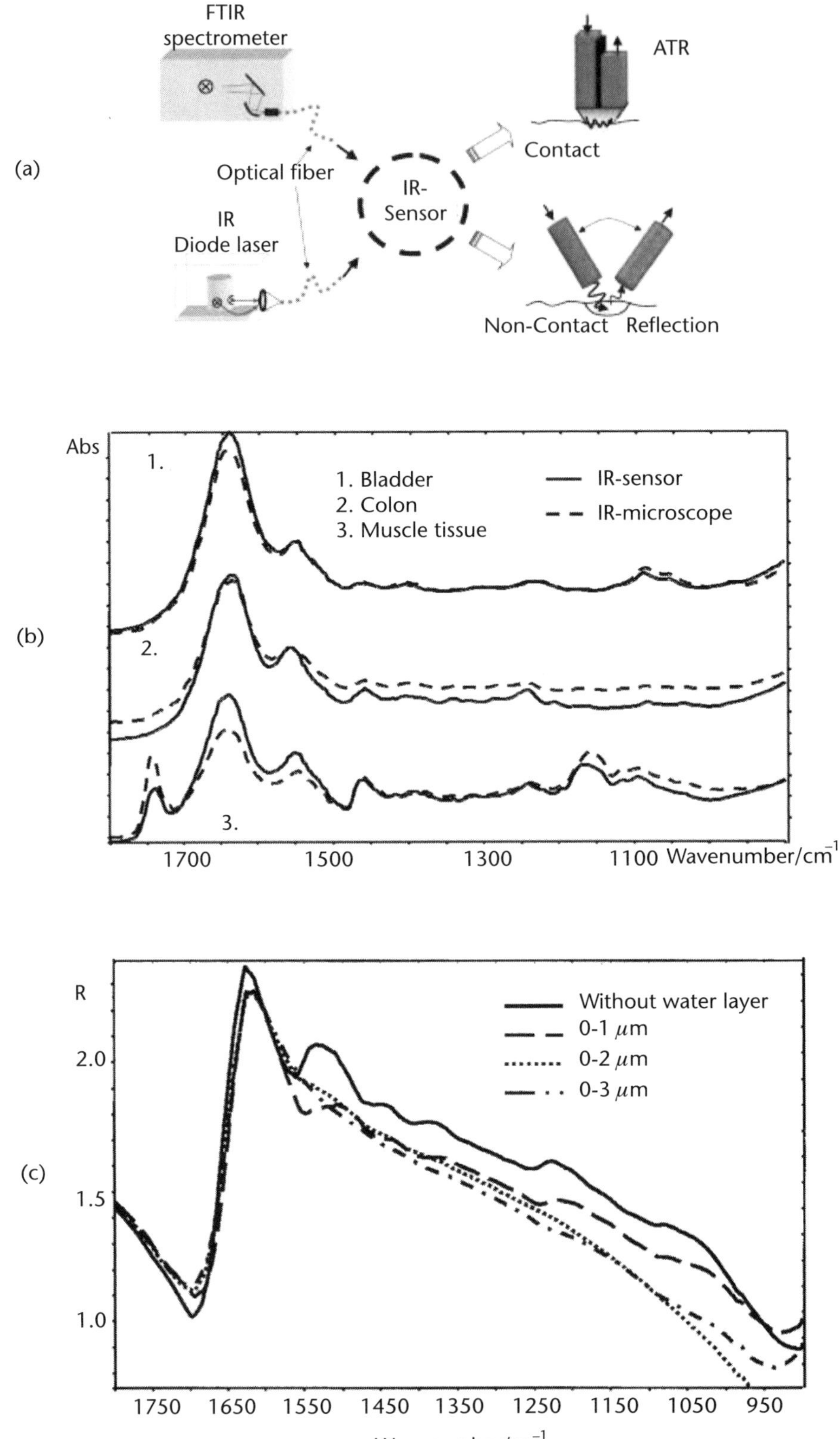

Figure 9.1 Fiber-optic spectroscopy for medical detection. (a) Typical layout of a spectroscopy system; (b) experimental spectroscopy for samples of tissues, measured with FTIR-ATR; (c) IR-reflectance spectroscopy for muscle tissues with increasing water contents. (From: [3].)

a backreflection from the measurand. Raman spectroscopy [2] is also used for the analysis of biomarkers; in Raman scattering, the shift in backreflected Raman wavelength is detected with a FTIR system.

The analysis of spectroscopic response is a weaker approach, compared to FOS; the reason is that spectroscopy is a label-free technique that does not detect a single constituent but obtains a response that depends on the whole mixture. Cross-sensitivities are often the main source of uncertainty in spectroscopy. Sensors, instead, using SPR approaches with aptamers immobilized on the sensor surface, have the capability to bind to specific biomarkers, and for this reason they are a more powerful approach. The task for researchers is to develop FOS technologies that build on spectroscopy applications already established, but have the capability to identify a biomarker without the impact of cross-sensitivities.

References

[1] Utzinger, U., and R. R. Richards-Kortum, "Fiber Optic Probes for Biomedical Optical Spectroscopy," *Journal of Biomedical Optics*, Vol. 8, No. 1, January 2003, pp. 121–147.

[2] Lin, K., et al., "Real-Time In Vivo Diagnosis of Laryngeal Carcinoma with Rapid Fiber-Optic Raman Spectroscopy," *Biomed. Opt. Expr.,* Vol. 7, No. 9, 2016, pp. 3705–3715.

[3] Bindig, U., et al., "Fibre-Optic IR-Spectroscopy for Biomedical Diagnostics," *Spectroscopy,* Vol. 17, 2003, pp. 323–344.

CHAPTER 10

Fiber-Optic Sensor Networks

10.1 Introduction

In the previous chapters, we described the working principles of several FOS configurations, focusing on the operation, sensitivity, fabrication, and which measurand(s) are accessible for each sensing configuration. In this chapter, we focus on the hardware part, which for FOS corresponds to the interrogation of the sensors. An interrogator is a device that accomplishes the goal of delivering the light in a fiber-optic cable to the sensing point(s), and collect the reflection or transmission components with a suitable detector. Depending on the working principle utilized for the sensing probe, the interrogator can operate with a broadband or laser source and, respectively, collect the output with a spectrometer (or similar) device or with a photodetector. All hardware is embedded in a single box, which serves as the optical hardware for the system.

The design of an interrogator is essential not only because it determines the performance of the system as much as the sensing principle and the transducing effects, but also because it allows bringing a technology from the laboratory to the market. Experimental research is typically carried out with bench-top instrumentation; commercial interrogators instead must respect different requirements that are derived from each application, and include:

- *Cost:* The cost of the interrogator (as a whole box, including all its components) must be limited in accordance to the specific application.
- *Size:* The size, weight, and shape of the interrogator depend on each application; typical requirements are to fit the size of a medical instrument rack or cart or to have a portable device.
- *Power consumption:* Low-power consumption is required for portable or battery-powered devices.
- *Recording speed and format:* The speed of data recording for each channel must be compatible with the application, and as a rule of thumb >10 times faster than the fluctuations of the measurand; for example, to measure a blood pressure pattern with variations of 0.1 second, the interrogator response time must be <10 ms for an accurate detection.

While in the 2000s the majority of commercial and research works were oriented to the working principles, with little emphasis on the hardware, in the latest 7-year period we are witnessing a progressive standardization of the interrogation hardware into few categories: intensity-based systems (compatible with any IBS principle), white light setups and scanning-source setups (compatible with FBG, FPI, SPR, and their combinations), and distributed sensors. In addition, a plurality of commercial systems has been industrialized. In 2016, the IEEE Sensors Council discussed standard proposals for FOS hardware.

In this chapter, we will detail the main interrogator architectures that are compatible with the sensors described earlier in the book, including hands-on expertise. This chapter will help readers to assemble their own interrogator or selecting the device closer to the application-driven specifications and is therefore aimed at a very practical but fundamental task for FOS systems. First, we will provide the description of the most standard forms of interrogators, providing also insights and photographic views of self-assembled interrogators as a guidance for readers; aspects such as sensor multiplexing and compensations are also discussed with practical outlook. Later, we will discuss and compare commercial instruments, highlighting main performance and differences.

10.2 Amplitude Detection Methods

10.2.1 FBG and Fabry-Perot Interferometers

Intensity-based sensors based on glass fiber exploit a very simple principle of operation and are among the most inexpensive fiber optic systems. However, some arrangements need to be performed in order to effectively operate in real-case scenarios. Two reference systems for an effective IBS sensing operation, based on glass fibers, are shown in Figure 10.1.

In Figure 10.1(a), the setup operates with sensors in transmission-mode. The source is a laser, operating at a single wavelength (λ_L), and finely controlled by a thermo-electric controller (TEC). It is essential to operate with a TEC in order to maintain a stable output power and a fine control of the emission wavelength. The laser needs to be isolated, and the insertion of an additional isolator (>20 dB) is recommended. In the figure, the sensing system is designed to provide several compensations: the sensor is connected to the branch (3), connected with a lead-out connector to the interrogator, and through a lead-in connector. This setup analyzes variations of the transmission spectrum of the sensor, at the laser wavelength λ_L. The main weakness of this setup is that the outer spectrum depends also on power and wavelength fluctuations of the optical source, which needs to be compensated. For this reason, the branch (1) is directly connected to the receiver, and its output is independent on sensor readout; sensor (2), internal to the device and having a reference spectrum equal to the sensor under test can be used to compensate slight variations of λ_L. Each channel is connected to a receiver, based on a fiber-coupled photodiode (PD), followed by a transimpedance amplifier (TIA). For a stable operation, it is highly recommended to use a TIA with reverse bias (typically –3V to –5V), followed by a TIA based on a low-noise operational amplifier with transimpedance gain $\leq$ 100 kΩ. Data acquisition (DAQ) is then performed for each

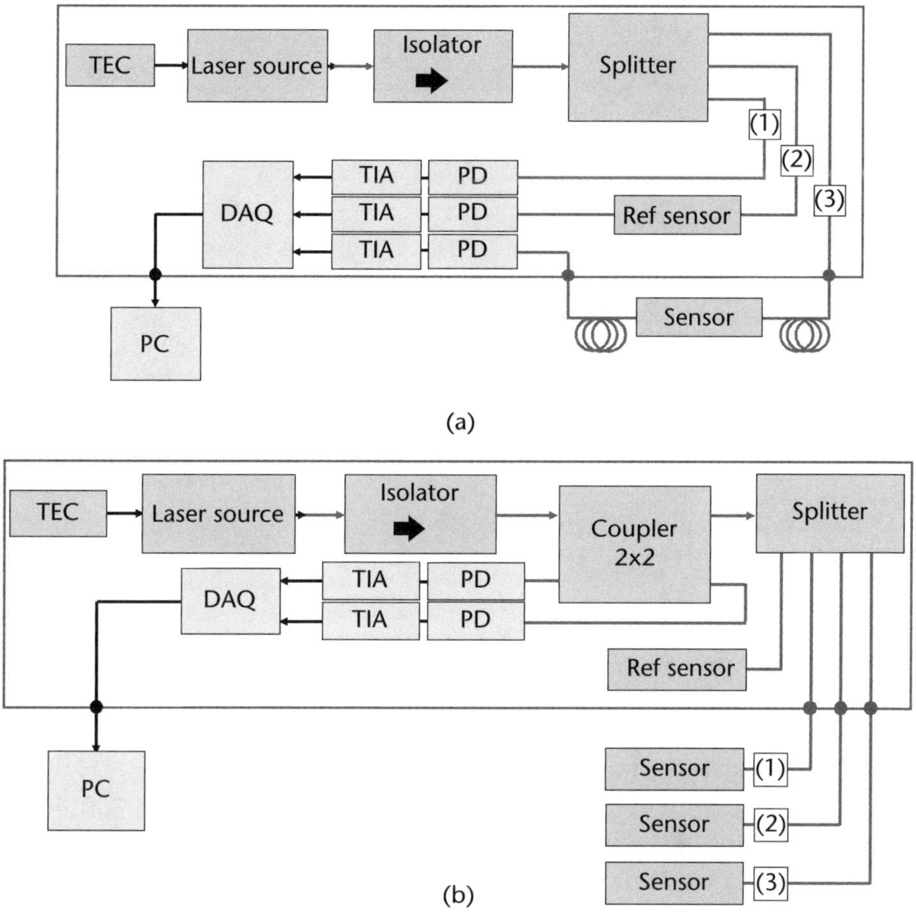

Figure 10.1 Two reference setups for IBS sensing for operation with FBG, FPI, or similar sensors, based on laser source. (a) Transmission-mode operation; (b) reflection-mode operation. The interrogator instrument encloses all components within the rectangular box, with lead in-out connectors represented as dots.

channel, converting to digital signals readable and suitable for postprocessing with a PC. In transmission-mode operation, the interrogator box includes all fiber-optic components and detection circuits. Often, a switch $1 \times N$ is used to multiplex different channels. In this configuration, the interrogation box can mount two sets of N connectors (lead-in and lead-out).

In biomedical applications, often it is preferred to operate in reflection mode, in order to have a single fiber propagating light to the sensor, and collecting its spectrum. The configuration is shown in Figure 10.1(b) [1] follows this principle, visualized for an array of 3 sensors. A 2×2 coupler is used to decouple the source and the detector(s) sides. One branch of the coupler is used to collect the power of the source, compensating for its fluctuations; the other branch is dedicated to the sensing array.

Figure 10.2 shows a photograph of the IBS interrogator developed at Politecnico di Torino, Italy, with off-the-shelf component. The laser source is a 1,550-nm diode (Alcatel Lucent, 1,550 nm), included in a 14-pin butterfly mount. The laser is controlled by a low-cost TEC (Teamwavelength DTC0520). A 3-port circulator

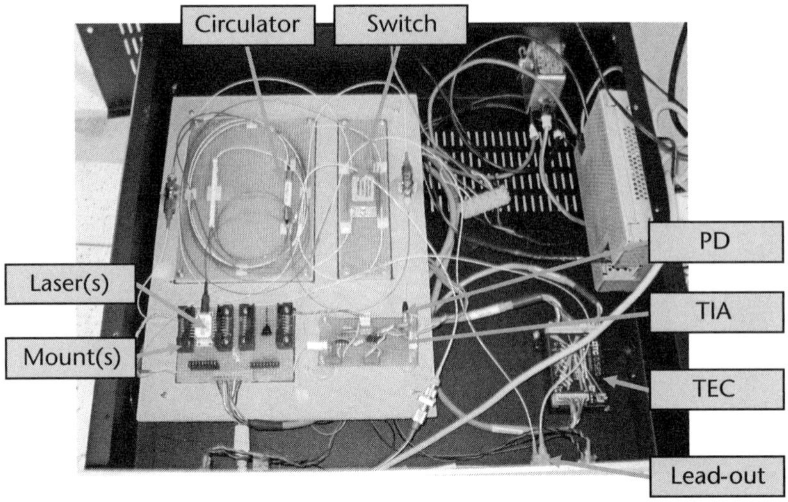

Figure 10.2 IBS interrogator developed at Politecnico di Torino, with breakdown of each component.

is used, rather than a coupler, to decouple input and output arms. A 1 × 2 switch (Sercalo MEMS 1 × 2), controlled by the acquisition card, is used to commute between two output channels. Detection is performed with a fiber-coupled InGaAs pin photodiode (JDSU EPM, responsivity 0.9 A/W), followed by a TIA and amplification circuitry. All DAQ and control are managed with a DAQ card (National Instruments, NI DAQ-6009, 5 kS/s) controlled via LabVIEW software. In this system, the TEC plays a significant role because it allows to have a fine adjustment of the emission wavelength, tuning the sensing point to the maximum sensitivity, or linearity, working regions (as described in Figure 3.3).

10.2.2 Plastic Fiber-Based Systems

IBS-based detection systems are the golden standard for POF sensors, whereas the simplified sensing approach substantially requires detecting the output power, dependent on losses along the fiber link [2, 3]. The choice of plastic, rather than glass, fibers allows a huge simplification of the setup: a simple LED can be used as light source, frontally aligned to the launch fiber. A simple voltage regulator can be used to stabilize the output power from the LED. The remainder of the setup resembles Figure 10.1(a). Often, a multiwavelength is system, implemented with an array of LED operating at different wavelengths, and a set of spatial filters, in order to obtain different working channels, or compensate for power fluctuations. Figure 10.3 shows the picture of an advanced POF interrogator, developed at Politecnico di Torino, Italy. It consists of 8 separate channels, each having a red LED coupled to the launch fiber through an alignment stage, and a detector based on a Si photodiode with multiple electrical amplification stages. Each channel connects to one standard POF (PMMA step-index 980/1,000 μm) sensor through a pair of connectors. DAQ is performed within the instrument allowing advanced functionalities, such as online data storage and remote control of the sampling time and gain of each electrical amplifier.

Figure 10.3 LoCK, a POF interrogator assembled at Poltecnico di Torino. (Courtesy of FCO group; http://www.fco.delen.polito.it/Spin-off_products.html.) The interrogator detects 8 channels, both having lead in/out connector pair, and embeds remote access features via GSM.

From the system point of view, the operation with POF sensors, particularly when light travels through an air path, is hampered by stray light; POF collect a large amount of light from the surrounding environment, which adds to the detected light at the end of the sensing system. In order to remove stray light, a simple and low-cost arrangement is to modulate the light source, and demodulate it at the receiver. This way, we ensure that disturbances, which are not modulated at the same frequency, are rejected during demodulation. A simple and effective way is to use the lock-in amplifier (LIA) technique implemented via software, described in [2]. In this setup, the LED is modulated with a carrier included within 1 kHz and 2 kHz, high enough to reject low-frequency components, but low enough to be detected with a DAQ card. The DAQ receives both the output of each channel, and the modulation signal, and computes the fast Fourier transform of the convolution of each output and the modulation. LIA ensures that the system keeps operating even when the modulation carrier fluctuates.

10.3 White Light-Based Setup

10.3.1 Setup and Instrumentation

The most popular and standardized arrangement for the interrogation of FOS is the white light based setup, which allows measuring the spectrum of an array of sensors. This approach has 3 key advantages over IBS: the measurement of spectral variations is much more resilient to power fluctuations of the optical source and each component in the system; several sensors can be combined into one array and interrogated at the same time; the use of a broadband source, rather than a laser, poses no problems of isolating backreflections and thus has a better signal-to-noise ratio (SNR). In contrast, this approach results in a more expensive and encumbering setup and the acquisition speed is limited. In the last half-decade, progresses in diffraction gratings and CCD have made spectrometers more affordable and compact, and resulted in a significant cost reduction for white light setup, making them a standardized approach. The measurement of optical spectra is a working principle compatible with any type of FBG, FPI, and SPR sensors, and their combinations.

A typical white light setup for interrogation of FBG, FPI, and other sensors that operate with spectral variations is sketched in Figure 10.4. A broadband light source, with stable output spectrum, is used as input; typically a fiber-coupled SLED with a control board is employed. A 1 × 2 coupler decouples the input light from the backreflections, and a 1 × *N* optical switch is used to multiplex different channels, typically 2 to 8. The detector is a spectrometer or a FBG analyzer (FBGA). SLED and FBGA are required to have identical bandwidth, covering the entire spectrum of the sensing arrays. Data acquisition is performed with the FBGA, by collecting and processing spectral information. Currently, most spectrometers operate either via USB, with a nominal data transfer rate of 1 to 100 Hz, or via RS-232 with detection speed up to 5 kHz. However, performances in terms of detection speed are often far from the nominal value, due to the need to transfer raw spectral data and the need to apply signal processing techniques in real time to track FBG or FPI spectral shifts.

Figure 10.5 shows a white light setup assembled by Politecnico di Torino, Italy, with off-the-shelf components, operating around 1,550 nm. A SLED source (Exalos, EXS2100, 60-nm bandwidth, Gaussian spectrum), with its control board (Exalos EBD5020), is used to couple white light into the system. A 3-port circulator is used to route SLED light into the output channels, and collect their backreflections. A switch 1 × 4 (Sercalo, SW 1 × 4 MEMS) is used to multiplex 4 output channels, each of them connected to a mating sleeve mounted on the front of the enclosing box. The detection is performed with a FBG analyzer (Bayspec FBGA, 2-pm resolution, 60-nm bandwidth, 512-pixel spectral acquisition). A DAQ card (National Instrument, NI-6009 USB) performs the tasks of data acquisition from FBGA and control of the 1 × 4 switch by selecting the digital output (4 bits) of its MEMS controller.

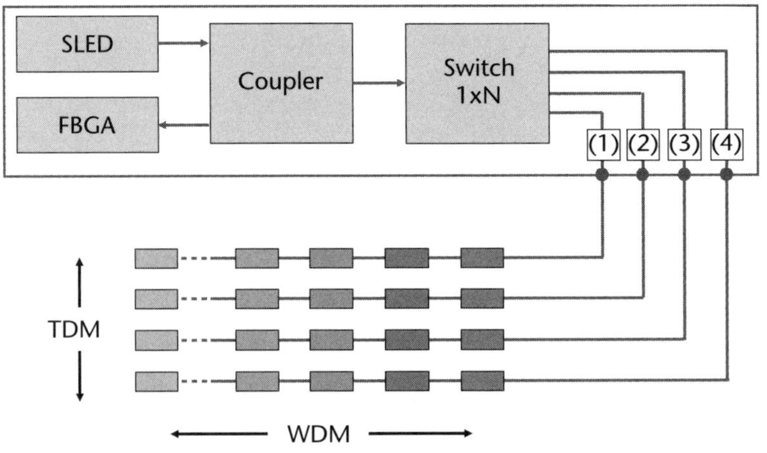

Figure 10.4 A standard white light setup for the interrogation of FBG sensors, arranged in a multichannel configuration with time and wavelength division multiplexing.

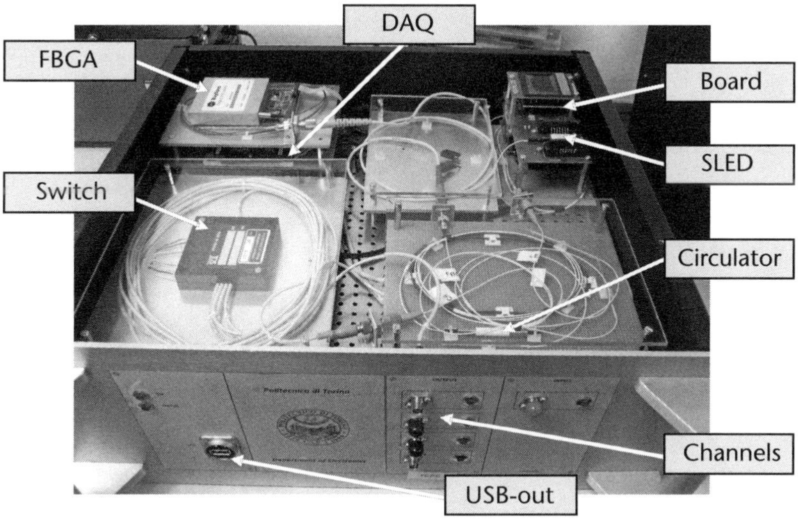

Figure 10.5 Photograph of a white light interrogation unit for FBG detection assembled at Politecnico di Torino.

10.3.2 Time/Wavelength Division Multiplexing

As in Figure 10.4, time-division multiplexing (TDM) and wavelength-division multiplexing (WDM) are the two underlying principles for FBG interrogation allowing high-capacity sensing (i.e., the detection of a plurality of sensing points). In TDM a $1 \times N$ switch is used to commute between N different channels: in this configuration, the active channel switches by changing the configuration of the MEMS circuit driving the switch. Typical configurations employ a 1×2, 1×4, or 1×8 switch. Most MEMS switches have a nominal switching time of 0.3 to 1 ms; however, the switch is controlled synchronously with data acquisition and the whole electrical circuitry, and this reduces the effective switching time to ~20 ms. In practice, TDM is an effective practice for the measurement of slowly changing parameters, <10 Hz, such as temperature variations, refractive index variations, and is still compatible with detection of blood pressure or intracranial pressure provided that the switch control is well synchronized. Often, the geometry of the sensing probe requires having a plurality of fibers and TDM is the correct approach to address this task.

WDM principle is the standard for inline FBG interrogation, whereas several FBGs are inscribed on the same fiber and interrogated altogether. WDM can apply to any sensor (FBG, TFBG, SPR) that has a spectrum tightly confined within a short bandwidth, and transparent to wavelengths outside of the bandwidth. As seen in Chapter 5, the FWHM bandwidth of an FBG is typically ≤0.4 nm, and sensitivities are 10 pm/°C and 1 pm/$\mu\varepsilon$; thus, a spacing of >1.5 nm between an FBG and its neighbor is correct for most biomedical applications, with the except of hyperthermia. In addition, recent advances in FBG sensing allow fabricating arrays of FBGs having identical spacing between each adjacent wavelength pair, and at the same time having the same physical distance between the center of each FBG in the fiber link. In this arrangement, we observe an equivalence between each FBG wavelength and its position, with respect to every other FBG; geometrically, it means having a one-to-one correspondence between the position of one sensor in the optical fiber and its wavelength in the FBG array. For FBG sensing networks, the combination of TDM and WDM allows operating networks with a high capacity, equal to $M \times N$ given that M is the number of sensors in each array and N is the switch size. Assuming that 2-nm spacing between each FBG is required and that standard interrogators operate with a bandwidth of 40 to 80 nm, up to 20 to 40 sensors can be inscribed on a single channel, and using TDM it is possible to interrogate up to 80 to 160 sensors using a 1×4 switch. As in Figure 10.4, in most practical implementations of FBG arrays, N identical arrays of M FBGs (each one identified with a color) are used. Although not reported yet in scientific works, T/WDM can apply also to TFBG and SPR sensors, and partially to CFBG. All these sensor types have a bandwidth broader than an FBG, and thus the number of sensors per array can be low. Whenever WDM is used, it is always necessary to obtain a neat separation between each spectral component, particularly when the spectral sampling is coarse, in order to be able to completely isolate the spectrum of each sensor without any cross-interference.

10.3.3 Cepstrum Division Multiplexing

WDM is not effective for FPI sensors, since their spectral envelopes are broadband, as for any resonant cavity. However, it was shown in [4] that a plurality of FPI sensors can be interrogated on the same array, similarly to FBG sensors. The main requirement is that the cavity length of each sensor is different, and thus each sensor can be uniquely identified by its cavity length. Since most FPI sensors used in biomedical engineering are fabricated as low-finesse interferometers, their spectrum resembles a sine-wave when sampled over a bandwidth of 40 to 80 nm, with each of the waves having a different period dependent on FPI cavity length. Thus, it is possible to discriminate FPI spectra by applying the Fourier transform on the optical spectra. This approach is labeled cepstrum-division multiplexing (CDM), referring to *cepstrum* as the spectrum of the (optical) spectrum. In the next chapter, we will show an effective algorithm for CDM-based high-capacity sensor tracking applied to tens of FPIs.

10.4 Scanning-Source Interrogators

The principle of operation of scanning-source interrogators is similar to white light setup, and aims at decoding an array of sensors (using WDM, and seldom TDM) by measuring their spectrum. The main difference is that, while in a white light setup no scanning element is employed and the detection is carried out with a device capable of discriminating the whole spectral envelope, in this setup the spectrum is sampled through using a light source that sweeps the whole wavelength bandwidth. The wavelength scanning process can be implemented in two ways, either by filtering a broadband source with a narrow bandpass filter (usually a Fabry-Perot resonator tuned by acousto-optics), or by using a tunable laser having a narrow linewidth and rapidly scanning all wavelengths. The latter method is more modern, thanks to the latest developments in oscillators.

The setup in Figure 10.6 is the typical system currently in for scanning lasers-based interrogators, arranged for multichannel interrogation. The light source is a laser with narrow linewidth, externally tuned by means of an oscillator. The laser

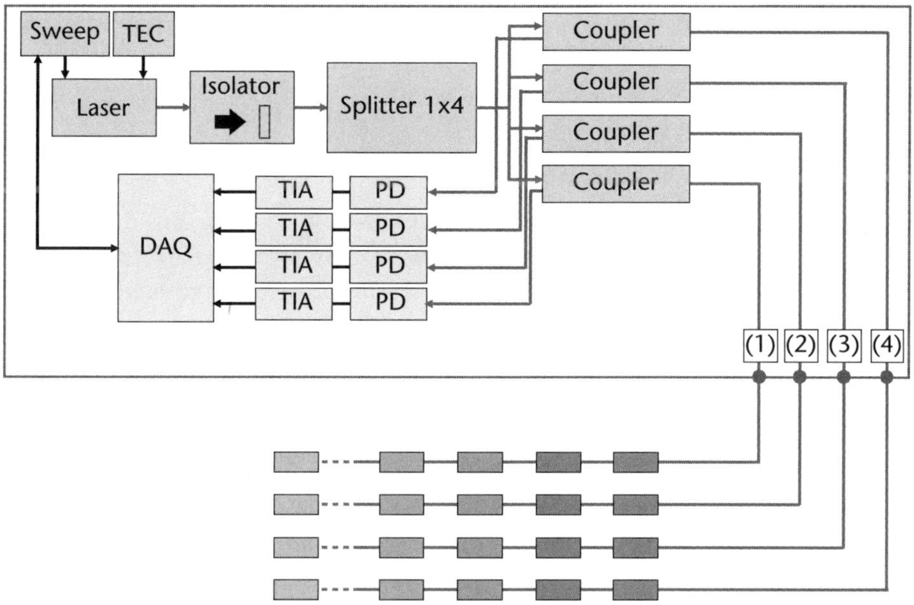

Figure 10.6 Schematic of a laser-scanning interrogator, arranged for 4-channel interrogation.

is controlled by a TEC to emit a constant power throughout the whole wavelength range, while the oscillator provides a linear-scanning function that covers the entire bandwidth of the sensors; current tunable lasers can cover a bandwidth up to 160 nm around 1,550 nm. Since the system operates with reflective element, it is essential to isolate the laser from backreflections, through the use of an isolator and angled connectors on the fiber link. In order to avoid using a $1 \times N$ switch, which in this setup can be hard to synchronize with the laser sweep function, it is preferable to physically separate each channel by using a $1 \times N$ splitter and N 1×2 couplers or circulators. The output of each coupler is then connected to the lead-out channel, while the second input branch is then acquired with a photodetector. In this system, the DAQ plays a key role, since it needs to control the sweep function of the laser and detect the TIA output in a synchronous way, therefore reconstructing the reflection spectrum point-by-point.

Albeit both setups achieve the same mission of measuring the optical spectrum of a sensor array, white light and laser-scanning setups differ in implementation and performances. White light setups are cheaper, usually by 2 to 4 times, and operate continuously without the need for synchronization; however, limitations in size of the diffraction element and CCD array limit the resolution (in terms of spectral distance between each pixel) within 40 to 160 pm. Also, these systems are typically faster, as spectrometers can transfer data to the PC at 5 kS/s nominal speed with RS-232 protocol, or via USB. In contrast, scanning-laser systems can detect spectra with narrow resolution, typically 1 to 10 pm, which allows appreciating all details of the spectrum, as required in precision sensing. However, the main drawback is that systems have a low scanning rate (1 to 10 Hz) typically; this is due to the fact that the laser must sweep the entire range of operative wavelengths. Fast-scanning systems have been developed, but they typically do not achieve a narrow wavelength sampling and the laser wavelength tends to slightly differ at each scan, requiring a complex synchronization. The dominant noise is due to the laser power fluctuations, and isolation is required to reduce laser noise.

10.5 Multiparametric Sensors

A key feature of the white light setups is the possibility to combine different sensing units, interrogating multiparametric probes in a straightforward way. To date, the most popular example of multiparametric probe combines an EFPI probe, fabricated on the tip of a glass fiber, with an FBG. This dual-sensing approach enables two key features, useful for advanced sensing techniques: (1) it allows the measurement of both pressure and temperature; and (2) it allows compensating the temperature cross-sensitivity in EFPI-based sensors. Figure 10.7 shows an example of a dual EFPI/FBG probe, fabricated by the University of Limerick, Ireland, and reported in [5]. The spectral analysis shows that the narrow spectrum of the FBG can be isolated and analyzed separately from the EFPI broadband envelope, which is detected by analyzing the variations of the right part of the spectrum.

The EFPI/FBG concept can be extended to larger sensor networks, and this approach is expected to create substantial breakthrough in biomedical engineering in the near future, whereas the need for miniaturized probes and the several cross-sensitivities require maximizing the sensing capacity of a single optical fiber. The operation with multiple sensors of different typologies can be expressed in a standardized matrix format. Given N sensors, each estimating variations of one biophysical or biological parameter Δp_n, $n = 1, ..., N$, the interrogation setup detects the spectrum of the overall sensing array, and estimates the spectral variations hereby labeled Δx_n. In the previous example of an EFPI/FBG probe, Δp_1 is the pressure variation (ΔP) while Δp_2 is the temperature variation (ΔT), while Δx_1 is the EFPI cavity length variation due to the pressure (and/or temperature) effect (ΔL) while Δx_2 is the FBG wavelength shift ($\Delta\lambda$). Now we can relate the array $\underline{\Delta p} = [\Delta p_n]$ with the array $\underline{\Delta x} = [\Delta x_n]$ through an $N \times N$ matrix, which contains all the sensitivities:

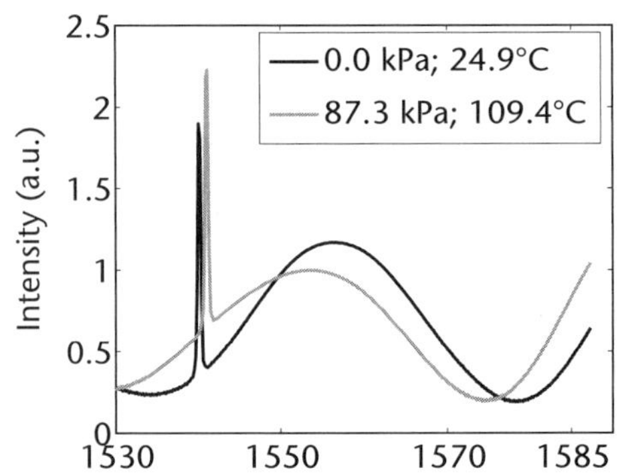

Figure 10.7 EFPI/FBG probe, fabricated at the University of Limerick, Ireland, and reported in [5]. The chart shows the spectrum recorded on a white light interrogator, for two values of temperature and pressure.

$$\begin{bmatrix} \Delta x_1 \\ \Delta x_2 \\ \vdots \\ \Delta x_N \end{bmatrix} = \begin{bmatrix} k_{11} & k_{12} & \cdots & k_{1N} \\ k_{21} & k_{22} & & \\ \vdots & & \ddots & \\ k_{N1} & & & k_{NN} \end{bmatrix} \begin{bmatrix} \Delta p_1 \\ \Delta p_2 \\ \vdots \\ \Delta p_N \end{bmatrix} \rightarrow \underline{\Delta x} = \underline{\underline{K}} \times \underline{\Delta p} \tag{10.1}$$

Each term k_{ij} contains the sensitivity of the jth sensor to the ith parameter. In the most ideal configuration, $\underline{\underline{K}}$ is a diagonal matrix (i.e., each sensor measures only one parameter with zero, or negligible, cross-sensitivities); however, in practical implementations, this is often not achieved. Each element of the matrix $\underline{\underline{K}}$ needs to be estimated through a calibration procedure, which can be performed by changing one of the parameters of the array $\underline{\Delta p}$ and setting to zero all the others; in the case of FBG sensors, manufacturers usually provide strain and temperature sensitivity. Since the interrogator detects $\underline{\Delta x}$ through spectral analysis, we can estimate the parameters by simply reversing $\underline{\underline{K}}$:

$$\underline{\Delta p} = \underline{\underline{K}}^{-1} \times \underline{\Delta x} \tag{10.2}$$

It is important to validate that the matrix $\underline{\underline{K}}$ is not singular, making sure its determinant is significantly higher than zero; this occurs when the sensitivity is good enough to guarantee that each parameter can be detected by at least one sensor with sufficient accuracy on each sensing range, and that all sensors respond differently to the parameters under test. The operation described in (10.1) and (10.2) is effective only when all sensors exhibit a linear sensitivity, or can be linearized with satisfactory precision. Most FPI and FBG sensors, in all their typologies, exhibit a linear behavior and therefore can be framed in a multiparametric probe.

The probe described in Figure 10.7 is, to date, the best example of dual probe, thanks to its capability to measure pressure with the EFPI probe and compensate for temperature variations. The probe has been manufactured by starting with a commercial FBG, and fabricating an EFPI sensor in proximity of the FBG using the all-glass probe method described in Chapter 7. The EFPI has pressure sensitivity $k_{11} = 1.60$ nm/kPa, and temperature cross-sensitivity $k_{12} = -0.0519$ nm/°C; the FBG has temperature sensitivity $k_{22} = 10.70$ pm/°C and negligible cross-sensitivity ($k_{21} \approx 0$ pm/kPa). Since one element of the matrix is close to zero and sensitivity values are significant, the matrix $\underline{\underline{K}}$ is clearly not singular, making it a robust sensing technique even considering tolerances in calibration and spectral detection.

10.6 Distributed Sensing Units

Distributed sensing systems, rapidly emerging also for medical applications, require the measurement of one parameter (typically temperature or strain) along an optical fiber. The two setups at the foundation of distributed sensing units, suitable for biomedical engineering, are the optical frequency-domain reflectometer (OFDR) and the microwave photonic interrogator (MWP). The main structure of an OFDR, based on Rayleigh backscatter measurement and with spatial discrimination capability, has been developed and patented by Froggat and Duncan [6] and Soller et al. [7]. The system is sketched in Figure 10.8. The light source is a tunable laser and is connected through a 90/10 coupler to two branches of an interferometer. The upper arm is the measurement part of the setup: a 3-dB splitter separates one feedback arm, connected via a polarization controller (PC) to a polarization beam splitter (PBS), while the upper arm connects to the fiber under test (FUT). The two output of the PBS are detected with a photodiode and amplifier and processed with a DAQ. On the bottom side, the trigger signal is generated with an interferometer that induces a delay line (τ_{tr}). This setup has the capability to measure the complex reflection spectrum of the FUT point-by-point, separating the reflectivity and the group delay.

As shown in [6] and explained in Chapter 6, it is possible to expand this technique to strain and temperature sensing: in presence of a fiber with no highly reflective element, Rayleigh scattering constitutes the dominant reflection component at the laser wavelength. Its pattern has a quite chaotic spectrum, which however linearly shifts with strain and temperature with coefficients ~1 pm/$\mu\varepsilon$ and ~10 pm/°C, similarly to FBGs. Also, it is possible to use the same OFDR setup for detection of shape deformations, in 3-D, using a fiber with 3 or more cores as the FUT. This setup is at the base of distributed shape sensors. In the OFDR method, performances are tightly constrained by the trade-off between spatial resolution, acquisition time, and accuracy; the three parameters are altogether dependent.

The MWP approach allows detecting one or two hot spots, and localizing them within a fiber link. This approach can be used for the detection of temperature hot spots (amplitude and position), typical in hyperthermia applications, or a localized and intense strain such as in gastroscopic diagnostic. The typical setup, consolidated by [8], is sketched in Figure 10.9. In this configuration, a narrowband source is used, having emission wavelength matching the array of FBG under interrogation. However, it is preferable to use a SLED followed by a bandpass filter (BPF), rather than a laser, since the system is capable of detecting hot spots having distance significantly higher than the coherence length, and therefore an incoherent source is

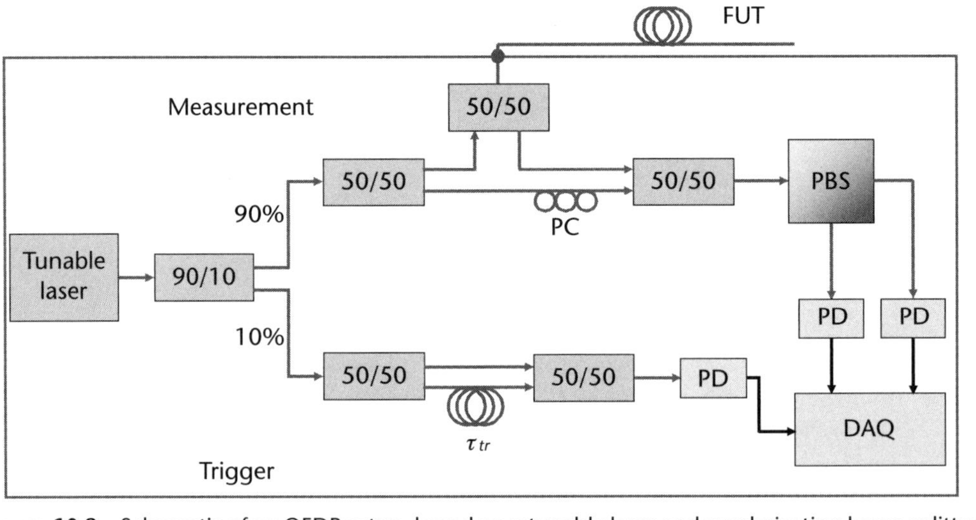

Figure 10.8 Schematic of an OFDR setup, based on a tunable laser and a polarization beam splitter, capable of acquiring the reflection spectrum of the fiber under test for every distance.

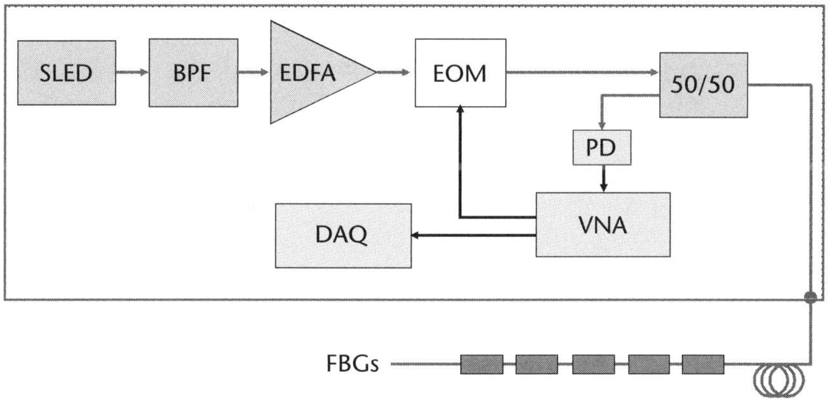

Figure 10.9 Schematic of a MWP setup for interrogation of an FBG array or an all-fiber grating, as proposed in [8]. The system is based on a VNA for signal detection and modulation.

required for narrow spatial resolution. Since the SLED power is narrowly filtered with a BPF having FWHM < 0.5 nm, an amplification stage with an EDFA may be needed. The output at the end of the amplifier is modulated with a microwave signal, which can directly be generated through a vector network analyzer (VNA). Typically, the microwave signal consists of a radio-frequency tone swept from 10 MHz to 10 GHz. Through a coupler or circulator, the swept signal is sent to the FBG array under test. The output, at the coupler side, is photodetected and sent to the VNA, which displays the S_{21} parameter, which is then acquired to estimate the hot-spot location. In this configuration, it is possible to interrogate an array of FBGs, or an all-fiber grating, having the BPF central wavelength matched with the FBG reflection peak. Due to the high reflectivity of the grating(s), most of the input signal is reflected at the initial section of the FBG array. However, a local change of temperature in a spot event placed at a certain point along the grating will produce a local Bragg frequency shift. When this occurs, besides the main reflected signal generated at the initial section of the grating, a second reflected signal is produced at the point where the hot spot is placed. In this way, the presence of a hot spot results in a two-tap MWP filter and it is possible to determine the location of the hot spot. Overall, the MWP setup is an excellent tool to interrogate sensing lengths up to several meters (which can be also located at remote distance from the interrogator by means of fiber patchcord), and its resolution is mainly limited by the VNA; it does not provide the full-range sensing data (temperature, strain, shape), but it is a fast system and it is also not subjected to the same constraints between resolution, accuracy, and sampling time typical of OFDR.

10.7 Commercial Devices

For FBG sensors, devices implementing amplitude-detection methods can be self-assembled as in Figures 10.2 and 10.3. It is desirable, for end-users, to assemble the device in house because each wavelength of the laser has to be matched to the Bragg wavelength of each FBG, and thus the FBG sensors have to be designed specifically for the available laser source, and vice versa. In contrast, this approach is well established for all Fabry-Perot based sensors, and particularly for EFPI pressure probes: in this case, thanks to the broadband spectrum of low-finesse EFPIs, it is relatively simple to arrange the system since virtually any wavelength can be chosen as sensing point. Commercial devices that implement amplitude-based detection, with an arrangement similar to Figure 10.1(b) operating in reflection and allowing for compensation, operate with the EFPI probes described in Chapter 7. The three most important systems are:

- *FISO (Canada, www.fiso.com):* The FISO signal conditioning systems operates in single-channel mode up to 15 kHz, or 105 Hz for dual (differential) detection. The system takes as input up to 2 probes and can perform temperature compensation. Probes are based on multimode fiber (62.5/125 μm). Several modules can be used for multipoint pressure measurement. The system is compatible with FISO FOP-M probes.
- *Opsens (Canada, https://opsens.com):* The Opsens system is entirely optimized for the fractional flow reserve application (Section 15.3). The system is based on an EFPI probe (Optowire) connected to the interrogator (Optomonitor), based on a bench-top instrument reading the fiber-optic probe and a portable monitor for display of pressure values, FFR estimate, and significant pressure data.
- *RJC (US, http://www.rjcenterprises.net):* RJC has two general-purpose systems, one bench-top (Model 650) and one OEM (Model 645), for single-probe interrogation up to a 200-Hz rate, with USB acquisition. Other systems, such as Model 600, are portable (handheld) and perform detection, processing and data visualization on a monitor, where pressure pulses are displayed together with average pressure and its repetition rate; these are intended for clinicians' use. Instruments are compatible with RJC probes.

In terms of FBG sensors, several white light setups laser-scanning interrogators are commercially available implementing exactly the principles in Figures 10.4 and 10.6, respectively, as well as C/L-band spectrometers (often referred as FBGA, FBG analyzers) which integrate the only detection device. Table 10.1 shows a list of the main commercial devices, which include FBG interrogators, both benchtop and portable and with multiple output channels, as well as spectrometers and FBG analyzers which can be used to assemble a custom-made interrogators, particularly useful when sensors data need to be processed in situ. Most instruments operate on bandwidth ranging from 40 nm to 100 nm, although up to 160 nm can be covered. Spectrometer-based systems sample spectra on 256 or 512 pixels, while laser-scanning instruments operate at 1 to 5-pm scanning resolution. Each instrument is completed with software, which perform FBG wavelength peak calculations and/or returns the raw spectral data; the Micron Optics systems can also perform FPI tracking, but only for high-finesse interferometers. In order to operate with other sensors (such as TFBG, CFBG, SPR, or their combinations) in real time, it is necessary to develop proprietary software that reads the output spectra.

In terms of distributed sensors, whereas the emergence of architectures based on OFDR, all-grating fibers, and microwave photonics, the commercial offer for high-density distributed sensors, having attractiveness for biomedical engineering,

Table 10.1 List of Main Commercial FBG Interrogators, Analyzers, and Spectrometers, and Their Performance

Manufacturer	*Instrument*	*Type*	*Bandwidth*	*Channels*	*Speed*	*Resolution*	*Precision*
Micron Optics	sm125–200	FBG interrogator (W)	1,520–1,580 nm	1	1 Hz	Undisclosed	±1 pm
Micron Optics	sm125–700	FBG interrogator (W)	1,510–1,590 nm	4	5–10 Hz	Undisclosed	±0.5 pm
Micron Optics	si255 standard	Interrogator (L)	160 pm	4–16	1 kHz	~5 pm	±1 pm
Micron Optics	si255 high speed	Interrogator (L)	80 pm	4–16	5 kHz	~5 pm	±2 pm
Bayspec	FBGA/ FBGA-IRS	FBG analyzer	1,525–1,565 nm	—	5 Hz or 5 kHz	~78 pm	±1 pm
Bayspec	FBGA/ FBGA-IRS	FBG analyzer	1,510–1,590 nm	—	5 Hz or 5 kHz	~156 pm	±1 pm
Bayspec	Wavecapture	FBG interrogator (W)	1,525–1,565 nm	1 or 4	5 Hz or 5 kHz	~78 pm	±1 pm
Bayspec	Wavecapture	FBG interrogator (W)	1,510–1,590 nm	1 or 4	5 Hz or 5 kHz	~156 pm	±1 pm
Ibsen Photonics	I-MON 256 USB	Spectrometer	1,525–1,570 nm	—	6 kHz	~175 pm	±0.5-5 pm
Ibsen Photonics	I-MON 512 USB	Spectrometer	1,510–1,595 nm	—	3 kHz	~165 pm	±0.5-5 pm
Ibsen Photonics	I-MON 256 HS	Spectrometer	1,525–1,570 nm	—	35 kHz	~175 pm	±0.5-5 pm
Ibsen Photonics	I-MON 512 HS	Spectrometer	1,510–1,595 nm	—	17 kHz	~165 pm	±0.5-5 pm
FBGS Intl.	704D	FBG interrogator (W)	1,525–1,565 nm	4	500 Hz	~78 pm	±1 pm
FBGS Intl.	804D	FBG interrogator (W)	1,510–1,590 nm	4	500 Hz	~156 pm	±1 pm
FBGS Intl.	FAZT I4	Interrogator (L)	1,528–1,568 nm	4	4 Hz /4 kHz	1 pm	±1 pm
Fibersensing	FS22	Interrogator (L)	1,500–1,600 nm	1, 4, or 8	1 Hz	5 pm	±1 pm
Fibersensing	FS42	Interrogator (L)	1,500–1,600 nm	4	1 Hz	5 pm	±1 pm

Type: W = white light interrogator and L = laser-scanning interrogator. Resolution = maximum spacing between two adjacent wavelengths sampled by the instrument. Precision = accuracy on FBG wavelength estimation, using software on-board of the instrument. (Web sites: http://www.micronoptics.com, www.bayspec.com, www.ibsen.com, www.fbgs.com, www.fibersensing.com.)

is limited to LUNA Technologies (www.lunainc.com) and the recent system commercialized by FBGS International and Sensuron (www.sensuron.com) based on all-grating fiber. The main performances of each interrogator (OBR4600, ODiSI, RT150, RTS125) are described in Table 10.2; it is hard to evaluate the exact performance of each system, since each performance rate depends on all the others. Systems can operate either with a standard single-mode fiber, or more recently with an all-grating fiber that increases the local reflectivity; shape sensors require a multicore fiber to discriminate the shape direction. All systems are available in multichannel option, and the sensing length can typically be adjusted even at remote distance, such that it is possible for the end-user to correlate each measurement output with the position of the fiber under test in a flexible way. Generally, most

Table 10.2 List of Main Commercial Distributed Fiber-Optic Sensing Systems

Manufacturer	*Instrument*	*Sensor*	*Detection*	*Length*	*Resolution*	*Accuracy*	*Speed*
Luna	OBR4600	Standard SMF	Losses	30–70m	10–20 μm	±0.1 dB	0.3–3 Hz
Luna	OBR4600 + module	Standard SMF	Temperature/ strain	30–70m	0.01–1 cm	±0.1°C / ±1 $\mu\varepsilon$	0.3–3 Hz
Luna	ODiSI-B HD	Standard SMF	Temperature/ strain	10m	1.3 mm	±25 $\mu\varepsilon$	1.25 Hz
Luna	ODiSI-B HS	Continuous FBG	Temperature/ strain	5m	1.3 mm	±35 $\mu\varepsilon$	23.8 Hz
Luna/Intuitive	—	Multicore fiber	Shape	2.5m	~2.5 cm	10 mm (radius)	-
FBGS Intl.	RT150	Continuous FBG	Temperature/ strain	100m	1–100 mm	—	100 Hz
Sensuron	RTS125	Continuous FBG	Temperature/ strain	13m	6.3 mm	±0.15°C/±1.2 $\mu\varepsilon$	100 Hz

Performances are rated from datasheet and hands-on experience, since OFDR systems operating on short length pose a trade-off between total sensing length, spatial resolution, measurement accuracy, and detection speed. Length refers to the sensing length per channel. The LUNA shape-sensing technology has been acquired by Intuitive Surgical, and is reported for its 2013 available data.

systems provide inferior spatial resolution in online measurements, with respect to raw data acquisition and storage, thus the minimum theoretical spacing between each measurement point can often be achieved only after data postprocessing.

10.8 Conclusions

In this chapter, we illustrated the hardware building block of a fiber-optic sensing system; each type of interrogator can work with a different sensing principle, and implement advanced sensing features such as compensation and multiparametric sensing. The hands-on details and the review of commercial systems will help readers to identify and assemble an interrogator that matches the specifications (performance, cost, size), as an essential task towards building an FOS system.

References

[1] Tosi, D., M. Olivero, and G. Perrone, "Low-Cost Fiber Bragg Grating Vibroacoustic Sensor for Voice and Heartbeat Detection," *Applied Optics,* Vol. 47, No. 28, 2008, pp. 5123–5129.

[2] Perrone, G., and A. Vallan, "A Low-Cost Optical Sensor for Noncontact Vibration Measurements," *IEEE Transactions on Instrumentation and Measurement*, Vol. 58, No. 5, 2009, pp. 1650–1656.

[3] Abrate, S., R. Gaudino, and G. Perrone, "Step-Index PMMA Fibers and Their Applications," *Current Developments in Optical Fiber Technology*, 2013.

[4] Tosi, D., "Simultaneous Detection of Multiple Fiber-Optic Fabry-Perot Interferometry Sensors with Cepstrum-Division-Multiplexing (CDM)," *IEEE/OSA Journal of Lightwave Technology*, Vol. 34, No. 15, 2016, pp. 3622–3627.

[5] Tosi, D., et al., "Fiber Optic Combined FPI/FBG Sensors for Monitoring of Radiofrequency Thermal Ablation of Liver Tumors: Ex-Vivo Experiments," *Applied Optics*, Vol. 53, No. 10, 2014, pp. 2136–2144.

[6] Froggatt, M. E., and R. G. Duncan, "Fiber Optic Position and/or Shape Sensing Based on Rayleigh Scatter," U.S. Patent 7772541 B2, 2010.

[7] Soller, B., et al., "High Resolution Optical Frequency Domain Reflectometry for Characterization of Components and Assemblies," *Optics Express*, Vol. 13, No. 2, 2005, pp. 666–674.

[8] Ricchiuti, A. L., et al., "Long Weak FBG Sensor Interrogation Using Microwave Photonics Filtering Technique," *IEEE Photon. Technol. Lett.*, Vol. 26, No. 20, 2014, pp. 2039–2042.

CHAPTER 11

Interrogation Software

An essential part of underestimated importance of an OFS system is the software that takes as input the optically detected signal (typically, an optical spectrum) and returns the estimation of the parameters to be measured. In this section, we will focus on spectral detection, and we will provide algorithmic tools for analyzing the spectra of FBG and EFPI, as well as some hints for analysis of TFBG and CFBG. As the literature on spectral detection software is broad, we will narrow down our analysis to a few, well-effective techniques.

11.1 FBG Tracking Direct Methods

The goal of FBG tracking is to estimate the FBG wavelength shift from the reference position, which is directly related to temperature and strain values (or to other parameters, in case a transducer is used), by simple spectral analysis. Since the wavelength grid of an optical interrogator is limited, usually within 10-pm to 156-pm sampling width (using the instruments for spectral detection listed in Chapter 10, Table 10.1), and the detector is noisy, accurate FBG tracking must be performed in order to retrieve temperature or strain with competitive accuracy. In order to provide a worst-case scenario, in this chapter we assume a wavelength grid $\delta\lambda$ = 156 pm, corresponding to a 9-bit quantization on the 1,520–1,600-nm wavelength range; and we quantize the spectrum amplitude $R_M(\lambda)$ over 16 bits within the range 0 to 1 (i.e., $\Delta R = 1/65{,}536$). These parameters correspond to the I-MON-USB interrogator from Ibsen Photonics, but are in agreement with most commercially available interrogators. Figure 11.1 graphically shows the proposed direct techniques for FBG tracking, and all three methods are listed in Appendix B.

11.1.1 Centroid

An effective technique is the estimation of the centroid of the FBG spectrum, defined as [1]:

Table 11.1 List of Parameters Used in This Chapter

Parameter	*Description*
λ	Wavelength (discretized: λ_i, $i = 1, \ldots, N$)
λ_B	Central wavelength of the FBG, to be estimated
λ_E	Estimated central wavelength of the FBG
$\delta\lambda$	Wavelength grid of the FBG/EFPI interrogator (here set to 156 pm)
kL	Strength coefficient on FBG
$R_M(\lambda)$	Measured reflection spectrum of the FBG, with $\Delta\lambda$ shift applied
$R_{ref}(\lambda)$	Reference reflection spectrum of the FBG, in zero-shift condition
ΔR	Quantization step for spectrum amplitude detection
F	Digital frequency (discretized and included within 0 to 1 range)
$\Im(f)$	Cepstrum, discretized as $\Im(f) = [\Im(f_1), \ldots, \Im(f_N)] = [\Im_1, \ldots, \Im_N]$
M	Toeplitz matrix underlying cepstrum, N N size
ξ	KLT eigenvalue string, in ascending order: $\xi = [\ \lvert\xi_1\rvert < \lvert\xi_2\rvert < \ldots < \lvert\xi_N\rvert\ ]$
H	Eigenvalue filters, $h = [h_1, h_2, \ldots, \lvert\xi_N\rvert]$

$$\lambda_E = \frac{\sum_{i=1}^{N} R_M(\lambda_i) \cdot \lambda_i}{\sum_{i=1}^{N} R_M(\lambda_i)} \tag{11.1}$$

whereas the spectrum is sampled on the $i = 1, \ldots, N$ points grid identified by $\delta\lambda$. A correct appliance of the centroid tracking, however, must also estimate with the same centroid approach also the FBG spectrum in reference condition, in order to correctly estimate the wavelength shift, since the centroid overestimates the central wavelength.

11.1.2 Bandwidth Tracking

This method is applied in most FBG interrogator software, available for customers, and is based on two steps. At first, the spectral grid is artificially expanded by oversampling the spectrum $R_M(\lambda)$, in order to remove the granularity due to the coarse spectrum sampling; in MATLAB, this task is accomplished with the resample function, which applies a polyphase quadrature digital filter to artificially expand the input signal. In the new spectrum, now densely sampled, we set a threshold value R_{TH}, and we consider all the wavelength values λ_{in} such that $R_M(\lambda_{in}) > R_{TH}$, such that the new set of wavelengths λ_{in} represents the inner part of the FBG bandwidth. The wavelength is then estimated as

$$\lambda_E = \min(\lambda_{in}) + \frac{\max(\lambda_{in}) - \min(\lambda_{in})}{2} \tag{11.2}$$

In most commercial software, the 3-dB bandwidth is computed, and therefore $R_{TH} = 0.5 \max(R_M)$. This method is typically used for static sensing [2].

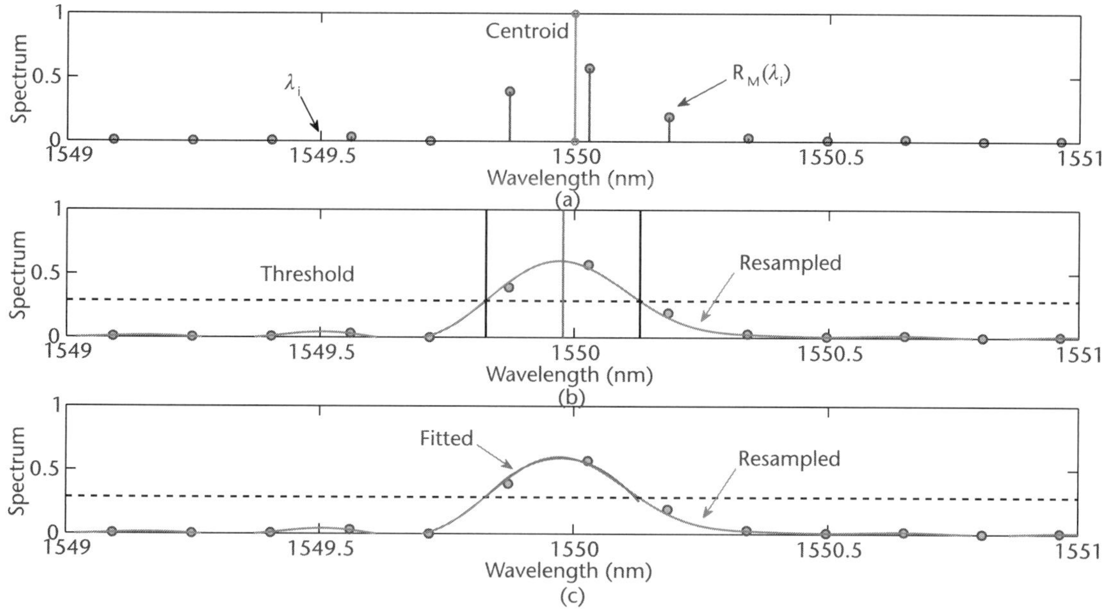

Figure 11.1 Direct and effective approaches for FBG wavelength estimation, applied to a simulated 1,550-nm FBG. (a) Centroid algorithm: all values of the spectrum, for each wavelength, are used to compute the centroid of the spectrum). (b) Bandwidth tracking: the spectrum is first resampled, with a 157:1 polyphase quadrature digital filter, the 3-dB bandwidth is then estimated, and the FBG wavelength is estimated as the central point of the values over threshold. (c) Polynomial fitting: the spectrum is first resampled and threshold is estimated, as in (b); then the inner part is fitted with a second order polynomial.

11.1.3 Polynomial Fitting

Similar to the previous method, we define a threshold value R_{TH}, and the spectral portion $R_M(\lambda_{in}) > R_{TH}$. We assume that this spectrum is can be approximated as a second-order function [3]:

$$R_M(\lambda_{in}) \simeq a_2\lambda_{in}^2 + a_1\lambda_{in} + a_0 \tag{11.3}$$

where the estimate of the coefficients a_2, a_1, a_0 can be easily performed with a quadratic regression. Then we can estimate the central wavelength, as the algebraic maximum of (11.3):

$$\lambda_E = -\frac{a_1}{2a_2} \tag{11.4}$$

A similar method was proposed by Chen et al. [4], whereas instead of applying a polynomial fit, a Gaussian fit was applied to the spectrum logarithm, reducing the estimate to a second-order polynomial fit. The solution for estimation of peak wavelength is the same as in (11.4).

All these techniques have the advantage of a fast computation, below 1 ms, and perform a direct estimation that operates on the measured spectrum, without additional requirements. In the case of fluctuating power levels, the estimator can still operate, since both centroid and thresholding operations are compatible with any power level. Centroid operation has privileges grating with broad spectrum (but no side lobes) with high $1 < kL < 2.5$, although it is effective with any FBG. Bandwidth or fitting techniques are mainly limited in performance by the resampling technique. Overall, these methods allow achieving a detection accuracy of 1 to 4 pm (depending on the signal-to-noise ratio) in most working conditions, even with a coarse wavelength grid of 156 pm.

11.2 EFPI Direct Tracking

For EFPI sensors demodulation based on the white light system, different algorithms need to be employed depending on the cavity length; it is therefore important to develop the EFPI architecture having in mind the demodulation algorithm, which is also adjusted for the sensitivity to pressure (or temperature) variations, and on the intended usage of the spectral window of the interrogator.

11.2.1 Short-Cavity EFPI

All-glass structures having the smallest size in terms of fiber and capillary are typically operating with a short cavity length (20 to 40 μm) which corresponds to a broad spectrum. In this condition, a good detection method makes use of the entire spectrum available, usually framing one period of the EFPI spectrum for the processing; alternatively, short-cavity EFPI sensors usually have a low sensitivity, close to 1 nm/kPa and therefore require using all the spectral information to extract the cavity length with satisfactory accuracy.

Due to the coarse wavelength sampling, and the lack of a narrow, resonant, spectrum like an FBG, we need to exploit the amplitude of the spectrum as the encoding information. An effective approach considers a wavelength grid $_i$, $i = 1, \ldots, N$, that contains one period of the EFPI spectrum (in reference condition). The grid is then split between the left part, of indexes $1, \ldots, N_h$ and the right part of indexes $(N_h + 1), \ldots, N$ [5]. We can then compute the parameter

$$\gamma = \frac{\sum_{i=1}^{N_h} R_M(\lambda_i)}{\sum_{i=N_h+1}^{N} R_M(\lambda_i)} \tag{11.5}$$

which is derived by dividing the sum of all of the left part of the spectral samples, by the sum of the right part of the spectral samples. An approximate choice of $N_h \approx N/2$ is recommended, but the algorithm works with most arbitrary choices of the half point. Figure 11.2 shows the implementation of this tracking method on a simulated EFPI: after selecting two intervals, the γ function is computed, and has an almost linear trend for small variations of cavity length. In Figure 11.2(b), it is shown that γ has a linear and sensitive behavior, with variations of cavity length in the order of 10 nm; considering that typical short-gap sensors have sensitivity on the order of 1 nm/kPa, this validates the effectiveness of the algorithm over a pressure range of 10 kPa, which is in agreement with typical pressure variations in medicine. The technique is also robust to power variations, since γ is a normalized

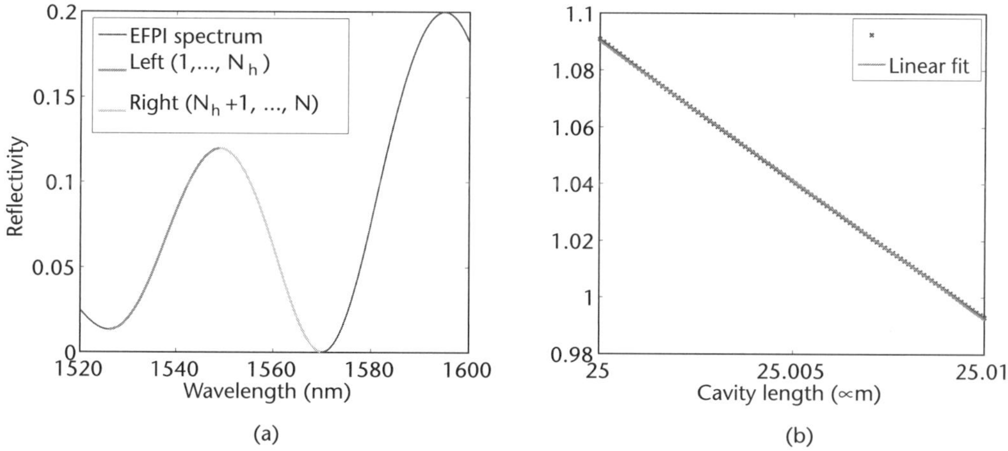

Figure 11.2 EFPI short-cavity tracking, for an EFPI sensor with a 25-μm air gap. (a) The spectrum is processed, selecting the left and right parts by choosing the correspondent indexes; then γ is estimated. (b) γ as a function of cavity length, for 0–10-nm cavity length variation. The trend is almost linear, with a sensitivity of –9.78 × 10–3 nm–1.

function, whereas wavelength drifts of the source needs to be compensated as they affect the estimation.

11.2.2 Long-Cavity EFPI

A long-cavity EFPI (0.1–1-mm Fabry-Perot cavity length) contains several fluctuations of the spectrum within the wavelength range of the interrogator, and they have a higher sensitivity, thanks to the use of flexible diaphragms on ferrule mounts, easily topping 100 nm/kPa. In this case, (11.5) is not as effective as tracking the sensor spectrum with the Fourier transform method. For low-finesse sensors, characterized by a spectrum similar to a sine wave, a good approach is to compute the Fourier transform of the EFPI spectrum:

$$\Im(f_k) = FFT\left[R_M(\lambda)\right] = \sum_{n=1}^{N} R_M(\lambda_n) e^{-j2\pi k\frac{n}{N}} \tag{11.6}$$

which is rapidly computed in MATLAB by the fast Fourier transform (FFT) function [6]. Since we compute the spectrum of an (optical) spectrum, it is reasonable to define the term $\Im$ as the cepstrum. Since several periods of the EFPI spectrum fall into the spectral window of the interrogator, the EFPI cepstrum will be represented as a steep peak in the digital frequency axis f (N points within 0 to 1), and variations of EFPI cavity translate into ceptral shifts. Figure 11.3 shows the cepstrum estimated as in (11.6), after removing the DC component, applied to a simulated EFPI sensor with cavity length varying from 500 μm to 525 μm. The cepstrum appears as a single peak, clearly emerging over the floor, and its position changes as a function of the cavity length. The right chart shows the inner cepstral shift, due to cavity length variation; the cepstral peak shifts linearly with the cavity length, with a sensitivity of 7.2×10^{-5} μm^{-1}. Upon identifying the narrow ceptstral region of interest, the peak position can be estimated with centroid or fitting methods, as previously identified.

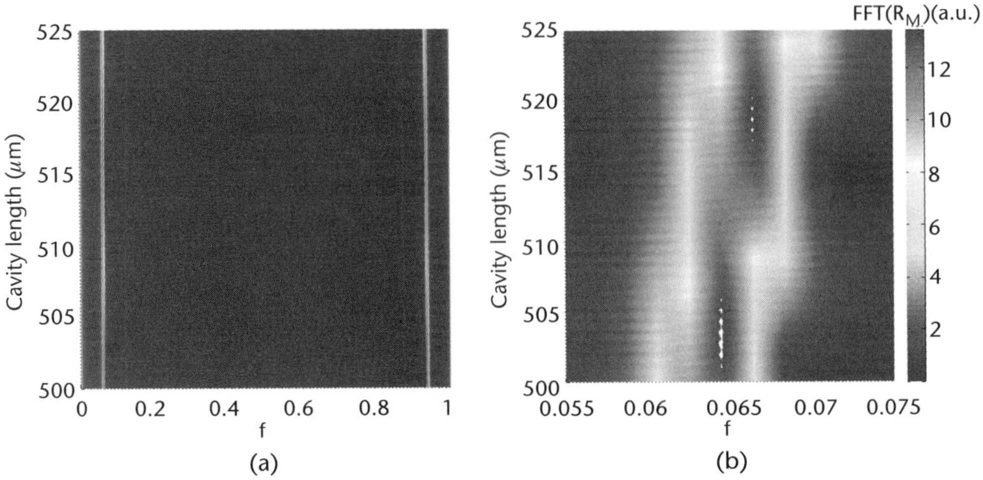

Figure 11.3 EFPI long-cavity tracking, for a simulated EFPI sensor with 500–525-m air-gap. (a) The cepstrum, estimated by applying the FFT to the EFPI optical spectrum. (b) The cepstral peak region.

11.3 Direct Karhunen-Loeve Transform

The Karhunen-Loeve transform (KLT) was applied for the first time to FBG tracking [7] and subsequently perfected and extended to EFPI and EFPI/FBG combined sensors in [8]. The KLT requires additional computation, and requires a more complex implementation, but allows achieving unmatched precision sensing. In the first step, we perform the FFT of the optical spectrum, as in (11.5), obtaining the cepstrum array $[\Im_1, \ldots, \Im_N]$. In a second step, we expand the cepstrum into its Toeplitz matrix $\underline{\underline{M}}$:

$$\underline{\underline{M}} = \begin{bmatrix} \Im_1 & \Im_2 & \Im_3 & \cdots & \Im_N \\ \Im_2 & \Im_1 & \Im_2 & \ddots & \vdots \\ \Im_3 & \Im_2 & \Im_1 & \ddots & \Im_3 \\ \vdots & \ddots & \ddots & \ddots & \Im_2 \\ \Im_N & \cdots & \Im_3 & \Im_2 & \Im_1 \end{bmatrix} \tag{11.7}$$

which has identical values on each diagonal, in a symmetric way.

Then the KLT is performed. The KLT takes the matrix $\underline{\underline{M}}$ as input and identifies an orthonormal basis $\underline{\underline{V}}$ over which it can be transformed. As demonstrated in Maccone's paper, dealing with the extraction of small signals buried into noise for deep-space communications [9], we can apply the singular value decomposition (SVD) as a special case for the KLT. This arrangement has two advantages: it is simple to implement, since most commercially available programs has the SVD routine based on the Cholesky decomposition, and it allows us to analyze the eigenvalues. The SVD decomposes $\underline{\underline{M}}$ in the following way [8]:

$$\underline{\underline{M}} = \underline{\underline{V}} \times \underline{\underline{D}} \times \underline{\underline{V}}^{-1} \tag{11.8}$$

where $\underline{\underline{D}}$ is a diagonal matrix containing all the eigenvalues of $\underline{\underline{M}}$ on its main diagonal, and $\underline{\underline{V}}$ is its corresponding orthonormal basis that contains on its lines the eigenvectors. The matrix $\underline{D}$ contains the N eigenvalues; as $\underline{\underline{M}}$ is a Toeplitz matrix, all its eigenvalues are real. The eigenvalue string is called ξ and is sorted in ascending order: for the moment we focus only on the top-rank eigenvalue ξ_N, neglecting the first $(N - 1)$ eigenvalues.

Figure 11.4 shows how the KLT applies to precision FBG detection, by plotting the highest rank eigenvalue ξ_N, as a function of FBG wavelength shift, for different grating strengths; the image is adapted from [8]. We observe that despite the coarse wavelength sampling, there is a sensitive variation of ξ_N even for small values of grating wavelength shift. The function is also periodic, with period equal to $\delta\lambda$: so we need to first estimate the correct half-period of $\xi_N(\Delta\lambda)$, using for example the centroid method, then refine the wavelength tracking by reversing this function to $\lambda(\xi_N)$. Data shown in [7] show that it is possible to achieve accuracy of 0.1 to 1 pm with relatively typical SNR values (42 to 52 dB), and it is possible to improve accuracy up to 0.01 pm with SNR > 62 dB, all despite the coarse sampling (156 pm).

Similarly, the KLT can also be applied to EFPI tracking, estimating Fabry-Perot cavity length. Figure 11.5 shows the variation of the top-rank eigenvalue, as a function of the variation of cavity length. Again, we observe a periodic function with period of 740 nm, and an extended (~600 nm) linear region. In order to reduce processing time, the chart is obtained by downsampling the spectrum (i.e., selecting one spectral sample, each 5 samples). We observe that for small variations of cavity length (0 to 1 nm), the response is almost linear. For typical values of SNR, higher than 40 dB, we obtain accuracy of 0.1 nm on the cavity length estimation.

The KLT can also be applied to FBG/EFPI combined sensors, used for pressure and temperature detection, as explained in [8]. The key asset is the analysis of the eigenvalue string ξ and how it relates to each sensor. In a dual sensor, the FBG dominates the reflectivity, since it has a highly reflective and narrow spectrum; and for this reason, it can be very well represented by the highest rank term ξ_N. The EFPI term is instead well represented by eigenvalues having lower indexes, in this case ξ_{N-2}. For small values of FBG wavelength shift and EFPI cavity length variation, which is the precision sensing context that we want to explore, we can almost

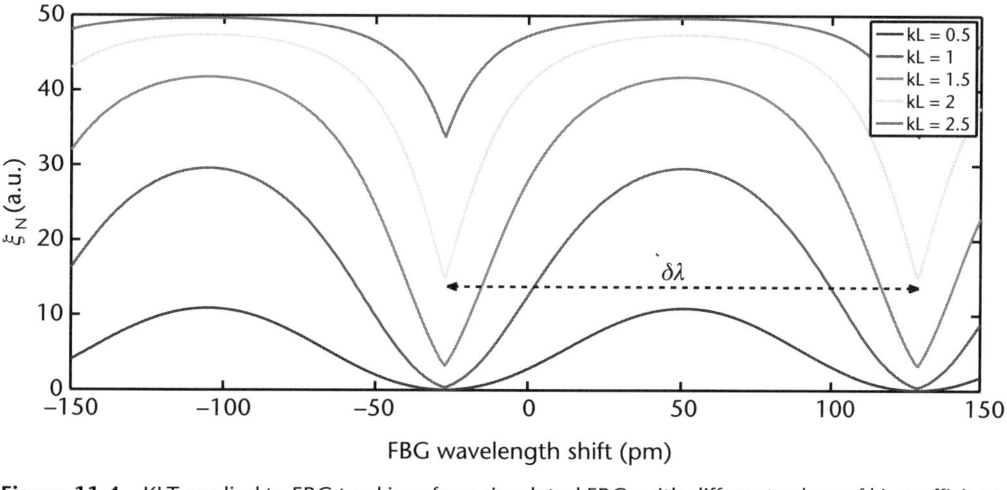

Figure 11.4 KLT applied to FBG tracking: for a simulated FBG, with different values of kL coefficient: the chart reports the maximum eigenvalue ξ_N for each value of the wavelength shift. (Chart adapted from [8].)

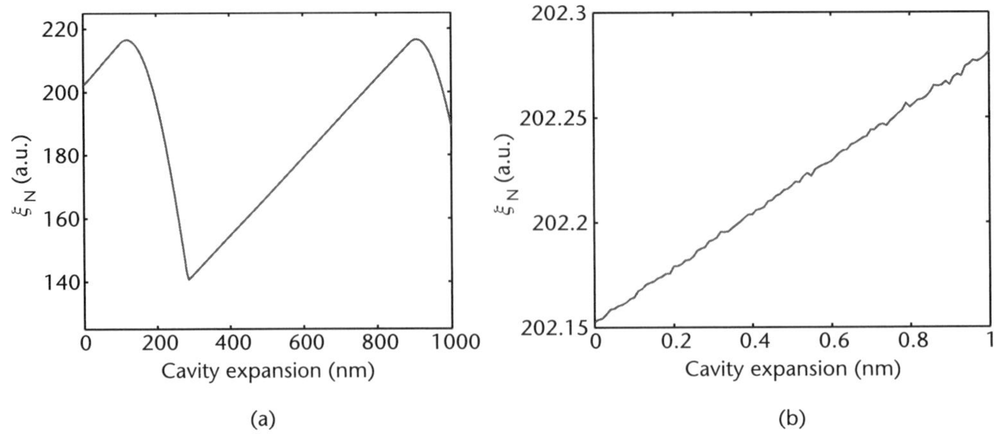

Figure 11.5 KLT applied to EFPI tracking: (a) for a simulated EFPI cavity having a gap length of 30 μm and a diaphragm length of 2 μm, the chart reports ξ_N as a function of cavity expansion; and (b) zoom on short cavity length variation. (Chart adapted from [8].)

perfectly separate the contribution of the FBG by analyzing the term ξ_N, and the contribution of the EFPI sensor by analyzing the term ξ_{N2}.

11.4 Multi-EFPI Systems

A modern trend in sensing is the interrogation of multiple sensors. For FBGs, this is done by means of WDM, isolating each FBG into a spectral slice. For EFPI sensors, since they have a broadband spectrum, this task cannot be performed in the spectral domain. As previously described, the FFT allows detecting a sensor in the cepstral domain, making the Fourier transform of the optical spectrum. Despite fast and simple, this approach is limited to few EFPIs, due to the frequency resolution limitations of the FFT. However, we can improve this approach by replacing the FFT with a better frequency estimator, not limited anymore in resolution. This approach consists in substantially shifting to a cepstrum-division multiplexing (CDM) approach.

At first, we take the spectrum of the EFPI combination, and we resample it (using the previously described method), in order to expand the number of points. Then we use the Capon estimator to perform cepstrum estimation. First developed in 1969 [10] as an equalizer for radar, is a good solution. Unlike the FFT, in the Capon algorithm the digital frequency is arbitrary and can be defined by the user at arbitrary choice, hence removing the limitation constraints typical of the FFT. For each digital frequency f, the Capon algorithm computes the optimum digital band-pass filter, defined as the filter that has a fixed unitary gain at f, and the minimum output variance. The power spectral density is then obtained as:

$$\Im(f) = \frac{1}{\underline{a}^H(f)\underline{\underline{R}}_{RR}^{-1}\underline{a}(f)} \tag{11.9}$$

where $\underline{R}_{RR}$ is the covariance matrix of the (resampled) optical spectrum, and $\underline{a}(f) = [1, \exp(-j2\pi f), \ldots, \exp(-j2\pi f(T-1))]'$ is the steering vector; $H(\bullet)$ is the Hermitian operator. A compact version of the Capon algorithm is included in Appendix B.

Figure 11.6 shows how it is possible to use the CDM to separate EFPI sensors. An array of 12 EFPI sensors with a cavity ranging from 50 μm to 3,425 μm is processed with the Capon estimator, estimating the power spectral density (PSD) in the digital frequency domain f. Thanks to the arbitrary choice of f, we can include 3,000 points in the frequency axis for low frequency (0 to 0.03). All 12 sensors appear as a peak in the cepstrum, and they can be separated in the cepstrum domain. The position of each peak can be then estimated with the methods in Section 11.1, obtaining the cavity length information. With this method, our numeric investigation shows that it is possible to interrogate up to 39 EFPI sensors with a cavity length within 40 μm and 3,600 μm.

11.5 Tilted and Chirped FBG

11.5.1 TFBG

The transmission spectrum of a TFBG is characterized by a combination of resonant cladding modes, a ghost mode, and the main FBG dip. In order to extract the information (particularly, the refractive index of the outer medium) from a TFBG, the analysis of ghost modes has to be performed. The combination of side shifts and amplitude changes can be effectively detected using a short-term Fourier transform (STFT); in MATLAB, this is performed by *stft* function. This approach is often used for time-frequency analysis of audio signals. Substantially, we divide the TFBG spectrum in blocks, each of size B, and we perform the FFT of each block. This way, we are able to split the analysis between modulus and phase of each FFT component. When a ghost mode that is nearly refractive-index independent falls within the STFT window, we observe nearly no change. When a cladding mode that exhibits

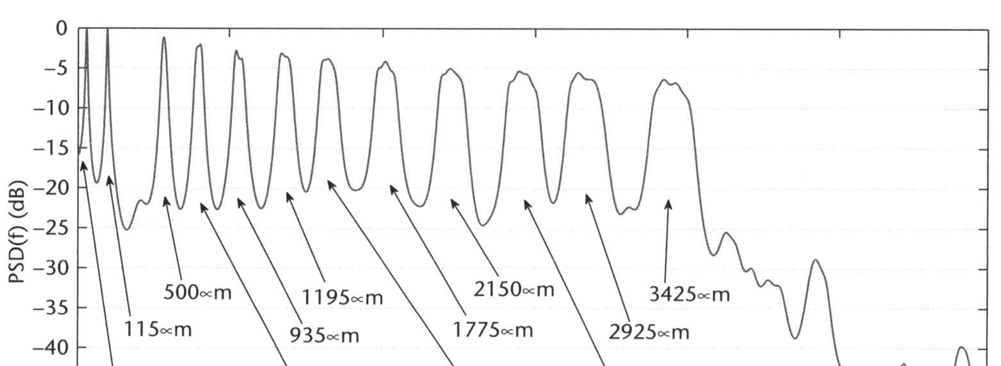

Figure 11.6 Cepstrum of an EFPI array, composed of 12 simulated EFPI sensors having cavity length of 50 μm, 115 μm, 500 μm, 700 μm, 935 μm, 1,195 μm, 1,445 μm, 1,775 μm, 2,150 μm, 2,560 μm, 2,925 μm, and 3,425 μm.

an amplitude change falls in the STFT window, we observe an amplitude variation at the low frequency (DC component). The STFT analysis is an effective investigation method for TFBG interrogation, as it allows framing the spectral window of interest and its dependence on refractive index.

The application of STFT to TFBG sensing is a recent topic; Figure 11.7 shows a simulation output, where the STFT output is reported for a standard TFBG having 1,550-nm Bragg wavelength and 8° tilt angle. For this specific TFBG, we do not observe large variations of the DC component. However, the STFT returns several peaks in the spectrum; some of these peaks are nearly independent on the refractive index, while two of them exhibit a visible amplitude change when refractive index is varying. The inner peak, corresponding to the sample 33, exhibits a steep amplitude change even for refractive index change of 0.04; compared to Figure 5.10, which shows the TFBG spectrum change with refractive index variations, the STFT is able to magnify a small change of the optical spectrum into a relatively large variation of the output. This work shows the premises for an STFT-based TFBG decoding that can break the current hardware limitations.

11.5.2 CFBG

To date, a method for interrogation of CFBG that extracts the temperature or strain spatial profile, as a function of CFBG spectrum, has not been determined. Tosi et al. [11] showed an approximate method for steep monotonic temperature change. Alternatively, engineers often track the grating bandwidth to estimate thermal/strain gradient or the spectral derivative to localize peak points, neglecting the whole spatial distribution. The current approach under investigation is based on the CFBG model of Chapter 5, as a chain of FBG: in the reference condition (i.e., no temperature applied, the CFBG is modeled as a chain of FBGs in static condition). When a temperature pattern is applied, the CFBG spectrum changes. The temperature pattern is reconstructed by identifying the temperature that, applied to each FBG of the chain, minimizes the root mean square error between the measured spectrum and the reconstructed spectrum. So far, this approach has been limited to Gaussian and super-Gaussian temperature spatial patterns, identified by 3 and 4 parameters, respectively. This way, we can collapse a nonlinear optimization problem with

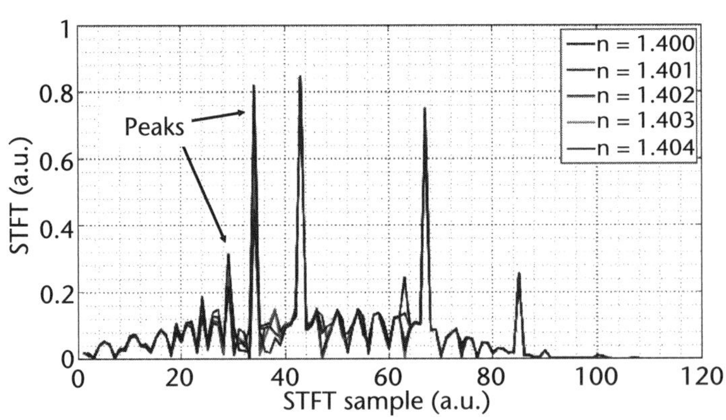

Figure 11.7 STFT of a simulated TFBG sensor, with a 1,550-nm Bragg wavelength and tilt angle of 8°, and sampled on a 156-pm wavelength grid; the STFT is applied on 120 blocks with size B = 60.

unsustainable computational complexity to the identification of few parameters. The limitation is that this approach works only for a predefined temperature distribution profile; however, this approach is effective in some applications, like thermal ablation, where temperature spatial patterns have a quasi-Gaussian shape.

11.6 LabVIEW Implementation

The code proposed in this section has been implemented in MATLAB and included in Appendix B, including the scripts for FBG and FPI tracking, KLT, and the Capon estimator. In most commercial and prototype systems, the data acquisition (DAQ) part is handled in LabVIEW software, thanks to its capability to interface with spectrometers and the easiness to design a graphic user interface suitable for prototypes and customers. In the releases since 2012, MATLAB code can be imported into LabVIEW: [12] showed a tutorial very useful for importing Mathscripts into LabVIEW functions and operated them as standard LabVIEW nodes. We recommend in particular familiarizing with Section 11.3, which relates to the SVD operation. This approach is recommended for research and development, since MATLAB allows computational tasks (particularly KLT and Capon operations, based on matrix algebra), but handling user-interface and data acquisition and display is the core asset of LabVIEW.

11.7 Conclusions

Software is an essential part of the whole FOS sensing system and is as important for the overall performance as hardware and sensing unit design. The task accomplished by software is to convert the sensor(s) output (typically the optical spectrum) into the variation of the measurand(s). The application of signal processing techniques that allows magnifying small signal variations or the extraction of weak signals buried into noise is a growing field in FOS science.

The future trend of research on FOS will progress in finding new software methods to interpret complex optical spectra and to isolate the contribution of each sensor from a spectral envelope. From the system point of view, this corresponds to standardizing the hardware block (which is essential to maintain FOS as a relatively low-cost and flexible unit) and developing new software capable of decoding any sensing element: this shift of complexity from the hardware to the software is a trend that aims at expanding FOS market attractiveness, as software (and its computation platform) costs a small fraction of the hardware required to operate sensors. To date, only standard FBG arrays are widely covered by commercial devices, but recent works are showing traction in FBG chains, FPI (with CDM capabilities), multiprobes, CFBG and TFBG, and a number of complex configurations.

References

[1] Atkins, C. G., M. A. Putnam, and E. J. Friebele, "Instrumentation for Interrogating Many-Element Fiber Bragg Grating Arrays," *Proc. SPIE*, Vol. 2444, February 2005, pp. 257–267.

[2] Kinet, D., et al., "Fiber Bragg Grating Sensors Toward Structural Health Monitoring in Composite Materials: Challenges and Solutions," *Sensors*, Vol. 14, No. 4, 2014, pp. 7394–7419.

[3] Ezbiri, A., S. E. Kanellopoulos, and V. A. Handerek, "High Resolution Instrumentation System for Fiber-Bragg Grating Aerospace Sensors," *Opt. Commun.*, Vol. 150, Nos. 1–6, 1998, pp. 43–48.

[4] Chen, W., A. Vallan, and G. Perrone, "Performance Assessment of a Fast Temperature Sensing System Based on Bare FBGs and Fast Spectrum Analyser," *Proc. 1st Int. Electron. Conf. Sens. Appl., Sciforum Electronic Conference Series*, Vol. 1, 2014, p. c004.

[5] Poeggel, S., et al., "Recent Improvement of Medical Optical Fibre Pressure and Temperature Sensors," *Biosensors*, Vol. 5, No. 3, July 13, 2015, pp. 432–449.

[6] Jiang, Y., "Fourier Transform White-Light Interferometry for the Measurement of Fiber-Optic Extrinsic Fabry-Perot Interferometric Sensors," *IEEE Photonics Technology Letters*, Vol. 20, No. 2, January 15, 2008, pp. 75–77.

[7] Tosi, D., "KLT-Based Algorithm for Sub-pm Accurate FBG Tracking with Coarse Wavelength Sampling," *Photonics Technology Letters*, Vol. 27, No. 20, 2015, pp. 2134–2137.

[8] Tosi, D., "Advanced Interrogation of Fiber-Optic Bragg Grating and Fabry-Perot Sensors with KLT Analysis," *Sensors*, Vol. 15, 2015, pp. 27470–27492.

[9] Maccone, C., "Advantages of Karhunen-Loeve Transform over Fast Fourier Transform for Planetary Radar and Space Debris Detection," *Acta Astronautica*, Vol. 60, 2005, pp. 775–779.

[10] Capon, J., "High-Resolution Frequency-Wavenumber Spectrum Analysis," *Proc. IEEE* Vol. 57, No. 8, 1969, pp. 1408–1418.

[11] Tosi, D., et al., "Fiber-Optic Chirped FBG for Distributed Thermal Monitoring of Ex-Vivo Radiofrequency Ablation of Liver," *Biomedical Optics Express*, Vol. 5, No. 6, 2014, pp. 1799–1811.

[12] Halvorsen, H. -P., LabVIEW Mathscript, http://home.hit.no/~hansha/documents/labview/training/LabVIEW%20MathScript/LabVIEW%20MathScript.pdf.

CHAPTER 12

Standards for Medical Sensors

12.1 Main Standards

The compliance to international and well-established standards is a main requirement for the application and the commercial exploitation of any medical device, including the majority of the sensors described in this book and all sensors operating in vivo through a self-contained medical device or associated with a medical device. The overall list of medical standard encompasses several regulations and is specific to each application and positioning. In this chapter, we focus on the most important standards, limitedly to the biomedical fiber-optic sensing devices and their applications. At the international level, the legislation in terms of standards is only partially unified: most European countries and Canada refer to International Organization for Standardization (ISO), whereas the U.S. Food and Drug Administration (FDA) is based on different standards, still similar to ISO. In this chapter, we will introduce the main standards currently in use for medical devices, showing the references and the breakdown of main items to be addressed. Figure 12.1 shows a flowchart [1] that engineers and sensor specialists can use as a reference tool to certify biomedical sensors according to FDA protocols.

12.2 ISO 10993

Submissions for approval of medical devices by regulatory agencies require that biocompatibility assessment be conducted to assure safety of the device and its materials throughout long-term operation. The main point of reference is the standard ISO 10993 "Biological Evaluation of Medical Devices" [2], which embraces a series of standards, reported in Table 12.1, for evaluating the biocompatibility of medical devices, and is to date the main standard for medical devices that operate in vivo. The ISO 10993 standard mainly pertains to devices making contact with the skin, mucosal membranes, devices operating in blood vessels, or in contact with tissue or bones (such as endoscopic and percutaneous devices), and implantable devices. The purpose of the standard is to protect patients from undesired effects and to serve as a framework for selecting tests to evaluate biological responses and the test protocols are designed in order to minimize the requirements for animal tests. As a

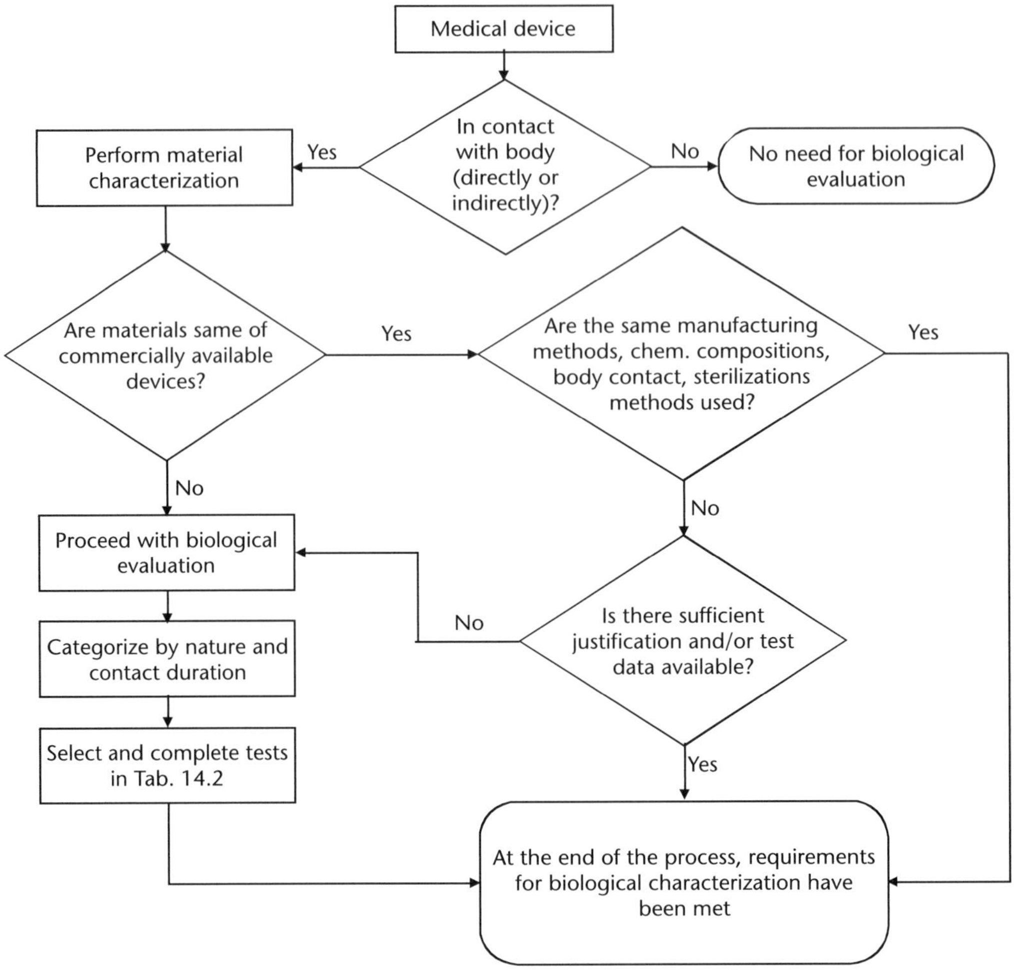

Figure 12.1 Flowchart for biological evaluation of medical devices, in accordance with FDA and ISO 10993 standards.

main focus, the ISO 10993 defines the table (Table 12.2) for the evaluation of the biocompatibility test, for different levels of exposure (less than 24 hours to permanent contact). This is reported also in the Blue Book Memorandum #G95-1 [3] that is in use by most medical device companies and is currently used by the FDA for the evaluation of medical devices in the United States [4].

Tables 12.1 and 12.2 are also the main reference points, limiting to FOS technology. FOSs are placed in vivo through percutaneous, endoscopic, or other insertion methods, through catheters; it is therefore necessary to evaluate the biological composition of each catheter. A good practice that avoids having to run the whole cycle in Figure 12.1 is the use of a commercial catheter, enclosing the fiber sensor within it; however, this poses a severe limitation to the FOS advantages, since the size of the overall device would be 10 to 50 times the size of the optical fiber. A custom-made catheter would have to go through the Table 12.2 evaluation of materials in order to evaluate the biological reactions to catheter insertion.

Table 12.1 List of ISO 10993 Standards

Part	*Date*	*Standard Content*
1	2009	Evaluation and testing in the risk management process
2	2006	Animal welfare requirements
3	2014	Tests for genotoxicity, carcinogenicity, and reproductive toxicity
4	2006	Selection of tests for interactions with blood
5	2009	Tests for in vitro cytotoxicity
6	2007	Tests for local effects after implantation
7	2008	Ethylene oxide sterilization residuals
8	2001	Selection of reference materials
9	1999	Framework for identification and quantification of potential degradation products
10	2010	Tests for irritation and skin sensitization
11	2006	Tests for systemic toxicity
12	2012	Sample preparation and reference materials
13	1998	Identification and quantification of degradation products from polymeric medical devices
14	2001	Identification and quantification of degradation products from ceramics
15	2000	Identification and quantification of degradation products from metals and alloys
16	1997	Toxicokinetic study design for degradation products and leachables
17	2002	Establishment of allowable limits for leachable substances
18	2005	Chemical characterization of materials
19	2006	Physico-chemical, morphological, and topographical characterization of materials
20	2006	Principles and methods for immunotoxicology testing of medical devices

12.3 ISO 13485

The ISO 13485:2006 [5] specifies requirements for a quality management system where an organization needs to demonstrate its ability to provide medical devices and related services that consistently meet customer and applicable regulatory requirements, and it caters to the medical device industry. The ISO 13485 is designed to be used by medical device companies throughout the whole device life cycle, from the initial conception, manufacturing and quality assurance, distribution, decommission, and disposal, including the aspects of storage, installation, and maintenance. The FDA and the European Union Medical Devices Directive 93/42/EEC refer to the ISO 13485 for medical devices, including sensors, probes, and their catheters. ISO 13485 is similar in scope and intent to ISO 9001, Quality Management Systems [6], the widest purpose quality management system employed in most countries, but adds specific guidelines for medical devices and their use that are sought by the FDA. Remarkably, all the commercial FOS technologies embedded in medical devices and described in this book, particularly in Chapter 14, have obtained the ISO 13485 certification.

12.4 IEC 60601

The IEC/EN 60601 [7] is a collection of standards aimed at the evaluating the safety of medical devices from the electrical point of view. It is the main reference

Table 12.2 ISO 10993 Biocompatibility Test Standards

Device Categories			*Initial Evaluation*								*Supplemental Evaluation*	
	Body contact	*Contact duration*	*Cytotoxicity*	*Sensitization*	*Irritation or intracutan. reactivity*	*Systemic toxicity (acute)/pyrogenicity*	*Subchronic toxicity*	*Genotoxicity*	*Implantation*	*Hemocompatibility*	*Chronic toxicity*	*Carcinogenicity*
Surface	Skin	A	•	•	•							
		B	•	•	•							
		C	•	•	•							
	Mucosal membrane	A	•	•	•							
		B	•	•	•	0	0	•	0		0	
		C	•	•	•	0	•		0			
	Breached/ compromised surface	A	•	•	•	0						
		B	•	•	•	0	0	•	0		0	
		C	•	•	•	0	•		0			
External communic.	Blood path indirect	A	•	•	•	•				•		
		B	•	•	•	•	0	•	0	•	•	•
		C	•	•	0	•	•			•		
	Tissue/ bone dentin communicating	A	•	•	•	0						
		B	•	•	0	0	0	•	•		0	•
		C	•	•	0	0	0	•	•			
	Circulating blood	A	•	•	•	•		0		•		
		B	•	•	•	•	0	•	0	•	•	•
		C	•	•	•	•	•	•	0	•		
Implant	Bone/tissue	A	•	•	•	0						
		B	•	•	0	0	0	•	•		•	•
		C	•	•	0	0	0	•	•			
	Blood	A	•	•	•	•			•	•		
		B	•	•	•	•	0	•	•	•	•	•
		C	•	•	•	•	•	•	•	•		

Exposure times: A = limited exposure (≤24 hours), B = prolonged exposure (24 hours to 30 days), and C = permanent contact (≥30 days). Test protocols: • = FDA and ISO evaluation tests; 0 = additional tests for FDA.

for surgical, endoscopic, and percutaneous devices, published and updated by the International Electrotechnical Commission (IEC), and then adapted, with some modifications, to national standards such as the EN 60601 for European Union or UL2601 for United States. The IEC 60601 standard is based on a generic standard, which governs definition and common terms for all the specific standards (part 1:1), 8 collateral standards (parts 1:2-11), and 58 particular standards (parts

2:1-68), which regulate particular requirement for each clinical procedure or instrument. Table 12.3 lists the IEC 60601 standards that are particularly related to the FOS application and have been granted to the vast majority of the commercial fiber-optic devices and catheters described in the next sections of this book.

From a strategic point of view, FOSs have a huge advantage over other sensing technologies thanks to their electromagnetic compatibility. While it is true that only a few (but very significant) medical applications related to sensing instruments are strongly affected by electromagnetic irradiance and electrical design guidelines, historically the cost of obtaining clearance for such standards has been uneasy and costly for research centers and academic spin-offs, partly because of the lack of know-how. Consultants are often employed for obtaining IEC 60601 clearance, often at cost higher than research grants awarded for investigations. Alternatively, the vast literature and historical track that see optical fibers as passive, electrically, and electromagnetically immune devices, provide a strong track record that simplifies the obtainment of standards, opening up for a faster certification process.

12.5 Other Standards

Within the several other standardization protocols, Table 12.4 contains a list of additional standards selectively addressable to biomedical FOS, as the main point of

Table 12.3 IEC 60601 Standard for Safety of Medical Electrical Devices: List of the General, Collateral, and Particular Standards Related to Present and Future Biomedical FOS

Part	*Related Use in FOS*	*Standard*
1:1	General FOS sensors	Safety requirements for medical electrical systems
1:2	General FOS sensors	Electromagnetic disturbances: requirements and tests
1:6	General FOS sensors	Usability
1:9	General FOS sensors	Requirements for environmentally conscious design
1:11	Remote and wearable FOS (Chapter 20)	Collateral standard: requirements for medical electrical equipment and medical electrical systems used in the home health-care environment
1:12	Pressure probes (Chapters 15 to 16)	Requirements for medical electrical equipment and medical electrical systems intended for use in the emergency medical services environment
2:2	Ultrasound, thermal ablation and their probes/sensors (Chapter 17)	Particular requirements for the basic safety and essential performance of high-frequency surgical equipment and high-frequency surgical accessories
2:6	Microwave ablation (Chapter 17)	Particular requirements for the basic safety and essential performance of microwave therapy equipment
2:22	Laser ablation (Chapter 17)	Particular requirements for basic safety and essential performance of surgical, cosmetic, therapeutic, and diagnostic laser equipment
2:49	Cardiovascular, diagnostic (Chapters 15 and 16)	Particular requirements for the basic safety and essential performance of multifunction patient monitoring equipment
2:62	HIFU (Chapter 17)	Particular requirements for the basic safety and essential performance of high-intensity therapeutic ultrasound (HITU) equipment

Table 12.4 Other Standards Applicable to Biomedical FOS

Standard	*Definition*	*Reference*
ISO 80601-2-56:2012	Medical Electrical Equipment, Part 2-56: Particular requirements for basic safety and essential performance of clinical thermometers for body temperature measurement	Equipment and measurement
ISO 80601-2-71:2015	Medical Electrical Equipment, Part 2-71: Particular requirements for the basic safety and essential performance of functional near-infrared spectroscopy (NIRS) equipment	Equipment and measurement
IEC 61010-2-101:2015	Safety requirements for electrical equipment for measurement, control, and laboratory use – Part 2-101: Particular requirements for in vitro diagnostic (IVD) medical equipment	Electrical safety
IEC 62366:2014	Medical devices: Application of usability engineering to medical devices	Instrumentation use
IEC 61326-1:2012	Electrical equipment for measurement, control and laboratory use: EMC requirements Part 1: General requirements	Electromagnetic compatibility
CLSI EP5-A3:2014	Evaluation of precision of quantitative measurement procedures; approved guideline	Measurements
ASTM F981-04	Standard practice for assessment of compatibility of biomaterials for surgical implants with respect to effect of materials on muscle and bone	Biocompability
ISO 10555-1:2013	Intravascular catheters: Sterile and single-use catheters, Part 1: General requirements	Cardiovascular
ISO 10555-3:2013	Intravascular catheters: Sterile and single-use catheters, Part 3: Central venous catheters	Cardiovascular
ISO 10555-4:2013	Intravascular catheters: Sterile and single-use catheters, Part 4: Balloon dilatation catheters	Cardiovascular
ISO 10555-5:2013	Intravascular catheters: Sterile and single-use catheters, Part 5: Over-needle peripheral catheters	Cardiovascular
ISO 11138-1:2006	Sterilization of health care products: Biological indicators, Part 1: General	Sterilization
ISO 17664:2004	Sterilization of medical devices: Information to be provided by the manufacturer for the processing of resterilizable medical devices	Sterilization

reference for medical-device companies. A fully comprehensive list of all the main standards relevant to medical devices is provided by Health Canada [8].

12.6 CE Mark

The CE mark (European Conformity) is a mark directly applied by the manufacturer upon approval that declares that the product meets the regulatory standards for being traded within the European Economic Area (since 1985). Today it is a requirement for products that are designed to be industrialized in Europe or sold in European Union. In order to be able to apply a CE mark, manufacturers need to comply with the respective regulatory body. These are listed in a collection of standards that are verified by each notified body [9]. The most important standards for FOS refer to the 3 medical device classes (93/42/EEC – Medical devices; 98/79/EC

– In vitro diagnostic medical devices; and 90/385/EEC – Active implantable medical devices) and to electromagnetic compatibility (89/336/EEC – Electromagnetic compatibility). According to CE regulations, devices are separated in different classes of risk from class I (nonsterile, nonmeasuring) to class IIIb (high risk). Depending on the risk class, the different ISO standards previously described apply and need to be verified. The production of the device must also respect the quality process, and ISO 13485 is the golden standard for today's devices.

Upon meeting the correct standards for the appropriate risk factor, manufacturers must prepare a CE marking technical file, which is a comprehensive description of the device intended to demonstrate compliance with European requirements and justifying design choices. Both the technical file and the quality management systems then are validated by the EEC body; upon obtaining the documentation, the manufacturer can then prepare a declaration of conformity (DoC) and apply the CE mark to each product. Most companies can select an external consultant that can handle the communication with EEC bodies, upon ISO standards compliance.

To date, CE marking is for most medical device manufacturer an additional step that adds up to the previous ISO/EN standards (that are the true checkpoints towards devices certification). CE mark is a need for operating in European Union at the industrial level.

References

[1] Saliterman, S., *Fundamentals of BioMEMS and Medical Microdevices*, New York: SPIE Press, 2006.

[2] ISO 10993, *Biological Evaluation of Medical Devices; International Organization for Standardization*, Geneva, Switzerland, 1995.

[3] "Use of International Standard ISO 10993, Biological Evaluation of Medical Devices—Part 1: Evaluation and Testing," G95-1, Rockville, MD: Department of Health and Human Services, FDA, 1995.

[4] "Use of International Standard ISO 10993-1, Biological Evaluation of Medical Devices, Part 1: Evaluation and Testing Within a Risk Management Process," Guidance for Industry and Food and Drug Administration Staff, 2016.

[5] ISO 13485, "Medical Devices – Quality Management Systems – Requirements for Regulatory Purposes," International Organization for Standardization: Geneva, Switzerland, 2016.

[6] ISO 9001, "Quality Management," International Organization for Standardization: Geneva, Switzerland, 2015.

[7] International Electrotechnical Commission (IEC) 60601-1, "Medical Electrical Equipment, Part 1: General Requirements for Basic Safety and Essential Performance," Geneva, Switzerland, 2005.

[8] Health Canada, "List of Recognized Standards for Medical Devices," 2016, http://www.hc-sc.gc.ca/dhp-mps/md-im/standards-normes/md_rec_stand_im_norm_lst-eng.php.

[9] CE Marking, "List of Notified Body," http://www.ce-marking.org/list-of-notified-bodies.html.

CHAPTER 13

Protocols and Tools for Validation

In this chapter, we will discuss definitions and procedures for research and test of biomedical fiber-optic sensors, as well as ethical issues concerned with biomedical research.

13.1 Moral Norms

Two moral norms, introduced in [1], have remained relatively constant across the various stages of medical device research, including the technologies related to sensing and measurements. They are beneficence, which is the provision of benefits, and nonmaleficence, which is the avoidance of doing harm. Every device under investigation, in relation to diagnostic of diseases, treatment of known diseases, or monitoring of evolution of diseases, must build the case that the device effective supports the healthcare for a patient, and one does not harm the patient under any reasonable circumstance. The whole modern healthcare system has to revolve around these two principles, which are both intended and verified by certification institutes (such as the U.S. Food and Drug Association [FDA]), but also constitute the cardinal pillars on which the research in medical devices is performed and accepted by most institutions, international journals and publishers, and funding agencies.

13.2 Relevant Definitions

Biology studies conducted for medical and pharmaceutical devices pose attention to protocols that allow testing in vitro, ex vivo, in vivo, on phantoms, and in several conditions. Whereas biological protocols are established and follow a general mainframe [2], the study of medical sensors suffers generally from a lack of procedural paths on the technological side, and from the only recent implementation of the application-driven development introduced in Chapter 3. Still, several of the concepts from biological science can be converted to counterparts that are also suitable for most of the biomedical devices described in this book.

13.2.1 In Vivo

The ultimate goal of most biomedical FOS is to operate within a patient's body, through an endoscopic, percutaneous, laparoscopic, or arthroscopic insertion. In vivo studies refer to all the test and research practices that take place in a whole living organism, which can be either a human or an animal. In vivo tests usually represent the ultimate part of the test protocols, as they allow highlighting all the effects of the living organism on the sensors and the catheters used to package the devices. The regulatory apparatus for medical research poses severe limitations to the capabilities of performing in vivo tests, both on animals and on patients, and experimental practices must undergo a procedure that allows to clear ethical issues and the nonmaleficence principle. Each test in vivo poses strong concerns for the overall condition of the patient or animal under test, which needs to be safeguarded: factors such as invasiveness, bleeding, and potential malfunctioning of the measurement device must be accounted before the tests are enabled, documenting the minimal risk for the device to harm the organism.

13.2.2 Ex Vivo

Experimentation ex vivo takes place in a tissue obtained from a nonliving organism, which mimics the scenario for the real application. Ex vivo measurements can be performed on human parts, often obtained through cadavers with a consensus for the use, or from animal parts that are often available commercially. Ex vivo experiments allow one to usually test devices at earlier stages than in vivo tests, as they allow an easier execution of measurements and experiments, and when animal parts are available commercially, they do not require compliance to ethical issues. Tests executed ex vivo can be performed with minimal concerns for invasiveness, tissue bleeding, and harm; in measurement science, this is positive as it allows, for example, inserting additional sensors in the tissue as reference instruments, or destructively processing the tissue after the experiments. In terms of reproducibility, ex vivo analysis does not take into account blood perfusion (and, consequently, the amount of bleeding) and effects such as breathing or tissue movements. Insertion and removal of the sensing probe into the tissue is simplified and does not take into account the ultrasound or MRI guidance of the device to the operation point, but this also allows studying the effects of different catheterizations in terms of insertion.

13.2.3 In Vitro

For medical devices, the term in vitro starts to appear as an improper definition of tests that are not performed on tissues (human or animal), but rather reproduce the condition of real testing in the laboratory, for example, using phantoms or calibration machines that mimic the behavior of the human body during the measurement procedure. Such tests are essential at preliminary development of sensors: they allow calibrating and measuring the performance (accuracy, repeatability, response time) of each sensing units, but they also allow isolating and analyzing each detrimental effect acting on the sensors. According to the availability of anatomic models or phantoms, in vitro test can be preliminary and lead to a simple sensor characterization by using a calibration facility or can be advanced thanks to the availability of tissue-mimicking phantoms that excellently reproduce the human organism. As a general guidance, in vitro analysis produce an ideal test condition, useful for calibrating sensors, that build the premises towards deeper tests ex vivo and in vivo.

13.2.4 Phantom

From the imaging terminology, for the biomedical sensors discipline, a phantom is any model or a sample of tissue that replaces the human organism for the test and validation of medical devices. In experimental practice, phantoms can be chosen as ex vivo animal organs that best reproduce the human counterpart: examples are porcine liver and pancreas or rabbits' ocular for eye treatments. These parts are often obtained commercially and require refrigeration and storage in order to maintain the properties of the tissue. In alternative, phantoms can be manufactured in order to replicate human parts and their points of access, mimicking the behavior of human organs. The current technologies, particularly for additive manufacturing, allow fabricating phantoms exhibiting an excellent degree of matching with human organs in terms of behavior and materials. Custom-made phantoms offer significant features for biomedical engineering projects, as they allow testing devices with a high complexity without any ethical issue; from a scientific point of view, the credibility of results obtained on phantoms is increasing. Most physical phantoms are fabricated using acrylonitrile butadiene styrene (ABS) material [3], while radiation-prone phantoms can be fabricated with polylactic acid (PLA) [4].

13.2.5 Principle of 3Rs

Since 1959, Russell and Burch [5] formed a set of guiding principles to minimize the impact of animal testing in medical devices. As of today, this process is implemented with the principle of 3Rs, which is inherent in research projects administered under the European Union and is also widely popular as a principle for any research project that involves testing on animal parts.

- *(1R) Replacement*: Methods that avoid or replace the use of animals in research. Phantoms and anatomic models are well representatives of techniques usable to minimize the use of animals particularly during the earlier stages of technologies.
- *(2R) Reduction:* Strategies for obtaining the same of comparable amount levels of information, while reducing the number of animals involved in the tests. In medical devices, this criterion implies the overall definition of the research plan, such that only the minimum amount of tests can be performed, maximizing the amount of knowledge generated in each run.
- *(3R) Refinement*: Use of methods that minimize and alleviate the pain for the animals used in the tests. In the framework of preserving animal welfare, scientists must minimize the impact of each test on the animals, preventing pain and discomfort whenever possible. In the current research systems, scientists

are asked to comply to the 3R criterion in each proposal submission, and funding agencies extend the 3R criterion also to ex vivo animal phantom.

13.2.6 Ethical Issues

Research in healthcare is concerned ethical issues, which regulate and limit the possibility of performing research on humans, animals, and tissues. Every research that is concerned with any ethical issues undergoes an approval from an ethics committee, which reviews the protocols and methodologies for the operative part, as well as the procedures for use and disposal of devices, data collection and protection, protection of patients and devices, and every possible aspect that can substantially raise a warning for the safety, privacy, and comfort of any living organism involved in the study. As a reference, we report in Table 13.1 the ethics issues involved in research projects proposed and funded under European Union Horizon 2020 program, as of 2016, limiting to those aspects involving biomedical sensors. The main two aspects related to early-stage technologies involve the experiments on animals, which require justification and must undergo ethics approval, and tests on human tissues, obtained through cadavers, which require the approval of an ethics committee and the justification of the study. At a much further level, experiments that involve humans must be justified. Experiments involving external applications of nondangerous materials, such as smart medical textiles, are required to justify the nontoxicity and nonharmfulness of devices; in contrast, technologies for intervention on study participants, such as real-time, in vivo sensors, must be approved by ethics committee. Patients and candidates recruited for the study are also part of the

Table 13.1 Ethics Issues Relevant to Biomedical Sensors and Devices for Research Projects and Proposals

Section	*Topic*	*Ethics Issue(s)*
1	Human embryos/ fetuses	Does your research involve human embryonic stem cells (hESCs)?
		Does your research involve the use of human embryos?
		Does your research involve the use of human fetal tissues/cells?
2	Humans	Does your research involve human participants?
		Does your research involve physical interventions on the study participants?
3	Human cells/ tissues	Does your research involve human cells or tissues (other than from human embryos/fetuses [i.e., Section 1])?
4	Personal data	Does your research involve personal data collection and/or processing?
		Does your research involve further processing of previously collected personal data (secondary use)?
5	Animals	Does your research involve animals?
7	Environment and health and safety	Does your research involve the use of elements that may cause harm to the environment, to animals or plants?
		Does your research deal with endangered fauna and/or flora and/ or protected areas?
		Does your research involve the use of elements that may cause harm to humans, including research staff?

Source: Horizon 2020, Grant Applications template, Part A, 2016.

ethical issues, as researchers need to outline the protocol and authorizations for collecting informed consents, the selection criteria for recruitment, and the protocols for personal data collection, processing, and protection.

13.3 Research and Test Methodologies

Although a unified methodology for testing and validation of FOS cannot be achieved, due to the different technologies, invasiveness, and positioning, we can frame a set of generic guidelines that engineers and developers can follow to prototype devices; guidelines must take into account the application-driven design in Figure 3.5. At first, the initial step is to validate the sensor unit in vitro, using an ad hoc test chamber such as a thermo-climatic chamber, a precision strain bench, or a pressure chamber. This initial test must occur in ideal conditions and using a set of reference sensors having accuracy superior to the sensor under test; if the proposed sensor has a nominal accuracy better than any reference, an indirect measurement can be used. The test in an ideal condition allows at first obtaining the calibration function and therefore validating the behavior of the sensors and identifying its working range, its linearity, and the input/output relation. Then it allows compiling the datasheet, examining in detail all the performance rates such as response time, repeatability, hysteresis, and noise. In this analysis, it is important to study the cross-sensitivities. For the in vitro test, the sensor has to be unpackaged, in order to obtain the performance rate of the sensing unit unaffected by its catheter or package form factor.

If sensors are designed for external appliance and are biocompatible and not harmful, they do not raise any ethical issue and can be directly tested on human candidates; this is the case of smart textiles or sensors for gait analysis. In most cases, when sensors are designed to be used in vivo, the next step involves the test on phantoms. Prior to proceeding with the test, the phantom has to be investigated. An effective phantom, besides having characteristics that mimic the human organs, allows inserting additional measurement instruments to be used as reference for the sensor under test, allows investigating a posteriori the effect of interventions on the phantom itself, and also allows evaluating the measurement artifact introduced by sensor packaging. In addition, considerations related to cost and the 3Rs principle are part of the procedure. As a general guideline, ex vivo animal phantoms can be used in applications that involve a strong interaction between the probes and the tissues; examples are thermal ablation and surgical procedures (Chapters 17 and 19). In this case, the catheterization and the artifact strongly differ between each tissue, and this is the main limitation. Cardiovascular applications mainly involve the pressure measurement in a pulsatile system through an access point located far from the measurement point; in this case, anatomic models are preferred as they allow simulating the pressure pattern with a pulsatile pump. The preliminary study on the phantom has the goal of estimating the degree of reproducibility of its features compared to human organs, and the weaknesses that need to be evaluated with specific tests.

In the next step, sensors can be tested on phantoms, replicating the medical procedures or the diagnostic method on the selected model. In a first stage, sensors are tested against a reliable reference to show the correct functioning of the probe and the mismatch between the ideal calibration and the one in phantom. When a satisfactory result is obtained, developers should focus on the package of the sensor, estimating both its measurement artifact and its compliance with the specific application (size, guidance to location, radio-opacity, and other features). Most phantoms also allow making a big-data quantitative analysis of the results, evaluating on a large number of tests and running statistical analysis to identify performance rates. In this moment, the decision on whether using a commercially available catheter, adapting a commercial catheter to the specific use, or develop a new catheter sample, has to be made on the basis of phantom test. When necessary, particularly for understanding the tissue-device interaction, ex vivo tests can be performed on human tissues upon the approval of the ethics committee.

The next step is to bring the technology in vivo, and depending on the scenario and the inherent risks related to the technology, experiments can be conducted directly on patients or first on animals; at this stage, the research and development team must include clinicians with specific operational experience. This part of the experimentation must undergo approval of ethics committee within the clinics running the tests, and all protocols and methodologies chosen for the study have to be disclosed and approved. Besides the results of the previous analysis, which need to show the usefulness of the probe and its impact on tissues, ethics committees pose specific emphasis on several aspects, including: aspects of the impact on patients that cannot be covered with phantoms (such as amount of bleeding, risk of infections); positioning and validation of probe positioning; analysis of the risks involved in the procedures, and contingency plan; sterilization, storage, removal, and disposal of the sensors; action plan in case of sensor breaking, or catheter breaking; possibility to use redundant sensors to obtain additional reference, and their impact. In addition, aspects considered by ethical committees involve the recruitment of patients, and their data management. A general-purpose reference for insights on the approval of medical device research is provided by National Institute of Health, in the United Kingdom [6], covering the main aspects of medical device approval.

13.4 Investigational Device Exemption

Upon successfully accomplishing the chain of test, a medical device must apply to a premarket approval (PMA), or a premarket notification 510(k), which represents the foundation for the follow-up clinical study. The FDA set of guidelines classifies significant risk and nonsignificant risk devices [7, 8]. The significant risk device class includes those devices that present a potential for serious risk to health and safety of a subject; in the field of FOS treated in this book, this category includes most of the cardiovascular devices, which may have the potential danger due to their positioning in proximity of the heart in carotid vessels, thermal ablation devices due to their heat effect and insertion, devices paired to invasive surgical devices such as robotic surgery. According to [7], the process for significant risk devices require the initiator or sponsor to undergo both the FDA and the institutional review board (IRB) approval prior to initiation of a clinical study, and the investigational device exemption (IDE) approval is the key step. According to [8], the initiator has to submit the investigational plan and report of prior investigations to the IRB at each institution where the investigation is to be conducted for review and approval and select qualified investigators, provide them with all necessary information on the investigational plan and report of prior investigations, and obtain signed investigator agreements from them. Nonsignificant risk devices, which for FOS relate to many urodynamic miniature catheters, gastroscopy devices, and other force-sensing microcatheters, may be exempt from IDA and need to satisfy only the IRB requirements.

Thus, for several of the devices based on FOS and related to risky functions, an essential step is the obtainment of an IDE, which is a requirement to obtain the FDA. The reference page for IDE is [8]. Although the IDE is well set in the U.S. FDA regulatory scheme, similar steps are also needed in most healthcare governing boards. An approved IDE allows a device to be legally used for the purpose of conducting investigations of without complying with other requirements that would apply to devices in commercial distribution. Upon approval of IDE, the sponsor may initiate the recruitment of subjects for the study in agreement with the IRB; typically, approval is based on conditions (approval with conditions), or it is granted for a portion of the intended study in a multistep test that aims at organizing the study proposed by the investigator as a sequence of experiments in order to verify several hypotheses within the IDE (staged approval or staged approval with conditions).

Within clinicians and medical-device investigators, two typologies of IDE are recurrent, considering that medical devices (including FOS) at the IDE application level are generally unique, because they overcome the present state of the art, and are often combined with other devices that govern positioning and function, therefore depending on the physician's actions. At the lower level, feasibility studies are run within IDE, particularly in staged IDE: feasibility studies may provide the validation of hypothesis or support the upper level tests, and the number of patients involved in these studies is often 10 to 40. Feasibility studies are called by the FDA to verify mainly the safety of the device for the patient and whether the device under investigation affects the patient's comfort; also, they are used to estimate the real benefits of the device, understanding if the device is worth its implicit risk. At the upper level, pivotal studies are the endpoint for the IDE, as they are intended as the primary document for supporting the PMA. Pivotal studies are designed to demonstrate a reasonable assurance of safety, and the effectiveness of the device to perform the function it is designed for. For this reason, statistical analyses on patients are used in order to verify the function, and generally the number of subjects for the study is larger than in feasibility studies. When designing medical devices that address sensing challenges in vivo, and FOS technologies are well suited for this thanks to the advantageous sensing properties, it is essential to have in mind the IDE structure as it will be the main bridge between all the preclinical studies, which mainly involve the technology development and characterizations, and the premarket stage when a medical device is prototyped and commercialized.

13.5 Useful Tools

Engineers and scientists developing biomedical sensors may find, in this section, some useful tool for the simulation and test of biomedical devices, particularly looking at the applications that will be detailed in Chapters 15 to 19 and constitute the current and emerging fields for FOS sensors.

- *Cardiovascular simulators:* Several technologies exist currently on the market that allow simulating the cardiovascular apparatus, reproducing the functioning of the heart and its blood vessels, and producing a pulsating pressure system. Testing sensors and their catheterizations on a CV simulator is a good practice in biomedical engineering: it is relatively easy to obtain a reference value for pressure or temperature (usually returned by the device itself), and it is possible to study in the simulator all the effects due to catheter insertion and positioning, their related uncertainty, response time of the sensors, and multiple location monitoring. CV simulators are often commercialized as training tools for surgeons and reproduce the heart and blood vessels with a 1:1 scale. In preliminary research, they can be filled with water or contrast liquid; eventually, small quantities of animal blood can be infused in the CV simulator to reproduce the blood flow. Commercially available systems include the Vascular Simulations Replicator (Vascular Simulations, Stony Brooks, New York) and the Corosim (Mecora Gmbh, Aachen, Germany). A more refined tool is the Cardiopulmonary Patient Simulator (Laerdal, Wappingers Falls, New York).
- *PhysioBank waveform database:* The PhysioBank database is available online at https://physionet.org/physiobank/database/, is a large archive of physiological data. Besides clinical and imaging files, it contains a large and well-comprehensive collection of waveforms, including electrocardiogram (ECG/EKG), heart rate, respiration patterns, oxygen saturation, gait data, and neuroelectric signals. Data are available under the Open Data Common license. When needed, data can be digitized using software and used to feed pressure chambers, testers, or simulators with realistic data.
- *Thermal ablation phantom:* Thermal ablation (TA), as will be described in Chapter 17, is one of the main emerging applications for biomedical sensors and represents the harshest environment for the probes described in this book. This is because, together with the biocompatibility, positioning, and catheterization problem that characterize most sensors, in TA temperature and pressure spatial gradients are extremely steep, making it very hard to test and validate sensors. In this direction, several authors proposed tissue-mimicking phantoms that assume different output (usually, colorimetric variations) depending on the local temperature. In 2016, Mikhail et al. from the U.S. National Institute of Health [9] proposed a tissue-mimicking phantom, applied to radiofrequency ablation, that is able to simulate the temperatures obtained during the ablation process with a colorimetric approach. The phantom, chemically composed of deionized water and acrylamide, reacts to the thermal ablation dose and returns a color, in RGB scale, that is particularly accurate in the 52°C to 62°C range. Dabbagh et al. [10] proposed a tissue-mimicking phantom method, sensitive over 50°C and usable for radiofrequency and ultrasound ablation, which also has a colorimetric output, and can also be verified under Fourier transform infrared

reflectometry. These and similar phantoms are excellent tools for calibrating thermal ablation sensors.

- *Respiration actuators:* When operating in vivo respiration acts as a detrimental effect on sensors output, inducing strain on the probes. In order to improve the accuracy of research made with tissue-mimicking or animal phantoms, it is possible to take into account the breathing pattern (amplitude and rate) through actuators. The procedure reported in [11] can be used as a reference point for simulating the effect of breathing pattern on animal tissues, during temperature measurement. The tissue is positioned within two one-dimensional linear actuators (M-235.2DD Physik Instrumente Gmbh, Karlsruhe, Germany), which simulate the expansion and compression as a periodic function. Accelerations, velocities, and displacements are given as input to the motor controllers (C-862, Physik Instrumente). Actuators are controlled and interfaced via software (LabVIEW). A sine-wave is used to drive the linear actuators, having time-varying amplitude and frequency. A physically sound system for simulating nearly any breathing pattern is the ASL 5000 Breathing Simulator (IngMar Medical, Pittsburgh, Pennsylvania).
- *Anatomical model platforms:* In addition to the previous simulators, 3-D Systems (formerly Simbionix, Beit Golan, Israel) provides a set of educational simulators designed for clinicians in order to train on hands-on surgical procedures. These platforms usually are quite representative of the human counterparts and therefore can be suitable for the initial assessment of biomedical technologies. Available anatomical models include gastrointestinal and urologic apparatus, as well as training systems for endoscopic, laparoscopic, and arthroscopic surgery and a robotic surgery simulator.
- *Three-dimensional (3-D) printed phantoms:* Thanks to the advances in additive manufacturing, 3-D printing techniques start to appear as a useful tool for the fabrication of phantoms, for testing medical devices. Cloonan et al. [12] provided a fabrication process for blood vessels based on two photopolymerizable polymers. In this work, the authors started from a 3-D model of abdominal aorta, obtained through a set of 2-D tomographic scans. The model has then been scanned via software, using graphic mesh. Then, using the obtained 3-D reconstruction, a phantom has been fabricated using TangoPlus FLX930 and HeartPrint Flex polymers. From a commercial point of view, Tessonics (Windsor, Ontario, Canada) has started the research and development of 3-D-printed phantoms for educational use. The 3-D printed phantoms can merge the benefit of low-cost fabrication with tissue-mimicking phantoms, advancing the possibility of in vitro testing of sensors.
- *Laser ablation phantom:* FOSs have a strong application in laser ablation sensing, and Chen et al., at Politecnico di Torino, have reported the fabrication and functionalization of a laser ablation phantom [13]. A phantom for laser ablation must match the optical coefficient of absorption of the human tissue, at the laser wavelength (usually 1,064 nm). In this work, the phantom is realized with agar jelly, solidified and colored to match the absorption coefficient in human liver. The specific characteristic of the phantom makes it easy to analyze under infrared imaging, thus providing a methodology for generating and testing thermal patterns with strong spatial gradients.

References

[1] Enderle, J., and J. D. Bronzino, *Introduction to Biomedical Engineering*, Chapter 2, "Moral and Ethical Issues," New York: Elsevier, 2000.

[2] Lipinski, C., and A. Hopkins, "Navigating Chemical Space for Biology and Medicine," *Nature*, Vol. 432, No. 7019, 2004, pp. 855–861.

[3] Kairn, T., S. B. Crowe, and T. Markwell, "Use of 3D Printed Materials as Tissue-Equivalent Phantoms," in Jaffray, D., (ed.), *World Congress on Medical Physics and Biomedical Engineering, IFMBE Proceedings*, Vol. 51, Toronto, Canada, June 7–12, 2015.

[4] Alssabbagh, M., et al., "Evaluation of 3D Printing Materials for Fabrication of a Novel Multi-Functional 3D Thyroid Phantom for Medical Dosimetry and Image Quality," *Radiation Physics and Chemistry*, February 5, 2017.

[5] Russell, W. M. S., and R. L. Burch, *The Principles of Humane Experimental Technique*, London, U.K.: Methuen, 1959.

[6] U.S. National Institute of Health, "Approval for Medical Devices Research," March 2008, http://www.hra.nhs.uk/documents/2013/09/approval-of-medical-devices-research-version-2-april-2008.pdf.

[7] U.S. Food and Drug Administration, "Information Sheet Guidance for IRBs, Clinical Investigators, and Sponsors," 2006, http://www.fda.gov/downloads/RegulatoryInformation/Guidances/UCM126418.pdf.

[8] U.S. Food and Drug Administration, "Device Advice: Investigational Device Exemption (IDE)," 2015, http://www.fda.gov/MedicalDevices/DeviceRegulationandGuidance/HowtoMarketYourDevice/InvestigationalDeviceExemptionIDE/.

[9] Mikhail, A. S., et al., "Evaluation of a Tissue-Mimicking Thermochromic Phantom for Radiofrequency Ablation," *Medical Physics*, Vol. 43, No. 7, July 1, 2016, pp. 4304–4311.

[10] Dabbagh, A., et al. "Reusable Heat-Sensitive Phantom for Precise Estimation of Thermal Profile in Hyperthermia Application," *International Journal of Hyperthermia,* Vol. 30, No. 1, 2014, pp. 66–74.

[11] Cavaiola, C., et al., "Error of a Temperature Probe for Cancer Ablation Monitoring Caused by Respiratory Movements: Ex Vivo and In Vivo Analysis," August, 2016.

[12] Cloonan, A. J., et al., "3D-Printed Tissue-Mimicking Phantoms for Medical Imaging and Computational Validation Applications," *3D Printing and Additive Manufacturing*, Vol. 1, No. 1, 2014, pp. 14–23.

[13] Chen, W., et al., "Performance Assessment of FBG Temperature Sensors for Laser Ablation of Tumors," *2015 IEEE International Symposium on Medical Measurements and Applications (MeMeA)*, Turin, Italy, May 7–9, 2015. pp. 324–328.

CHAPTER 14

Sensor Catheterization

14.1 Characteristics of Medical Catheters

After a fiber-optic sensor has been conceptualized, developed, and calibrated, it must be packaged in a form factor suitable for its operation. In biomedical engineering, this usually involves the use of a catheter, which is a thin tube guiding one or more instruments into the body to the point of action; a catheter is required for cardiovascular, gastroscopic, neurologic, and urologic diagnostic and interventional procedures and most of the detection problems discussed in Chapters 15 to 18. Restricting our analysis to FOS sensors, the catheter must accomplish several tasks: guiding the sensor to the target position, either in a stand-alone format or together with a surgical device, protecting the inner fiber from the surrounding environment, while guaranteeing an effective sensing approach. The impact on the catheter on the sensor is not negligible, and indeed catheter and sensors coexist as an overall measurement instrument: it is possible that the catheter alters the performance of the inner sensor(s), but the design must minimize this effect.

In the following, we will highlight the main critical points for the design of a catheter suitable for in vivo operation and effective for accomplishing diagnostic tasks. We will also provide some reference to the commercial catheter to highlight current clinical devices devised for specific diagnostic tasks. In the next section, we will review several FOS solutions and their applicative implications.

14.1.1 Invasiveness and Size

One of the main aspects of a catheter is its invasiveness when it is applied in vivo. Invasiveness is directly related to the patient's status and comfort, during the whole procedure of diagnosis, surgery, or treatment, and in which sensors play a role. Minimizing the invasiveness is an essential task in modern surgery and can be accomplished by reducing the treatment length, pain, occurrence of complications, or posttreatment recovery length. From an engineering standpoint, the most immediate implication for sensors and their catheters is the need to minimize the size, intended as the size of the whole set of catheters employed for the measurement. Whether engineers prefer to measure catheter diameters in millimeters, clinicians use different units. For flexible catheters, usually the unit French (Fr) is employed:

the conversion is 1 mm = 3 Fr. For rigid metallic catheters, such as needles, the most popular unit is gauge (Ga), which is inversely proportional to the diameter size. Table 14.1 shows the conversion between mm, Fr, and Ga for typical catheter/needle sizes. Smaller catheter size is an important indicator for lower impact on the patients. In measurements performed in blood vessels, the catheter occupies a significant amount of space within the vessel itself, causing possible obstructions and altering the measurement; similar considerations can be performed for measurement performed in anatomic channels, such as for gastroscopy or urology, and in these applications, reducing the size contributes also to lower pain or discomfort. For penetrative packages, such as needles used in thermal ablation, lower size minimizes the tissue damages and bleeding.

14.1.2 Form Factor and Rigidity

When an FOS is placed inside a catheter, since the fiber is miniature and lightweight, the catheter itself defines the rigidity and mechanical resistance to penetration. From a measurement standpoint, urologic and gastroscopic diagnostic are similar to perform a measurement in air of fluid-filled channel, which is partially occupied by the catheter. For this task, it is necessary to have a flexible catheter, usually made of rubber or plastic compound, that bends and follows the flow of the channel without opposing a high resistance. The catheter can often be externally fixed to the patient. Cardiovascular diagnostics usually require the measurement inside a blood-filled pulsating vessel and require a slightly higher resistance to avoid catheter movements after positioning, whereas the bending of the catheter needs to be partially guaranteed to achieve the correct positioning. In this case, semiflexible architectures are preferred, with the catheter having the capability to penetrate follow the blood vessels but having some mechanical resistance to perfusion. Rigid structures, usually with metallic needles, are in use for penetrating into the tissue without altering the sensor(s) form factor. Neurologic sensors, such as for

Table 14.1 Conversion Chart Between Typical Diameters of a Catheter or Needle Form Factor in SI Units (mm) and Their Correspondent Values in Clinical Practice Evaluated for Soft Catheters (French, Fr) and Metallic Rigid Tubes or Needles (Gauge, Ga)

Diameter (mm)	*Catheter Size (Fr)*	*Metallic Tube Size (Ga)*
0.1	0.3	38
0.3	1	29
0.7	2	21
1	3	18
1.3	4	16
1.6	5	14
2	6	12

intracranial pressure measurement, are required to sustain the strength of subdural penetration; however, thermal ablation catheters maintain a penetrative shape but allow some degrees of bending along their extension. Both the catheter and the inner measurement device must be biocompatible.

14.1.3 Insertion

The catheter must support the clinician through the insertion process. It needs to easily allow the access to the entry point and provide the correct amount of resistance to the insertion while being able to move into the hosting tissue. For interventional procedures, the surgeon must be able to insert both the catheter containing the FOS probe(s), and the surgical instruments.

14.1.4 Positioning

The positioning of the measurement catheter is an essential feature of the whole diagnostic procedure. At first, the catheter must guarantee the correct positioning of the catheter in the correct measurement point, with respect to the tissue or the location under investigation. Then it is necessary to define the correct positioning of the optical fiber containing the probe with respect to the catheter itself; and it is required to maintain this positioning through the penetration. It is also necessary to design the catheter such as its position, and the position of the sensors, can be estimated by the clinician either with manual methods, or via imaging (X-ray, magnetic resonance imaging [MRI]). In some measurements, such as thermal ablation, the sensors need to be positioned relatively to an applicator, which itself needs to be positioned with respect to a tissue. The monitoring of catheter position can be supported by the use of radio-opaque markers, which are visible under X-ray. Finally, it is necessary then to safely remove the catheter at the end of the measurement.

14.1.5 FOS Protection and Functionality

The catheter needs to ensure that the fiber is correctly inserted in the sensing environment, while it is protected from mechanical stress. If a portion of the fiber is uncoated, such as in EFPI and in many FBG sensor configurations, the fiber tip is extremely fragile and the catheter needs to protect it; in case of a fiber breaking, the catheter needs to guarantee that no fiber elements can reach out of the catheter and remain in the body. For this reason, architectures based on distributed sensors, or draw-tower gratings that do not require removing the fiber protective jacket, are preferred. The catheter also needs to sustain the high-temperature in case of thermotherapy measurement. In contrast, the catheter must ensure that the inner sensor is effectively sensitive to the biophysical or biological phenomena. This is often achieved by means of holes on the side of the catheter structure that connect the sensor with its measurement point. Depending on the size of the holes and the whole catheter form factor, different performance in terms of response time and sensitivity can be obtained, as discussed in [1].

14.1.6 Metrologic Issues

When defining the sensor operation in real time, in vivo, it is necessary to understand that the catheter size, for the majority of applications, is not small or negligible with respect to its surrounding environments. This creates a number of metrologic issues that overall alter the behavior of the sensor in its specific context. In blood vessels, the catheter partially obstructs the flow, resulting in biasing pressure estimates. In thermotherapies, the catheter absorbs part of the heat, causing to overestimate temperatures. These effects, often referred to as *artifacts*, need to be minimized or, whenever not negligible, compensated with additional sensors. Other effects that need to be taken into account are fiber bending, which can cause optical extra-attenuations, cross-sensitivities (particularly in biological measurements), the influence of vital functions such as heartbeat and breathing pattern. Thanks to their small size and light weight, FOSs tend to minimize the impact of artifacts and improve the overall measurement quality, but we need to guarantee that the catheter structure is effective in preserving these attributes. Friction is also an issue in the catheter form factor, referring to both external friction (the correct transferring of forces exerted on the catheter tip to the sensing element), and internal friction (fiber-catheter contact forces that can act on the sensing element, particularly prestrained FBG).

14.1.7 Disposable Format and Sterilization

In the typical implementation, the catheter is a disposable medical device, used for the measurement task and disposed after its end. It connects to an interrogator, which can be separated or integrated in a medical device, depending on its features. It is essential to guarantee that the whole catheter cost, which includes the catheter itself, the inner FOS, and the fiber connector, is reasonable in comparison to the significance of the measurement task that it is designed to accomplish. The sterilization of medical catheters prior to their first usage can be accomplished by several methods, either physical or chemical, well outlined in [2]. The autoclave steam method is the main physical technique for sterilization, exposing the catheter to a combination of heat, pressure, and humidity; the method is efficient with current catheters, but may endanger thin layers (such as Fabry-Perot diaphragms or thin-film SPR coatings). Chemical methods are based on exposure to ethylene oxide, chlorine dioxide, and hydrogen peroxide in vapor or plasma form. Chemical methods that can be combined with physical ones work by binding to microbial or bacterial contaminants, providing a high level of removal without exposing the catheter to a strong mechanical stress. In alternative, radiation-based methods (gamma-ray, electron beam) can be used to sterilize the catheter and are quite effective on glass. In an alternative to a disposable format, the catheter can be sterilized after use. This operation is strongly discouraged, because in terms of medical standards catheters are considered to be critical devices as they come in contact with tissues and therefore require a complete sterilization cycle. This operation may alter the relative positioning of the sensors, as well as their capability to measure.

14.2 FOS Catheterizations

In the following, we will outline several state-of-the-art FOS catheterizations, describing how each catheter encloses the sensors and what tasks the probe fulfills. Our overview includes most of the latest works presented for pressure, temperature, and force, taking into account both research and commercial devices.

14.2.1 Urologic Pressure and Multipressure Catheter

The solution proposed by Poeggel et al. is a catheter that embeds EFPI pressure probes, in a single catheter, for pressure measurement in urology; initially, the system was deployed for a single probe [3], then was updated to host two fibers in the same probe [4]. The catheter is shown photographically in Figure 14.1 and described in [5]. In urology, the pressure catheter needs to be inserted in the patient's bladder, and record the bladder pressure in two different points in real time. During the examination, the bladder is filled with a saline solution, which is subsequently expelled through urination; thus, the measurement occurs in a fluid (substantially similar to water). The catheter has a threefold goal: insert the sensors from the external point to the bladder, through the urinary channel; protect the fibers, which are fragile as the EFPI building process require removing their outer jacket; and ensure, by means of three side holes, that the outer liquid penetrates the inner part of the catheter such that the EFPI can detect its pressure change. The solution adopted is described in [5]: a commercial silicone catheter (Vygon Nutrisafe, 5 Fr), a standard urologic catheter, has been used to house the sensors. The sensors are inserted in the catheter, one in proximity of the catheter tip, and one at a 1-cm distance; the distance is checked by illuminating each tip with red light. The catheter is terminated with a Y-junction that seals its structure, as shown in Figure 14.1(c). On one side, the junction leads to the external urodynamic device, which infuses the liquid to the patient's bladder; on the other side, the two fibers connecting to the EFPI run through the end of the junction, which is sealed with epoxy to guarantee that no liquid can leak out and that the sensor positions are fixed. This operation is done externally to the patient, as the catheter has 1-m length and therefore no epoxy comes in contact with any parts of the patient. The proposed work is an elegant solution that guarantees multisensor pressure detection in a single catheter,

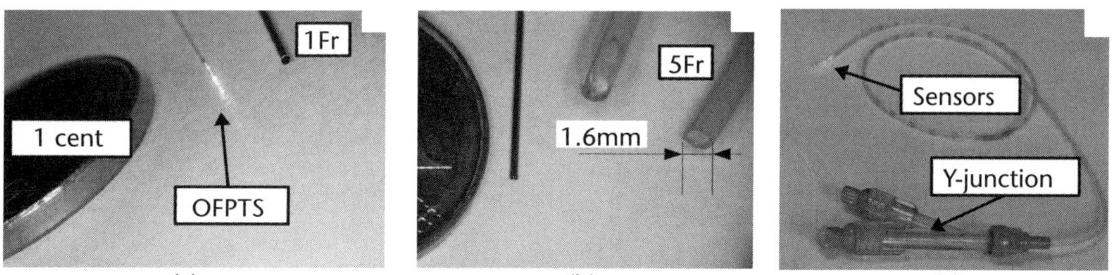

(a) (b) (c)

Figure 14.1 Urologic catheter embodying two EFPI pressure (and temperature) probes, labeled OFPTS. (From: [5].) (a) Single EFPI probe, compared to a Euro coin and 1 Fr tube; (b) two 5 Fr catheters for urology; and (c) a catheter containing two EFPI probes, one of which illuminated, and terminated with a Y-junction.

with dual function (measurement and liquid infusion); in addition, the fibers used for this design are bend-insensitive to guarantee that no bending attenuation occurs at the two tight bending points through the bladder in male patients.

14.2.2 Fiber-Optic Manometer

The work reported by CSIRO (Australia) shows the realization of an FBG-based manometer, which is a pressure-sensing device suitable for operation within the digestive apparatus [6]. In this application, the catheter is also a flexible element, but unlike the previous application it does not require having side holes on the catheter structure. The pressure-sensing mechanism is instead due to the pressure-to-strain conversion, due to the muscular activity involved in the swallowing: the action of the muscle causes a contraction of the esophagus, which converts to strain acting on the FBGs. The catheter developed at CSIRO has 9-Fr diameter and is nearly as invasive as commercial manometer; in the proposed works, measurement have been compared with Gaeltec (Dunvegan, Scotland) solid-state manometer. The catheter has a cylindrical shape and is sealed on the tip with a room-temperature-vulcanizing rubber plug, while at the external end (leading out of the patient) the catheter is terminated with an optical connector. In a further work, reported in 2009 [7], a catheter with an almost identical design has been applied in colonic analysis. For the measurement, a radio-opaque marker has been applied on the external side of the catheter to make it visible under X-ray.

14.2.3 Force-Sensing Device for Epidural Anesthesia

The work proposed by University of Sannio, Italy, by Carotenuto et al. [8] shows the construction of a force-sensing device, based on FBG, that records insertion force in real time as the device penetrates the epidural space where anesthesia is performed. During this procedure, a rigid and penetrative needle is inserted through the tissue, from the skin to reach the spinal cord, crossing different layers of tissues; as the needle travels through the different layers of tissues, it senses a different longitudinal force on its tip, which converts to a strain detectable by the FBG. The system is based on an FBG sensor integrated inside a Tuohy epidural needle, a metallic hollow hypodermic needle slightly curved at the end. The FBG probe is written with the draw-tower method. The FBG sensor has a 0.9-mm-thick eternal coating, in order to match the Tuohy diameter (18 gauge), while the fiber probe is locked to the needle by means of a standard clamping adapter. The probe emerges out of needle tip, of ~1 mm, in order to obtain a physical contact between the fiber and tissues during penetration. The tests, performed on phantoms, show that the force detected by the FBG shows an initial rise through the subcutaneous tissues, and decreases when the penetration reaches the intraspinous tissue, and then force rises a peak in correspondence of the dura mater, at the contact with the spinal cord, where the correct positioning is reached. The whole strain-sensing mechanism relies on the matching between the FBG coating thickness and the inner diameter of the Tuohy, allowing the FBG to be prestrained and thus sensitive to compressive and extensive strain.

14.2.4 Cranial Intraventricular Catheter

The most popular catheter for intracranial pressure (ICP) sensor is the Camino ICP Monitor (Integralife.com), which is typically used in the intraventricular location. For the ICP measurement, the catheter has a semiflexible format: the penetration through the skull is performed with a drill, which, starting from an incision, is capable to penetrate through the dura mater into the ventricular location [9]. The catheter is then positioned after the access point has been granted, thus it does not have to sustain the penetration itself. For ICP, the catheter has to accomplish two tasks: the first is the ICP measurement, which in the Camino sensor is performed with a two-fiber interferometer, while in more modern approaches, it makes use of an EFPI; the second task is the drainage of the cerebrospinal fluid, which is required to reduce the ICP. The dual operation is implemented with a single catheter, terminated with a Y-junction similar to Figure 14.1(c): one side of the catheter is connected to a fiber-optic connector, which is then sent to a detector, while the second side is connected to a draining terminal. In this structure, the catheter has holes on all sides to allow draining liquid to fill the catheter tip and being drained out, while the FPI measures the pressure.

14.2.5 Needle Thermotherapy Probe

The catheter proposed by Bio-Medical Campus of Rome (Italy) in [10] is suitable for the monitoring of thermotherapies. In this case, the process of catheter insertion is percutaneous: the sensor penetrates through the tissue and reaches the target point in the tissue where thermal ablation is performed. In this case, a rigid structure has to be used, as the temperature-sensing FBG probe must sustain the penetration in the tissue without breaking or losing its thermal sensitivity. In the first design described in the proposed work, the package is based on a surgical needle (21 Ga), with a hollow structure. One FBG, having 1-cm length, has been inserted inside the needle, and the fiber is uncoated in the FBG active zone. The FBGs were encapsulated into the needle through two different procedures. The first one employs epoxy adhesive, which ensures both mechanical and thermal contact between the FBG and the needle: the active portion of the fiber was immersed in a solution of dichloromethane, subsequently the epoxy adhesive Araldite 2011 was deposited on the uncovered fiber length and the fiber was finally inserted inside the metallic needle and rested for 24 hours. In this solution, the authors estimated a response time of 104 ms and a sensitivity of 27 pm/°C, higher than standard FBGs. The second encapsulation procedure employs thermal paste, which guarantees only thermal contact between the two elements. Here, a thermal paste (Down Corning silicone heat sink compound) was deposited on the active FBG length, and finally, the fiber was inserted inside the metallic needle and rested. In this solution, the authors estimated a response time of 99 ms and a sensitivity of 10 pm/°C, very close to the standard uncoated FBGs. After the initial investigation, the authors were able to conclude that the needle probe exhibits a measurement artifact due to strain acting on the FBG; strain occurs mainly due to the patient breathing at cyclical rate. For this reason, a further work was proposed in 2016 [11]: the structure of the needle has been updated, going to a 22-Ga format and expanding the sensing array to 6 FBGs. In this case, an algorithm was developed in order to estimate the artifact due to breathing, and tested in simulated environment and in vivo. The key for compensating the artifact is based on the plurality of FBGs, which are held at different temperature during the thermotherapy but sense a similar strain pattern due to respiration, due to the fact that the grating array is localized in a small portion of the tissue (2.8 cm of the total FBG array length). A similar package, based on a metallic hollow needle having 21 Ga size, has been used in [12] to map temperatures in thermal ablation with a pulsed radiofrequency applicator (Habib 4x).

14.2.6 Fiber-Optic Guidewire

A guidewire is a catheter that is used to navigate through a blood vessel to reach a target portion of the vessel where treatment or pressure monitoring is performed. Guidewires are based on solid steel package or (in modern design) with a nitinol core wires, which guarantees the best performance in terms of mechanical properties. Guidewires are pushed by the clinician from the external side to advance the wire into the vessels, and thanks to the semiflexible form taken by nitinol it is possible to steer the guidewire to navigate the vessel, maintaining its shape over time. Usually, the end portion of the guidewire is coated with a radio-opaque marker to make it visible under fluoroscopic imaging. The Optowire is a guidewire developed by Opsens (Canada) and embedding an EFPI pressure sensor and is, to date, the most advanced device for pressure measurement. The design is based hollow stainless steel cable, semiflexible, surrounded by a coil of nitinol material to improve the torque. Inside the hollow part of the design, an EFPI pressure sensor is introduced in proximity to the guidewire tip. Near the tip, the guidewire has a fence that allows the blood pressure to push the EFPI diaphragm and thus be recorded with the outer detector. The catheter has a 175-cm overall length, with the last 3.5 cm being coated with the radio-opaque marker, while the overall thickness is 1 Fr. The outer part of the catheter is terminated with an optical connector (proprietary design by Opsens). The design proposed by this design substantially converts the existing guidewire standard to a hollow format, suitable for hosting one FOS probe; however, it maintains the same problem of having a translated measurement, as EFPI sensor is positioned at 35 mm distance from the tip, thus not returning a localized measurement.

14.2.7 Biosensor Catheterization

To date, biosensors intended as real-time sensors monitoring the refractive index of biological analytes are rapidly emerging but the research is mainly limited to the technology and the functionalization. Most research works presented so far are limited by the fact that a proper catheterization, which allows fiber placing into tissue and detection in real time, has not been developed. We envision the combination of FOS biosensors with optofluidics-based catheter structures for detection of refractive index in fluids, whereas a catheter with judiciously sized holes allow analytes to penetrate to the inner part of the catheter and allow binding to the sensor structure, much like EFPI pressure probes in fluids require being immersed into the fluid under test while the catheter protects the fiber.

14.3 Perspectives

The catheterization of FOS, particularly when a complex transducing phenomenon occurs, is a key task for their in vivo usage, and system performances are strongly tied to the catheter form factor. FOSs have the undoubted advantage of miniature size, and future catheter developments must focus on microcatheters that allow exploiting this feature. From the measurement science perspective, it is important to evaluate the effectiveness of the catheter in preserving the fast response and decoupling mechanical stimuli from physical sensors. We envision the application of catheter concepts to biosensors, for in vivo operation, as the way to obtain an impact in real-time biological detection.

References

[1] Cooper, M. A., et al., "Comparison of Air-Charged and Water-Filled Urodynamic Pressure Measurement Catheters," Neurourology and Urodynamics, Vol. 30, No. 3, March 1, 2011, pp. 329–334.

[2] Maxim Integrated, Application Note 5068, "Sterilization Methods and Their Impact on Medical Devices Containing Electronics," 2011, https://www.maximintegrated.com/en/app-notes/index.mvp/id/5068.

[3] Poeggel, S., et al., "Fiber-Optic EFPI Pressure Sensors for In-Vivo Urodynamic Analysis," *IEEE Sensor Journal*, Vol. 14, No. 7, 2014, pp. 2335–2340.

[4] Poeggel, S., et al., "Differential In Vivo Urodynamic Measurement in a Single Thin Catheter Based on Two Optical Fiber Pressure Sensors," *Journal of Biomedical Optics*, Vol. 20, No. 3, 2015, p. 037005.

[5] Poeggel, S., et al., "Recent Improvement of Medical Optical Fibre Pressure and Temperature Sensors," *Biosensors*, Vol. 5, No. 3, July 13, 2015, pp. 432–449.

[6] Arkwright, J. W., et al., "In-Vivo Demonstration of a High Resolution Optical Fiber Manometry Catheter for Diagnosis of Gastrointestinal Motility Disorders," *Opt. Expr.*, Vol. 17, No. 6, 2009, pp. 4500–4508.

[7] Arkwright, J. W., et al., "Design of a High-Sensor Count Fibre Optic Manometry Catheter for In-Vivo Colonic Diagnostics," *Opt. Expr.*, Vol. 17, No. 25, 2009, pp. 22423–22431.

[8] Carotenuto, B., et al., "Lab in a Needle for Epidural Space Identification," *International Society for Optics and Photonics 6th European Workshop on Optical Fibre Sensors (EWOFS2016)*, May 30, 2016, pp. 99161E–99161E.

[9] Roriz, P., et al., "Review of Fiber-Optic Pressure Sensors for Biomedical and Biomechanical Applications," *J. Biomed. Opt.*, Vol. 18, No. 5, 2013, p. 050903.

[10] Polito, D., et al., "A Needlelike Probe for Temperature Monitoring During Laser Ablation Based on Fiber Bragg Grating: Manufacturing and Characterization," *Journal of Medical Devices*, Vol. 9, No. 4, December 1, 2015, p. 041006.

[11] Cavaiola, C., et al., "Error Caused by Respiratory Movements on a Probe for Distributed Temperature Monitoring During Cancer Ablation: Ex Vivo and In Vivo Analysis," *IEEE Sensors Journal*, Vol. 16, No. 15, 2016, pp. 5934–5941.

[12] Palumbo, G., et al., "Temperature Monitoring During Thermal Ablation on Ex-Vivo Organs by Fiber Bragg Gratings," *IEEE Sensors Conference*, Orlando, FL, October 30–November 2, 2016.

CHAPTER 15

Cardiovascular Sensors

15.1 Pressure Measurement

Cardiovascular (CV) sensors aim at performing pressure, temperature, or biochemical detection within blood vessels. The majority of CV measurements involve the detection of pressure, or pressure differential, in specific points of the blood vessels. Compared to traditional guidewires embedding MEMS [1], FOS-based catheters, particularly in EFPI configuration, have five major benefits: (1) small size of the fiber and its sensor tip, which allows a reduced invasiveness that translates into a smaller physical impedance to blood flow; (2) advantageous form factor, which allows placing a plurality of microsensors inside a single catheter, therefore enabling multipoint measurement, also it is important to mention the symmetrical shape of most FOS which allows better mechanical performances of miniature devices and prevents artifacts when rotating devices while facilitating navigation through tortuous paths; (3) measurement stability, thanks to the lower thermal detuning and long-term stability, resulting in a more accurate baseline measurement; (4) instantaneous time response, well below milliseconds and limited only by hardware; and (5) insensitivity to any electromagnetic interferences (EMIs) that makes the sensor suitable for use under magnetic resonance imaging (MRI) or in the presence of electrically disturbing surgical tools.

CV pressure measurement involves the detection of pressure pulses, directly corresponding to heart rate. The heart works as a pulsatile pump, characterized by a periodicity of contractions (systole) where blood is ejected from both ventricles and pumped into the circulatory system, and relaxations (diastole), where ventricles are refilled with blood, preparing the next contraction. The resulting waveform, depicted in Figure 15.1, is characterized by a steep pressure rise, having typical 40-mmHg amplitude and <0.2-second duration, until the peak systolic pressure is reached; then a relaxation phase is observed reaching the minimum diastolic pressure, with a dicrotic notch in correspondence of a waveform transition. The period of the waveform highly depends on the heart rate, ranging from 0.8 second to 1.2 seconds in relaxation, when most measurements are performed. The values of pressure indicated in Figure 15.1(b) [2] are indicative to show typical CV pressure values (always recorded in mmHg, 1 mmHg = 0.133 kPa) expected in patients with normal and hypertension condition.

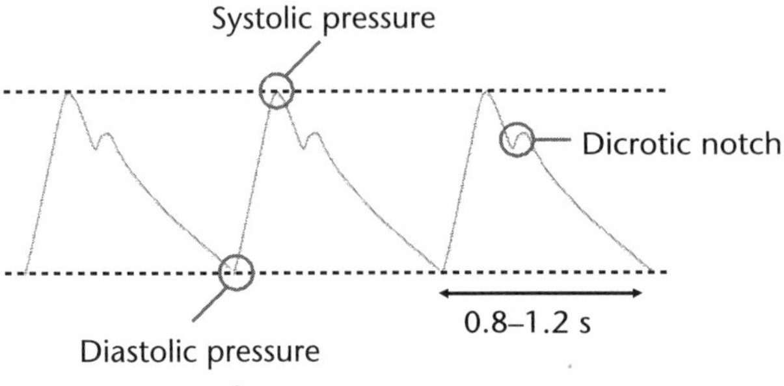

(a)

Pressure	Diastolic [mmHg]	Systolic [mmHg]
Normal	<80	<120
Pre-hypert.	80-89	120-139
Hypert. 1	90-99	139-159
Hypert. 2	>100	>160

(b)

Figure 15.1 (a) Typical shape of blood pressure pulses, highlighting the major regions of interest; (b) Typical diastolic and systolic pressure values, according to [2], for patients in normal, prehypertension, and hypertension conditions.

Standard blood pressure measurement is not an application of interest for FOS, since minimally invasive MEMS and external systems, such as Holter monitoring for blood pressure, are widely available and inexpensive. Instead, FOSs for CV diagnostic focus on several applications in which at least one of the aforementioned factors is essential, and where pressure is localized within a specific organ or medical device. Selective localization and high performance are the two call-to-action factors for FOS over other technologies. The wide market of CV sensors has attracted several companies, manufacturing EFPI sensors for specific CV applications: to the best of our knowledge, three companies are actively producing CV EFPI sensors, and are listed Table 15.1 together with their product names [3]. In the following, the main CV applications and their future evolutions are discussed, showing the role and metrologic properties of sensors.

15.2 Blood Pressure Measurement

Blood pressure monitoring in artery is a typical task of sensors, in vascular applications. This operation, when performed with a minimally invasive catheter, usually involves a guidewire or FOS. EFPI probes have often been used in blood pressure monitoring [4]. Most of proposed works based on FOS are based on the displacement sensor based on a bifurcated bundle, whereas the solution of the single-fiber EFPI discussed in Chapter 7 is more modern, accurate, and reliable. Thanks to progress in MEMS manufacturing, and the advances in noninvasive pressure monitoring, the measurement of blood pressure with a fiber-optic probe is confined to small niches.

A recent work based on EFPI probes and aimed at validating CV pressure monitoring has been proposed by Poeggel et al. [5] and described in Figure 15.2. The probe, developed by University of Limerick, Ireland, is based on an all-glass Fabry-Perot interferometer fabricated by fiber splicing and wet or dry etching of the glass

Table 15.1 Commercial Pressure Sensors Based on EFPI for CV Applications

Company	*FISO*	*Opsens*	*RJC*
Product name	FOP-MIV / FOP-M260	OPP-M25	Model 40
Min. pressure	–300 mmHg	–50 mmHg	500 mmHg (abs.)
Max. pressure	300 mmHg	300 mmHg	1,100 mmHg (abs.)
Resolution	<0.3 mmHg/0.1 mmHg	0.5 mmHg	<0.1 mmHg
Probe diameter (catheter)	0.55 mm/0.26 mm	0.25 mm	0.17 mm
Sampling frequency	250 Hz up to 15 kHz	250 Hz	1 kHz
Approved for	FDA, CE Mark	—	AAMI
Applied in	Intra-aortic balloon, fractional flow reserve	Fractional flow reserve	Aortic pressure

Source: [3].

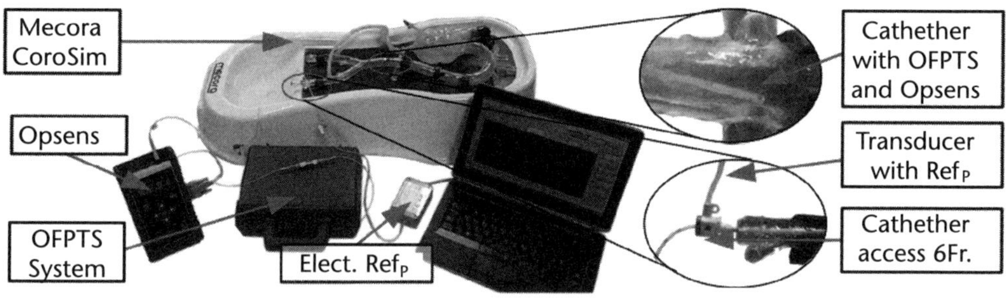

Figure 15.2 Description of the EFPI-based pressure probe and its cardiovascular analysis in a heart simulator, proposed by Poeggel et al. (From: [5].)

diaphragm. The sensor was packaged in a single catheter (Vygon Nutrisafe, 5 Fr) with a 1.6-mm outer diameter, together with a reference sensor (Opsens OPP-M). A heart simulator (Mecora CoroSim) was used to test the response of the sensor, in the time domain. A different approach has been implemented by Leitao et al. [6]: in a recent study, the authors reported the measurement of arterial pressure in vivo using two FOS configurations, including an FBG and a POF sensor, fabricated in a pen-shape form factor. In this case, pressure detection occurs externally to the patient, in correspondence of the carotid artery where the clinician records the blood pressure. The study of Leitao was used to record the entire blood pressure cycle, and subsequently perform a waveform analysis to estimate biophysical parameters.

From the commercial point of view, a recent system which has recently been developed and prototyped, is the Endophys Pressure Sensing Access System (Endophys Holdings, LLC, USA). The Endophys system was approved by the U.S. Food and Drug Administration (FDA) in 2014. It consists of a sheath for vascular procedure, containing an EFPI-based sensor (of undisclosed manufacturer) with a 0.33-mm diameter, positioned at the tip of the sheath. An external box contains the interrogators and a display. More information on the sheath format has been outlined in a recent patent [7]. The sheath is formed as a round catheter, with cylindrical shape, incorporating an EFPI probe positioned at distal location. The Endophys system is designed for long-term patient monitoring for patients with acute stroke, heart attack, or severe vascular conditions that involve an interventional catheterization to perform angioplasty and stent positioning. The sheath has already been applied for monitoring of pressure in the femoral artery.

15.3 Fractional Flow Reserve

Whenever, in a liquid-flow pulsating system, an obstruction occurs, a pressure differential is observed between the pre-obstruction and post-obstruction points; the higher the pressure differential, the stronger is the effect of obstruction. This concept, widely popular in fluid dynamic science, is increasingly applied in medicine, and it has been established as a golden standard for blood vessels obstruction diagnostics, under the name of fractional flow reserve (FFR). The FFR technique is well described by Pijls et al. [8] and is a well-established method for detection of stenosis (i.e., blood vessel restrictions or obstructions). Stenosis is often due to coronary artery narrowing caused by atherosclerosis, which promotes the formation of plaques inside inner artery walls. More recently, FFR is emerging as a diagnostic technique for ischemia. The advantage of FFR over other methods, usually based on intravascular ultrasound and CT coronary angiography, is that it provides a quantitative estimation of the obstruction effect and, thanks to its artifact, can provide a localization of the obstruction. Also, FFR is a robust method, based on normalized readings. The main disadvantage is the invasiveness, since (compared to imaging-based methods) a microcatheter is needed to insert a pressure probe.

The FFR measurement is performed by recording pressure in two points as in Figure 15.3: at first, the mean aortic pressure (P_a) is recorded prior to a potential obstruction, and represents the maximum achievable pressure. Then the distal pressure (P_d) is recorded at the distal point after the obstruction and represents the mean stenotic pressure that suffers from the effect of stenosis. The FFR is accounted as the ratio between the values: FFR = P_d/P_a. The FFR is a normalized metric that is independent on heart rate, contractility, or changes in blood pulsation. The maximum theoretical value of FFR is equal to 1, while values >0.95 are obtained in healthy patients. FFR < 0.75 is the threshold for diagnosis of ischemia, while for FFR > 0.8, no ischemia is found.

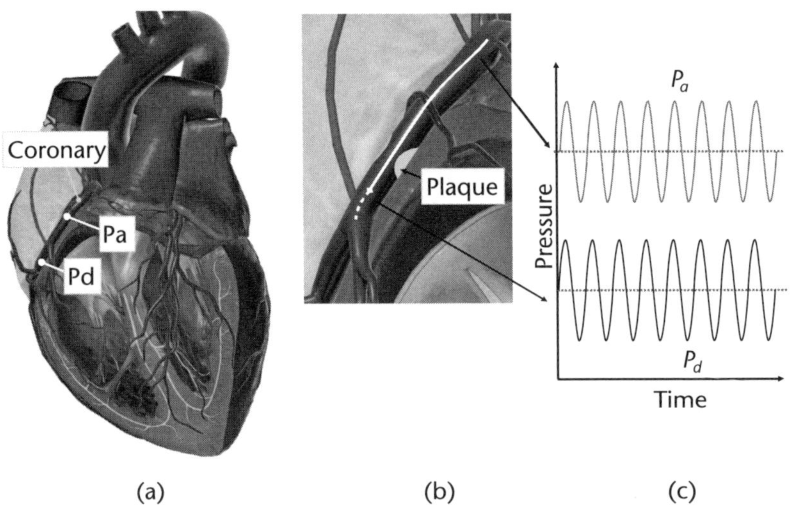

Figure 15.3 FFR measurement with a fiber optic probe. (a) Positioning of the catheter in the carotid aorta; (b) pressure recording and FFR estimation by sliding the probe into the distal position; and (c) sketch of pressure traces.

In the medical practice, a single pressure probe is used. The clinician first inserts the catheter in the coronary artery, through percutaneous angioplasty. As the catheter reaches the prestenosis condition, the blood pressure is observed, obtaining a chart similar to Figure 15.1, and the mean value of pressure is stored as P_a. Then the clinician progressively advances the pressure probe into the stenosis, and the (mean) distal pressure P_d is recorded, estimating the FFR in real time. As the probe advances in the stenosis, the distal pressure changes until the minimum FFR is reached. The minimum FFR is then accounted and used as a diagnostic metric.

FOSs have established as a tool for FFR measurement; in particular, OpSens Medical has industrialized a system for FFR measurement (OptoWire One) based on a nitinol catheter incorporating one EFPI probe, and a similar system has been developed by FISO Technologies (Canada; datasheet available at http://fiso.com/section.php?p=104). The catheter is based on a cable of a 175-cm length, terminated with an EFPI tip fabricated by OpSens, with an undisclosed design. The other side of the cable is terminated with a fiber-optic connector, based on a design patented in 2014 [9]. The catheter has a thickness of 0.36 mm, similar to the PressureWire Aeris (0.35 mm) that is the golden standard for vascular measurement. The system has 1-mmHg accuracy, and drift < 1 mmHg/hour. It would be difficult to omit to mention the *ACIST* Navvus rapid exchange FFR microcatheter which was launched around the same time and the *Boston Scientific* Comet pressure guidewire launched few quarters ago, but already seen as a flagship product for growth.

The measurement of FFR is an acquaintance, and as of 2010 it had become a golden standard for its simplicity and robustness. To date, FOSs are competing with MEMS-based guidewires in this application. Although FOSs have a smaller size, packaging is bulky and the size of 0.35 mm is the limitation to current catheterizations. The main advantage of FOS is the better stability in terms of temporal drift: EFPI probes and their interrogators achieve a stability better than 1 mmHg/hour, while guidewires tend to exhibit a more visible drift; the authors observed a drift of 6 mmHg/hour on a commercial guidewire. This effect is not negligible, although FFR measurements are relatively fast. However, the full potential of FFR, and generally multipressure measurements in CV diagnostic, is far from having been explored. In first place, the possibility of positioning multiple probes inside a single catheter is a unique feature made available by FOS and not commercialized yet. The possibility of detecting pressure in two points (aortic and distal) simultaneously, rather than at different times, can improve the accuracy of FFR by a significant factor. However, it is the possibility to use a multiprobe geometry, with simultaneous pressure recording in different locations, that offers imaginative applications: a similar geometry, with a catheter positioned in long-term placement, can give the possibility of calculating a plurality of pressure ratios, and visualize the effect of stenosis in each part of the plaque. This indication can not only be used as a metric for better diagnostic of ischemia, but also used during revascularization, which is the surgical procedure to reactivate blood perfusion in a body part that suffered ischemia. Notably, a technique similar to FFR will be discussed in the next chapter, for urologic diagnostics.

15.4 Heart-Assistive Devices

Heart-assisting devices are typically brought to action, when a patient's heart is failing or has a reduced workload. A popular therapy for assisting coronary perfusion was consists of an intra-aortic balloon (IAB), a popular yet simple device to mechanically increase perfusion through an external pump. The IAB pumping (IABP) counterpulsation therapy is a procedure for management of patients undergoing cardiogenic shock, unstable angina, post-cardiopulmonary bypass, or high-risk patients undertaking cardiac surgery. The principle of IABP is inserting a miniature balloon, fabricated in polyethylene and inflatable, with an inner pressure sensor that detects blood pressure. The balloon is positioned as in Figure 15.4: the surgeon inserts the balloon through the descending thoracic aorta, and distal to the left subclavian artery. The position is close to the heart and allows for both recording proximal blood pressure (and therefore with minimal delay with respect to the heart pumping), and is within a large aortic diameter to allow the balloon to easily extend and counterpulsate.

The IAB performs two tasks. At first, with the inner pressure probe, it records the blood pressure with a high sampling rate (100+ Hz), and instantaneous time response. Second, it can be inflated or deflated by means of an external injection of He. Both the inflation system and pressure readout are incorporated in a single instrument, usually positioned in a medical-grade rack near the patient's bed. The balloon inflates during diastole, when heart is relaxing and the minimum pressure is found, and deflates during systole, when the heart contracts and maximum blood pressure is observed. Figure 15.4(b) shows the IAB value proposition: in patients with reduced heart workload, low pressure (and particularly low-pressure differential) is observed. The pressure reduction during diastole calls for a high myocardial requirement for oxygen. The blood pressure pulse benefits from the effect of the IAB during the decreasing part: in correspondence of the dicrotic notch, the quick inflation of the IAB provides a steep pressure peak, higher than the peak pressure, and followed by a pressure decrease during the deflation period. Thus,

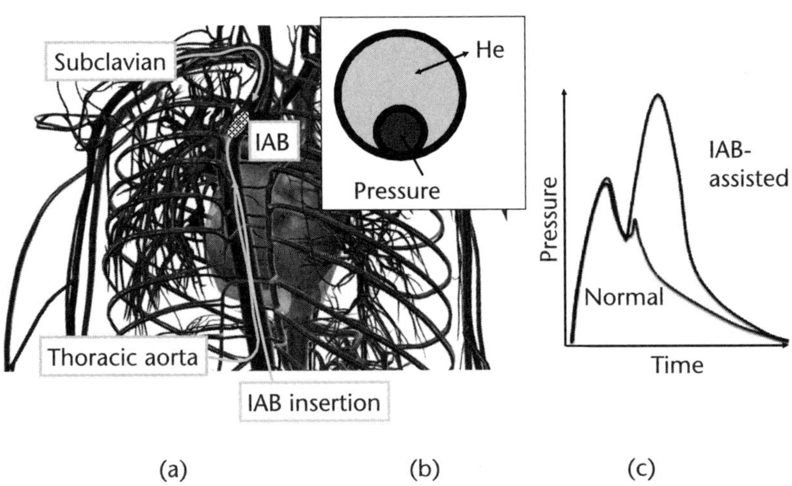

Figure 15.4 IABP therapy and IAB device. (a) Positioning of IAB device, in thoracic aorta. (b) Top view of the IAB section, highlighting pressure probe and He inflation zone. (c) Original and IAB-assisted blood pressure pulsation.

the IAB-assisted pressure pattern is characterized by two peaks per cycle, and the increased mean pressure reduces the requirement for myocardial oxygenation.

The role of the pressure probe is essential, and OFS have two critical advantages in this application. In first place, the small size is needed to minimize the mechanical impedance to blood flow. Because the unassisted blood pressure is weak, a millimeter-sized device produces a non-negligible effect and would significantly induce a pressure gradient. In the second place, an extremely fast response is needed, since the whole pressure reading and IAB inflation time needs to be significantly smaller, around 1/50th of the repetition rate, in order to guarantee a perfect synchronization of the IAB-induced pressure peak. A sensor with a slow response might be harmful for this application, since balloon inflation during the systole might create a highly pressurized cavity. Pressure probes for IABP or FFR systems have been commercialized by FISO Technologies (Canada), recently acquired by Nova Ventures (Boston, Massachusetts): the probe operation, also reported in [10], is based on an EFPI sensor having a 0.125-mm to 0.55-mm outer diameter and is based on a silicon or glass membrane working as diaphragm. The sensors were tested in vitro, on a Bio-Tek 601A waveform generator capable of reproducing a standard blood pressure. FISO Technologies probes are photographically shown in Figure 15.5. The figure shows the different architecture of EFPI probes from the manufacturer, showing the outer diameter of the whole probe. The left and center probes are based on an EFPI octagonal micro-optomechanical system (MOMS), with a silicon membrane structure and an outer diameter of 550 μm and 260 μm, respectively. The design on the right is more modern and minimalistic and foresees a tip-cavity with a thin glass diaphragm; this reduces the diameter to the size of the input fiber, requiring an ad hoc catheter to protect the sensor during operation; this architecture is compatible with multiprobe catheters, as described in Section 16.2. All sensors are positioned at the tip of a multimode fiber with standard cladding diameter. The FISO Technologies system is controlled with an interferometric compact detector, operating in visible wavelength. Several interrogators are commercially available (see www.fiso.com) with single or multiple parallel channels

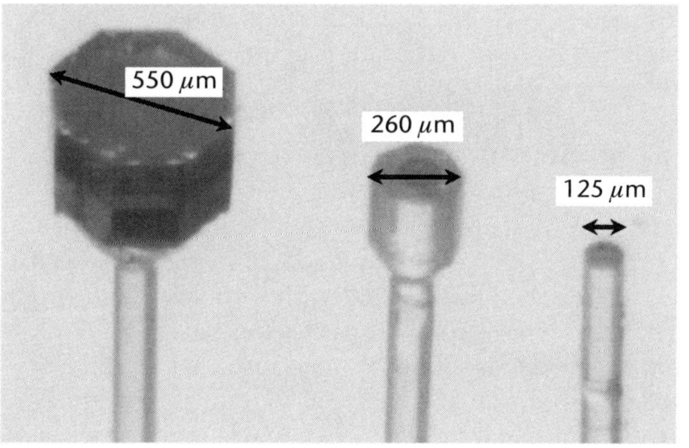

Figure 15.5 EFPI pressure probes, manufactured by FISO Technologies. Picture shown with permission of FISO Technologies (www.fiso.com), and [11].

with acquisition rate compatible with CV monitoring (typically 250 Hz for human, but also up to 15 kHz useful for small animal testing).

Another assistive device that is increasingly used in medical practice is the ventricular assist device (VAD). A VAD is an electromechanical pump used to compensate the function of a failing heart, and in some cases to completely replace the heart function. The VAD provides perfusion of blood from a lower chamber of the heart, and pumps it through the upper part, substantially replacing the effect of a failing ventricle. VAD can perform left (LVAD), right (RVAD), or dual (BiVAD) ventricle assistance. The VAD incorporates several parts: a flexible cable where blood flows, connected to the left (or right) ventricle where blood comes in, and to the aorta where it is pumped to the body; a cable connecting the inner part to the control unit, positioned externally; an external pump, which can be a continuous flow pump (typically used for long-term treatments) or a pulsatile pump that mimics heart pulsation; and finally, an external control unit that embeds all controls and a set of sensing data and alarms. Initially, VAD were externally applied to patients during hospital care, with connecting cables interfaced through the abdomen. More recently, several companies fabricated portable LVAD devices, in which the entire functions are comprised in a battery-powered wearable device; the Thoratec Heartmate II is the most popular implantable LVAD.

Although not a primary function, blood pressure sensing is a functionality of a LVAD. Zhou et al. [12] reported an EFPI-based pressure sensor to take the function of blood pressure monitor in an implantable LVAD; the probe is fabricated with a fiber-optic tip, facing a silicon mirror mounted on a parylene diaphragm of 5-μm thickness before etching. The bendable form factor of the diaphragm allows a very high sensitivity (326 nm/mmHg), obtained with a thickness of 300 μm. The EFPI probe offers continuous cardiac output estimation during the lifetime operation of the LVAD, and the key advantages of the fiber-optic probe, similarly to IAB, are the small size, low drift, and instantaneous response to pressure variations, much faster than any fluid-filled catheter. In 2015, Ruschen et al. [13] developed a methodology based on two fiber-optic pressure sensors, inserted through the inner catheter of a LVAD device (Impella CP, Abiomed, USA), to estimate the cardiac output of a LVAD-assisted heart: the methodology is based on estimating the left ventricular pressure inside the catheter, and the aortic pressure outside of the LVAD catheter, and estimating by means of nonlinear state-space analysis the cardiac output.

15.5 Blood Temperature Sensor

Although for CV diagnostics and minimally invasive procedures, pressure detection is the most popular approach, other works have been presented, recently, for thermal and biochemical sensing with intravascular approach. The work of Zou et al. [14] presents a temperature sensor, based on EFPI, and its application on coronary artery disease (CAD) treated via thermal angioplasty. The formation of plaques, which are often diagnosed via an FFR method, may force surgeons to interventional procedures, particularly when atherosclerotic or ischemic drug-resistant conditions are present. Angioplasty is the preferred approach to clinically remove plaques formed in the inner wall of coronary arteries, and two therapies are established. Balloon angioplasty is a method based on a rigid stent, with a shape somehow similar to an IAB but surrounded by a metallic wire cage; when the balloon is expanded, the stent inflates and removes the plaques from the inner walls. Lately, thanks to advances in solid-state lasers and optical fibers, this technique has been progressively complemented or replaced by thermal angioplasty, which makes use of a laser coupled into an optical fiber to ablate the plaque tissue in the specific location. In this framework, the temperature measurement can be performed with a miniature FOS. In [14] the thermal sensor was designed with an EFPI probe having sensitivity of 10.3 pm/°C, and fabricated within a Kapton tube terminated with a polymer bead. The temperature measurement performed in vivo on a swine shows small temperature variations in the blood, inferior to 5°C. The work presented by Zou et al. is indeed preliminary, but shows the premises for a FOS application that is still unexplored, particularly since the sensor was used as a control method for excessive blood temperature and not as an angioplasty efficacy monitoring: it may possibly to extend to multiplexed/distributed sensing following the latest results on thermal ablation (Chapter 18). Temperature probes as small as 200 μm based on GaAs technology and compatible with medical applications (accuracy of 0.5°C over a temperature range from 0°C up to 80°C with a resolution of 0.05°C) are available commercially from FISO Technologies.

15.6 Conclusions

Cardiovascular detection is one of the most mature application areas for FOS, particularly in aspects related to pressure sensing. Future applications will explore the localization of pressure values: thanks to miniature footprint and cabling size, it will be possible to use a plurality of EFPI sensors to detect pressure in multiple points, localizing obstructions with unprecedented precision but also introducing control elements in personalized drug delivery.

References

[1] St. Jude, PressureWire Aeris, "FFR Guidewire," 2013, https://professional-intl.sjm.com/products/vas.

[2] Chobanian, A. V., et al., "The Seventh Report of the Joint National Committee on Prevention, Detection, Evaluation, and Treatment of High Blood Pressure: The JNC 7 Report," *JAMA*, Vol. 289, 2003, pp. 2560–2571.

[3] Poeggel, S., et al., "Optical Fibre Pressure Sensors in Medical Applications," *Sensors*, Vol. 2015, 2015, pp. 17115–17148.

[4] Roriz, P., et al., "Review of Fiber-Optic Pressure Sensors for Biomedical and Biomechanical Applications," *J. Biomed. Opt.*, Vol. 18, No. 5, 2013, p. 050903.

[5] Poeggel, S., et al., "Recent Improvement of Medical Optical Fibre Pressure and Temperature Sensors," *Biosensors*, Vol. 5, 2015, pp. 432–449.

[6] Leitao, C., et al., "Optical Fiber Sensors for Central Arterial Pressure Monitoring," *Opt. Quant. Electron.*, Vol. 48, 2016, p. 218.

[7] Purdy, P. D., and S. J. Ferry, "Device Configured for Real-Time Pressure Sensing," U.S. Patent 20,150,289,816, 2015.

[8] Pijls, N. H., et al., "Measurement of Fractional Flow Reserve to Assess the Functional Severity of Coronary-Artery Stenosis," *New England Journal of Medicine*, Vol. 334, No. 26, 1996, pp. 1703–1708.

[9] Belleville, C., S. Lalancette, and A. Proulx, "Interface Connector Handle for Disposable Guidewire Optical Connection," U.S. Patent 0248021, 2014.

[10] Pinet, É., A. Pham, and S. Rioux, "Miniature Fiber Optic Pressure Sensor for Medical Applications: An Opportunity for Intra-Aortic Balloon Pumping (IABP) Therapy," *International Society for Optics and Photonics Bruges, Belgium-Deadline Past*, May 2005, pp. 234–237.

[11] Pinet, É., "Pressure Measurement with Fiber-Optic Sensors: Commercial Technologies and Applications," *21st International Conference on Optical Fibre Sensors (OFS21), International Society for Optics and Photonics*, 2011.

[12] Zhou, M. -D., et al., "An Implantable Fabry-Pérot Pressure Sensor Fabricated on Left Ventricular Assist Device for Heart Failure," *Biomed. Microdevices*, Vol. 14, 2012, pp. 235–245.

[13] Rüschen, D., et al., "Continuous Cardiac Output Estimation Under Left Ventricular Assistance," *IFAC-PapersOnLine*, Vol. 48, No. 20, 2015, pp. 569–574.

[14] Zou, X., et al., "Miniature Fabry-Perot Fiber Optic Sensor for Intravascular Blood Temperature Measurements," *IEEE Sensors Journal*, Vol. 13, No. 6, 2013, pp. 2155–2160.

CHAPTER 16

Diagnostics in Gastroscopy, Urology, and Neurology

16.1 Gastroscopy

Gastroscopy, or esophagogastroduodenoscopy (EGD), is a medical procedure aimed at the visualization and monitoring of the digestive apparatus. Most EGD procedures are based on small and general-purpose endoscope, equipped with a camera and a 8 to 12-mm probe mounted on the tip is used to render images of the pharynx, esophagus, stomach, and intestine; optionally, the endoscope can mount other tools, such as tweezers or forceps used to grab a small portion of tissue to be processed via biopsy for bacterial testing or pressure sensors typically mounted in a guidewire. Gastrointestinal disorders occurring in the digestive apparatus can be therefore observed with this tool, which combines imaging with moderate sensing and tissue manipulation capabilities.

In this scenario, OFSs offer several advantages over standard sensing technologies. In the first place, OFS (using FBG, or the most recent distributed sensors) allows a time-space analysis, whereas pressure, or strain, or a combination of the two values, is recorded with narrow spatial resolution. This feature is essential in solid-state manometers, which record the muscular activities associated with the transit through each stage of the digestive apparatus. In addition, the small size of optical fibers allows the minimum invasiveness. A miniature catheter reduces discomfort for the patient, allowing the use of nasal hole as entry point for the catheter, rather than the mouth, particularly for the analysis of the tract extending from the pharynx to the stomach, which may potentially drop the requirement for local anesthesia. The lower impedance to the flow of dense liquid makes the measurement more accurate, particularly within narrower sections.

To date, the most advanced systems for gastroscopy based on fiber-optic technologies have been documented by Arkwright et al., at the Commonwealth Scientific and Industrial Research Organisation (CSIRO) [1–4].

16.1.1 Sensing Element

Most gastroscopic sensing procedures require detecting pressure, in different traits of the digestive apparatus, during simple procedures such as swallowing or digesting food. The time-space analysis fits well multiplexed and distributed sensors; a good gastroscopy device needs to have a dense spatial resolution in order to render every small part of the digestive apparatus and a fast response time. Miniaturization is also an essential requisite for the whole catheter. The main task to achieve a successful sensor is pressure transducing; pressure is a measurand that can usually be detected with a tip-sensor, such as an EFPI. In this context, due to the need for a multiplexed (or distributed) detection, it is necessary to transduce pressure into strain that can be recorded with FBGs.

The first, and most significant, sensing element has been proposed in [1] and illustrated in Figure 16.1(a). Each sensing element is formed by a rigid substrate having a 3.6-mm length, and an FBG with a length of 3 mm. The transducing element is a flexible diaphragm; changes in surrounding pressure cause diaphragm to flex sideways, deflecting the FBG and therefore shifting the peak wavelength. The calibration performed in vitro shows that this setup achieves a sensitivity, defined as Bragg wavelength shift as a function of pressure variation from baseline, of 1.2 pm/mmHg, and therefore achieves accuracy close to 3 mmHg with a standard FBG interrogator. The final realization of the catheter embodies 32 FBGs, in a rubber catheter with 3-mm thickness (sufficient for nasal intubation); the spacing between each FBG is 1 cm.

An improved architecture has been documented in [2] and subsequently patented in [4]. Two fibers, each hosting 36 FBGs, are used to form the catheter as in Figure 16.1(b). The FBG has a 3-mm length, and the spacing between each FBG is 10 mm. The catheter reproduces a similar concept of the previous design, but each building block contains two FBGs, one fixed to the substrate, and one strained by the diaphragm flexing, achieving a total of 72 sensing FBGs. Fibers are woven into

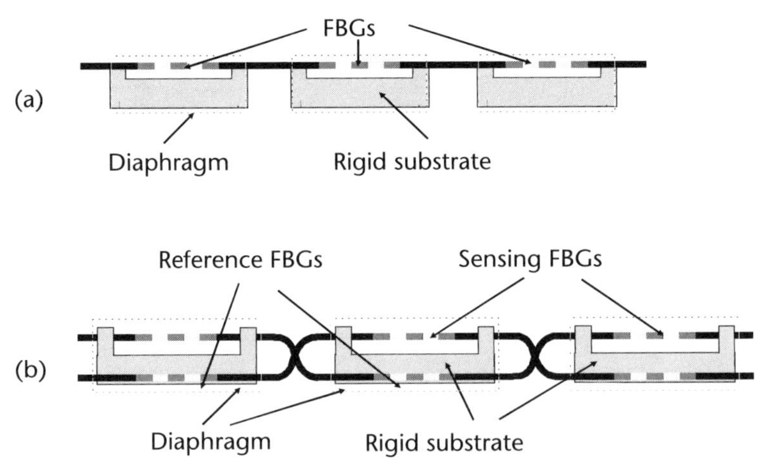

Figure 16.1 Sensing element for gastroscopy fiber-optic FBG sensors. (a) Sensing element transducing pressure to strain, built with a bendable diaphragm mounted on a rigid substrate, on which one FBG is fixed. The structure is repeated for each FBG, as in [3]. (b) Improved sensing element having a reference FBG, fixed in the substrate, and a sensing FBG, pressure-sensing, thanks to the diaphragm position. Fibers are interlaced and woven to form a catheter with separate sensing elements, as in [3], and patented in [4].

the substrate, with an interlaced structure that enhances the flexibility of the catheter. The diameter of each sensing block is 2.2 mm, while the catheter size is not disclosed. Each sensor is visualized under the X-ray.

16.1.2 Esophageal Detection

Arkwright [1] proposed the early-stage diagnosis of gastrointestinal motility, with the method in Figure 16.1(a) packaged as a high-resolution pressure manometer. Results have been obtained in vivo for a healthy volunteer, acquiring the pressure data in real time during a swallow. Figure 16.2 shows the main outline of this measurement technique. On the left side, we can observe the catheter positioning: insertion is performed as a standard nasogastric probe, through the nasal hole having a typical diameter of 2 to 3 mm, and therefore compatible with the 3-mm catheter. During a swallow, contractions of the muscles in the gastrointestinal tract, particularly towards the lower esophagus, result in a pressure change, spatially confined in tract of a few centimeters. The time/space chart, rendered in Figure 16.2(b) and obtained in Figure 8(B) and 8(C) of [1], shows a pressure peak that transits towards the lower part of the esophagus over time. A pressure peak appears within the middle section of the esophagus, having about a 30-mmHg value; this peak tends to enlarge as the contractions progress towards the stomach, reaching about 120 mmHg in the lower esophagus, and then reducing its amplitude. In the time/space domain, the pressure peak moves with rate of 5 cm/s, as recorded by the high-speed interrogator. The conclusion of this study shows the premises for an FBG-based manometer with resolution of 1 cm, fast acquisition, and allowing nasal intubation; with this tool, monitoring of peristalsis from the pharynx to the stomach can be achieved.

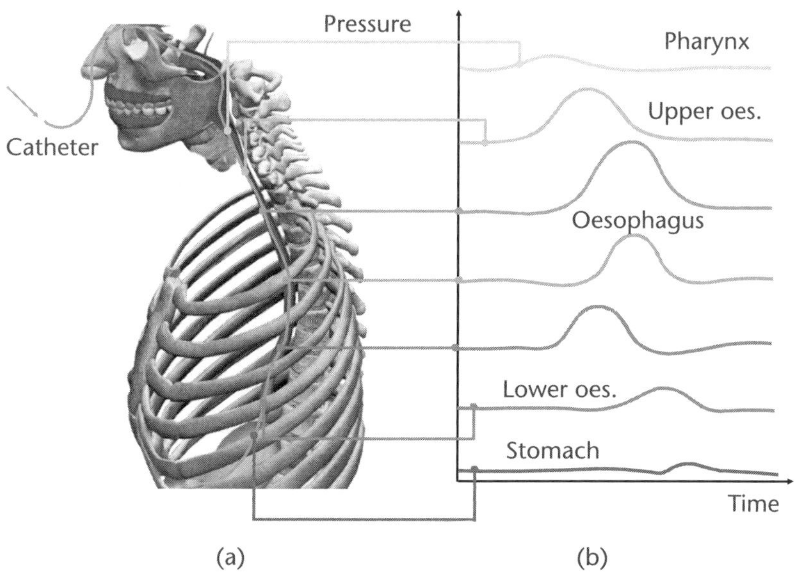

Figure 16.2 Upper digestive tract pressure, recorded between the pharynx and the stomach, through the entire esophagus; the result is described in [3]. (a) Positioning of the fiber optic catheter, inserted through nasal hole and reaching the stomach through the esophageal tract. (b) Rendering of pressure time-space analysis of a swallow, showing the transient of a pressure rise through the esophagus.

16.1.3 Colonic Analysis

In [3], the catheter having the structure of Figure 16.2(b) is applied in a long-term colonscopy (24 hours): the catheter is positioned through the rectum, through the whole colon, reaching almost the cecum. In this arrangement, 5 sensing points are located in the ascending colon, 11 sensing points in the transverse colon, 15 sensing points in the descending colon, and 5 sensing points in the sigmoid colon. In this example of colonscopy, the FBG-based catheter is aimed at the detection of propagating sequence (PS), defined as: sequence of 3 or more pressure waves recorded from adjacent recording sites in which the conduction velocity within that sequence lies between 1 and 12 cm/sec. Practically, the probe performs a space-time analysis as in the previous case, recording pressure (with a sensitivity of 1 pm/mmHg). In this architecture, tight fiber bends are imposed, particularly at the interface between the rectum and the rectum and sigmoid colon; the use of bend-insensitive fibers, such as ormoceramic fibers suitable for FBG inscription, facilitates the catheter operation. The FBG interrogator has two channels, switching at a 10-Hz rate and reading the pressure differential between each recording sensor and its corresponding reference.

The study is partially conclusive. FBG sensors allow achieving a high-resolution manometer, with a performance adequate for colonic analysis and straightforward interrogation. The catheter has the capability to perform an almost complete pan-colonic tracking of pressure waves due to colon contractions, and establishes as a high-resolution manometer. The architecture has been patented [4]. In contrast, the lack of reliable indicators for colonic disorders, obtainable through the space-time analysis, is still preventing to reference the measurement to a specific condition. It is expected that the availability of high-resolution fiber-optic manometers can push clinicians to obtain more markers on colonic data in the coming years.

16.1.4 Dual Sensing

In [3] a more advanced strain sensing was presented, based on FBG and with transduction of pressure to strain. The system improves the architecture of Figure 16.1(b) by providing a dual-sensing structure. A sensing element as in Figure 16.1(a) is used as circumferential sensor: contractions of the circumferential muscles of the gut distort the fiber sideways, hence causing a change in the Bragg wavelength. In contrast, for a longitudinal sensor having 2 FBGs fabricated on the same fiber and axially separated by a slider glued to the fiber, any relative motion of the wall past the catheter is picked up as a change in the reflected wavelengths of the two FBGs. Fibers are woven in an interlaced way as in Figure 16.1(b). In this configuration, a catheter hosting 5 circumferential and 5 longitudinal sensors was designed, with a 1.5-m total length. Experiments in vivo in rats and humans affected by functional gastrointestinal disorders have been reported, showing the functionality of the sensor.

16.1.5 Present and Future Perspectives

To date, FOSs for gastroscopy analysis have demonstrated the potential for improving over standard technologies for the space-time pressure analysis required to render the behavior of digestive apparatus. We foresee a further development

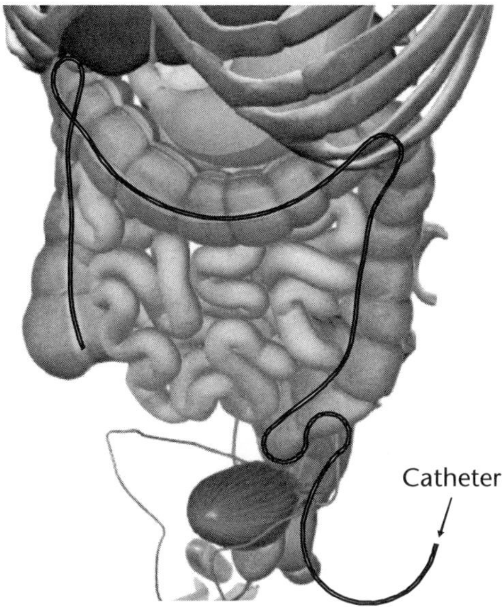

Figure 16.3 Positioning of the catheter described in [3], for colonic monitoring.

by improving the transducing technologies, and the sensing mechanism, in four directions.

1. *Transducer:* A major task in gastroscopy is the transduction of the traveling pressure within the digestive apparatus, into a strain that is easily detectable with technologies such as FBG or distributed sensors. The sensing elements proposed by CSIRO achieve this goal, but contribute to increase the catheter size to about or over a 3-mm diameter. This level of miniaturization allows using the nasal hole as access point for intubation, but still does not minimize the invasiveness for the patient. A further miniaturization of flexible diaphragm, or a further miniaturization of multifiber solutions as in Figure 16.1, will contribute to miniaturize the catheter, eventually reaching the portability criterion that can access continuous monitoring.
2. *Spatial resolution:* In parallel, a further narrowing of spatial resolution can be sought using distributed sensing. A 10 times increase, from 1 cm to 1 mm, can be achieved using current distributed sensors such as LUNA OBR4600. In addition, distributed sensing has the advantage of having an all-sensitive fiber, while an FBG array requires a perfect match between the FBG position and the diaphragm; otherwise, the sensor may be significantly decalibrated.
3. *Shape sensing:* Shape sensing can instead offer the opportunity to disrupt gastroscopic analysis. The monitoring of the deformation of the inner part of the digestive apparatus during swallowing, and during the entire digestive cycle, can be performed using shape sensors such as 4DSP/FBGS RT150. This is an early-stage task, whereas issues such as accuracy and positioning still have to be resolved.
4. *Multitool gastroscopy:* In clinical procedures, the space-time pressure analysis is typically combined with other features. The main ones are the

possibility of rendering the images of the inner part of the stomach, and grabbing with a microforceps a part of tissue for biopsy, often for detection of specific digestive bacteria, such as *H. pylori*. A well-thought design of a catheter can combine sensing with imaging on the tip; an example is [5]. SPR sensors, in addition, can combine with distributed sensors to measure specific analytes at the detection point.

16.2 Urology

Sensing in urology is a key task, particularly for urodynamic analysis [6]. This medical procedure aims at reproducing the natural functionality of the bladder, during its filling and voiding operations. In clinic, this examination is turned into a 10-minute to 30-minute procedure, whereas a saline solution is progressively infused in the patient's bladder at a constant rate until the peak stimulus to urinate is reached, and subsequently the patient is asked to stand up and urinate, emptying the bladder. Urodynamics is concerned with diagnosis of diseases in the lower urinary tract, particularly bladder outlet obstruction (BOO), incontinence, detrusor muscle diseases, and other conditions. In general practice, urodynamics is more targeted to BOO detection than other imaging-based techniques, such as fiber-optic cystoscopy, that aim at close imaging of the bladder walls and is used for bladder irritation, cancer, or blood traces in urine.

16.2.1 Urodynamics

The standard urodynamic diagnostic is based on a combination of uroflowmetry (that involves the measurement of flow and speed of urine in and out of the bladder), and pressure flow study (which is particularly concerned with the muscular activity related to the detrusor muscle) [7]. In addition, other tasks such as post-void urine analysis and leakage tests are performed. For BOO detection, the combination of uroflowmetry and detrusor muscle pressure is referred as the golden standard. During the examination, the patient's bladder is filled in with a saline solution, typically at a standard rate of several milliliters per minute, until the maximum urination stimulus is obtained. The patient is then asked to urinate in a calibrated container, which records in real time the urine flow (Q, in mL/s). The maximum void rate (Q_{max}) is often obtained during the first part of urination.

Simultaneously, the detrusor pressure (P_{det}) is measured. Since a direct measurement is not available, an estimate of P_{det} is performed by using two separate pressure sensors, one positioned in the patient's bladder and recording the bladder pressure (P_{ves}), and one positioned in the rectum and recording the abdominal pressure (P_{abd}). The detrusor pressure is then accounted as $P_{det} = P_{ves} - P_{abd}$. The golden standard of BOO detection is proposed in Figure 1 of [6], which shows a nomogram reporting Q_{max} on the x-ccordinate, and P_{det} at Q_{max} on the y-coordinate. This diagnostic approach is substantially qualitative: depending on the position of the P_{det} and Q_{max} values on the nomogram, the clinician can estimate whether a patient is potentially affected by BOO, which therefore leads to surgery. However, the large variability of Q_{max} values, the possibility that the maximum P_{det} value is obtained for smaller values of Q, and the natural enlargement of bladder due to aging, which increases pressure values, can contribute to a large number of inconclusive examinations. Using OFS, it is possible to improve this scenario, and in the following the main paradigms are discussed. In a [8], EFPI pressure sensors took the place of standard catheters, improving the accuracy and impedance to flow, necessary for an accurate detection. Following up, [9] proposed using a different technique, more similar to fractional flow reserve measurement, for detection of obstructions in the bladder. The promising results lead to new design criteria, which can lead to disruptive innovation in urology.

16.2.2 Traditional Diagnostic

Figure 16.4 shows the urodynamic diagnostic, performed with the traditional and differential pressure-reading method, showing the catheters positioning and their shape. In the traditional method, described in [8], two catheters are employed and positioned by the clinicians as in Figure 16.4(a). The first catheter, in Figure 16.4(b), records P_{ves}, it has 4-Fr (1.33-mm) thickness and 16-inch length and is sealed on the front side, while three side-holes (H) let the liquid inside the catheter (C). In order to detect pressure, an EFPI probe is fabricated as in Figure 16.4(b): the Fabry-Perot (FP) cavity is enclosed within a single-mode (SM) fiber and a multimode (MM) fiber serving as diaphragm, and it is sealed with a capillary (Cap). The abdominal pressure P_{abd} is positioned in the rectum; the rectal catheter has 4-Fr thickness and 16-inch length and is terminated with a balloon with a 1-cm diameter. Inside the catheter, the same EFPI structure is used for pressure sensing. Both catheters are fixed with medical Scotch tape at the outer end (outside of the patient's body). An

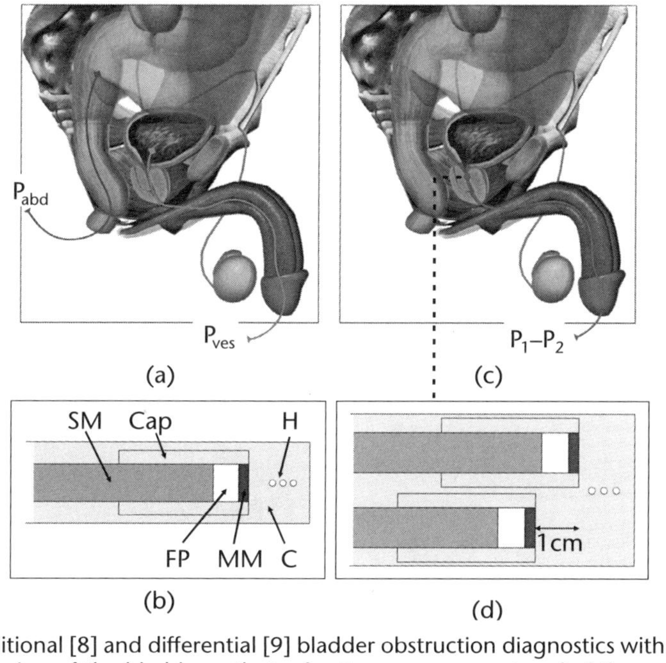

Figure 16.4 Traditional [8] and differential [9] bladder obstruction diagnostics with fiber-optic EFPI sensors. (a) Positioning of the bladder catheter for P_{ves} measurement and of the rectal catheter for P_{abd} measurement [8]. (b) Structure of both EFPI sensor, and embodiment in bladder catheter. (c) Positioning of the catheter for differential pressure measurement $P_{diff} = P1 - P2$ in the bladder [9]. (d) Structure of the bladder catheter, with 2 EFPI sensors positioned at a 1-cm distance.

interrogator (Ibsen I-MON-USB), connected to a switch (Sercalo 1x2) was used to commute between the two probes at 20 Hz, and the detrusor muscle pressure was then estimated by computing the difference between the two pressure values.

In [8], the first demonstration of fiber-optic based sensing versus traditional catheters was performed, in vivo on 7 patients. For each patient, two pairs of catheter were colocated by the clinician inside the bladder, and in the rectum. The first pair of catheters embedded the EFPI probes, and the second pair of catheters were the standard urology catheters, controlled by PICO2000 (Menfis Biomedica S.r.l.). Prior to the measurement start, all catheters were zeroed. Then the patient was asked to cough, producing a contraction of both bladder and abdomen, resulting in a sharp pressure pulse in all catheters, in order to verify the correct positioning. The patient's bladder was infused at constant rate by the PICO2000 until reaching 200 to 375 mL, through the bladder catheter controlled by the PICO2000. Finally, the patient was asked to urinate. All pressure traces, as well as uroflowmetry values, were recorded in parallel by PICO2000 and by the fiber-optic interrogator.

Figure 16.5 reports the P_{det} trace, recorded with the pair of EFPI sensors, for the duration of 70 seconds, during the urination phase following the saline solution infusion (375 mL). As the urination stimuli are initially high, we observe a high-pressure value, equal to 42 cmH_2O, close to 0.1 kPa; we note that in urology, the cmH_2O units for pressure are preferred over kPa and mmHg. Pressure initially floats, approaching 50 cmH_2O during the first phase of urination, and then gradually decreases, approaching a zero value. Other studies, performed on patients with potential BOO, show steep pressure peaks, and higher P_{det} values, even during the infusion part, as well as during urination.

16.2.3 Differential Diagnostic

The second study, presented in [9], extended the previous demonstration and provided the preliminary result for a differential pressure-based diagnostic. The underlying hypothesis is that, in case a bladder obstruction of any kind is observed, similarly to an obstruction of any fluid-filled capillary, a differential of pressure $P_{diff} = P_1 - P_2$ is observed between the pressure measured after the obstruction (P_1) and

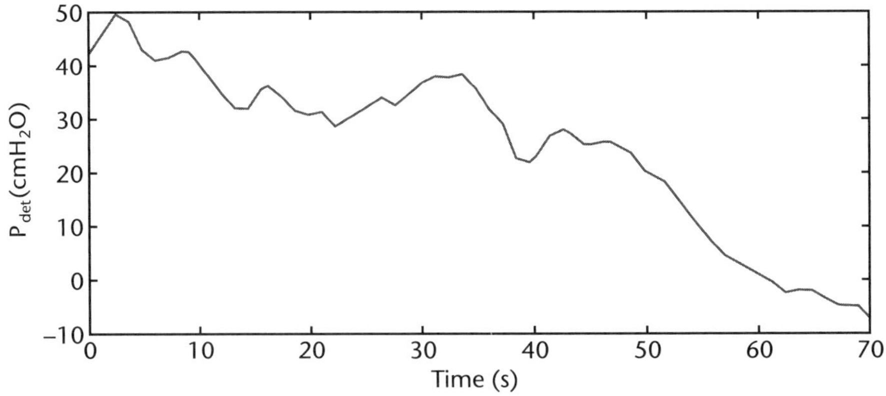

Figure 16.5 Detrusor pressure, estimated with a pair of EFPI sensors, for a male patient with no diagnosed bladder condition, during urination (0 to 30 seconds) and after bladder voiding (30 to 70 seconds).

the pressure measured before the obstruction (P_2). This assumption requires a pair of sensors, positioned pre-obstruction and post-obstruction, having a negligible impedance to the flow. In addition, accuracy is a key issue. In urology, small pressure values (typically 0 to 100 cmH_2O) are observed, and accuracy of 1 cmH_2O, typical of most catheters, is adequate to plot the nomogram of [6] with sufficient precision. Instead, when looking at pressure differentials obtained in a narrow spacing, an accuracy value of 0.1 cmH_2O is required to have a reliable estimation of P_{diff}.

Under these assumptions, the study of Poeggel et al. provided the first-ever differential pressure measurement in the bladder, for the attempt of improving the BOO diagnostic. The EFPI structure is the same of Figure 16.4(b), based on a biocompatible all-glass sealed structure having typical sensitivity ≥ 1 nm/kPa (≈0.1 nm/cmH_2O). A number of 8 sensors were fabricated with this structure, on bend-insensitive fiber to allow obtaining low losses when inserted in the urinary tract. The main difference with respect to the previous study is the catheter shape, as in Figure 16.4(d). Inside a commercial flexible catheter (Nutrisafe-2, 5 Fr – 1.6-mm thickness), 2 EFPI probes have been inserted, with a 1-cm distance between the two tips. The catheter is sealed on the tip and has side holes that allow the liquid penetration to the catheter tip. Alternatively, the catheter has a Y-junction that allows 2 connections: the first connector is sealed and allows the two fibers to leave the catheter and reach the interrogator positioned at remote distance; and the second connector is attached to the control machine (SmartDyn, Albyn Medical), which controls the infusion of saline solution, and records uroflowmetry. This setup, rather than measuring the detrusor muscle pressure, aims at having both probes positioned in the bladder, with no rectal catheter, and quantifying the P_{diff} value.

A full urodynamic examination, performed with the SmartDyn and the EFPI probes, has been performed in vivo on 4 patients (2 male, 2 female), at Federico II University of Naples, Italy. Figure 16.6(a) shows a picture of the urodynamic room, set up for patient analysis. In Figure 16.6(b), the pressure trace is reported for a male patient, reporting the value of both pressures estimated by each EFPI probe. In the first part of the chart, the pressure values are low, and tend to increase around 6 minutes and 20 seconds time, where a pressure pulse is observed only in $P_1$and not in P_2. In the last part of the trace, the patient is urinating. When the bladder contracts, a steep pressure rise is observed by both sensors; in this part of the trace, the pressure pattern is very similar for both pressure values.

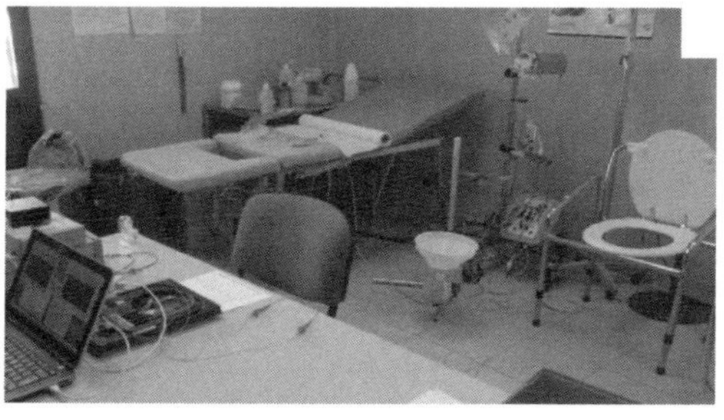

(a)

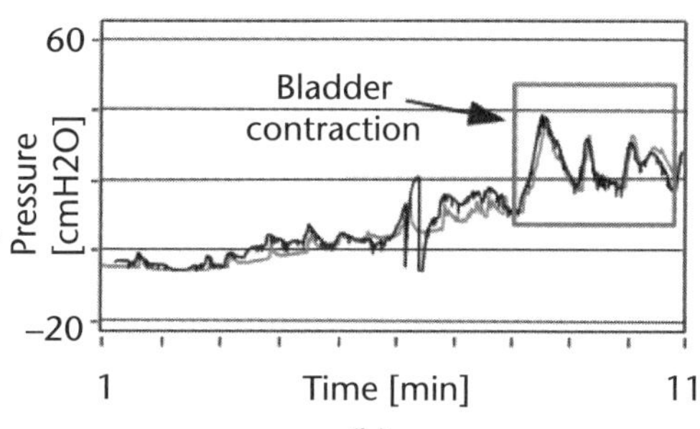

(b)

Figure 16.6 Differential pressure system for urodynamic measurement. (a) Photographic view of the urologic room. (b) Urodynamic measurement, performed with two EFPI probes (black curve = P1, gray curve = P2). (From: [10].)

16.2.4 Current Outlook and Future Prospects

The study of [9] sets the preliminary basis for the study of differential bladder pressure. This measurement is not available when using standard air-charged or fluid-filled catheters, as they have excessive size. The miniature size of an EFPI, instead, allows inserting several sensors in a single catheter (2 in the investigation, up to 7 in a standard 5-Fr urologic catheter): this enables exploring the different pressure traces, in different parts of the urinary tract. The preliminary investigation shows that the EFPI have excellent metrologic properties, in particular bend-insensitivity (a key factor, in male urinary tract), and precision sufficient to measure P_{diff} with ~0.2-cmH_2O accuracy. In order to gain significance from the medical point of view, clinicians need to explore the EFPI technology in a relevant number of patients, and find relevant metrics based on differential pressure that are unambiguously tied to BOO. As a proof, the investigation of Poeggel et al. has shown that: (1) differential pressure values are higher in male patients than in female patients, and this is in agreement with BOO statistics [6]; (2) pressure differential extends between 20 to 60 cmH_2O, with sensors positioned at 1-cm distance, and tends to rise for patients that appear to have more urination stimuli during the infusion; and (3) for the same patients, the values of P_{diff} appear to vary between urination and infusion phases.

It is expected that this diagnostic technique may replace the standard nomogram, particularly when reporting the charts suggested in [9], that is, reporting P_{diff} versus P_1 and the volume of infused/urinated saline solution versus P_{diff}. In this case, we may move from a qualitative diagnostic of BOO, which makes an indirect measurement, to a more quantitative diagnostic, directly measuring a parameter related to the amount of obstruction. The possibility of filling a catheter with multiple EFPIs is an asset: expanding the spatial resolution may enable the possibility to localize an obstruction in the urinary tract, besides detecting its effect.

16.3 Neurology

The most successful application of fiber-optic sensors in neurology is the intracranial pressure (ICP) measurement [11]; in this case, commercial technologies have been established as clinical procedures or are currently under clinical tests and validation. ICP measurement is essential in the diagnostic, and often treatment, of specific neurological disorders such as strokes, ischemia, hemorrhage, and traumatic brain injury. ICP measurement requires the detection of pressure in specific locations, inside or on the border of the skull. The review from Kawoos et al. [12] identified several techniques for noninvasive and indirect ICP measurement, usually with a subdural method, although this protocol is not clinical yet. In contrast, the invasive technique, which requires intraventricular or intraparenchymal pressure measurement, is the current gold standard [13] and is the best method for detection of traumatic brain injury and accurate diagnostic of potential strokes.

ICP measurement has been performed since the late 1960s, with intraventricular catheters containing simple pressure probes [14]. The procedure is invasive: after anesthesia and preliminary treatments, a small hole (~2-mm diameter) is drilled in the skull, to allow the penetration of an intraventricular catheter, containing a pressure transducer, in the brain lateral ventricle. Often, particularly in case of suspected stroke, the catheter is inserted in the parenchymal region, for a more difficult insertion. Figure 16.7 shows the positioning of the intraventricular pressure probe, within the lateral ventricle of the brain, which requires penetration through the skull, compared with the subdural and intraparenchymal positioning.

ICP measurement has been among the first examples of application of FOS to biomedical engineering: the Camino sensor has been in use for more than a decade as a clinical device [15]. The Camino technology (OLM IntracraData pressure monitoring kit Model 110-4B; Integra NeuroSciences Camino: San Diego, California) is based on a simple fiber bundle that makes use of a transmitting fiber, and a separate receiving fiber, as shown in Figure 16.8 [14]. The bundle is packaged in a catheter, with holes that allow liquid penetration. Inside the catheter, a bendable diaphragm is mounted: when pressure is applied, the diaphragm bends reducing the displacement between the input/output fiber and the target. This short displacement, in the order of hundreds of micrometers, acts as a Fabry-Perot cavity modulated by pressure inside the catheter. The diaphragm is highly flexible and reflective to improve performances. The structure of the Camino, well described in [14], has key advantages in the easiness of fabrication (particularly, thanks the current methods for precision alignment, and the possibility to embed the structure directly in the catheter), the high sensitivity and reflectivity of the diaphragm, and the low cost of input and output fibers, as no elements are inscribed or fabricated on the fiber tip. Despite the simple approach, the need for two separate fibers has the key drawback of increasing the overall probe size with respect to a single-fiber architecture, whereas in deep-seated ICP positioning miniature dimensions are needed.

For this reason, recent studies have documented the application of EFPI-based fiber-optic sensors for ICP measurement, using the last EFPI generation. In 2012, Murtha et al. [13] proposed an investigation (including a video of the procedure) of ICP measurement based on a commercial transducer (420LP, SAMBA Sensors, Sweden; SAMBA technology has been acquired, more recently, by FISO Technologies, Canada). The use of a single fiber has the key advantage of reducing the catheter size: the previous generation of EFPI-based tip sensor had a diameter of 35 to 0.55 mm, while the current generation achieves a 0.125-mm diameter size. The miniature size allows also biomedical studies, in rats, aimed at optimizing the positioning of the pressure transducer relatively to the brain. The study has been

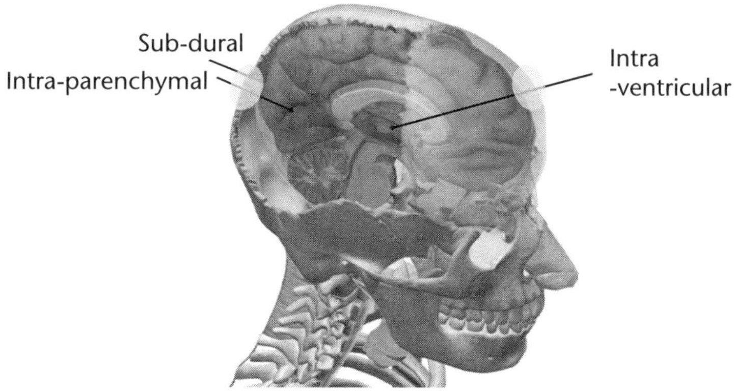

Figure 16.7 Positioning of ICP probe, for intraventricular, intraparenchymal, and subdural measurement.

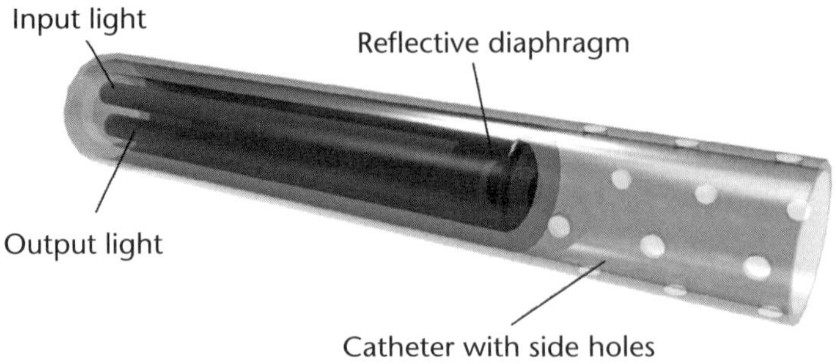

Figure 16.8 Schematic of the Camino sensor. (From: [14].)

performed in vivo on a rat, monitoring the ICP during normal operation, and during abdominal compressions (that induce an artificial pressure rise, useful to validate the catheter operation).

The study shows that the ICP trace is visible and follows a clearer path than in fluid-filled catheter. In the ICP trace, pressure has a baseline value, usually within 1 to 20 mmHg; in [13] the baseline value was found to be function of the skull enclosing, with fluctuations up to 2.5 mmHg when skull holders were changed. The pressure trace exhibits a periodic pattern, with approximately 1-Hz period, and a good identification of the inner details has been demonstrated by EFPI probe. Currently, most diagnostic techniques are based on the baseline ICP value [11]; usually, high values in proximity of 20 mmHg are found to be harmful and particularly appear as indicator of brain traumas. The analysis of the details of the pressure trace, instead, is often not performed as a clinical procedure: the reasons are that fluid-filled catheters (including the Camino) often have a slow response, which prevents discriminating all the pressure pulses, and also the bigger active size tends to average out fast pressure fluctuations. Similar to the previously discussed applications, biomedical engineers and clinicians should exploit the advantages of size and response of fiber-optic sensors, perform tests on a broad number of phantoms or samples, and identify reliable metrics that enable advanced diagnostic features, unavailable with the present generation of sensors.

16.4 Epidural

Epidural is one of the most popular loco-regional technique for the operative, postoperative, and chronic pain management. Epidural consists in the injection of a drug, anesthesia, or contrast agent in the epidural space of the spinal cord. In clinical practice, epidurals are performed with a Tuohy needle, a metallic hollow needle that penetrates through the skin and the interspinous ligament, identifies the epidural space that lies within the ligament and the dura mater of spinal cord, and injects the liquid required for the treatment. The Tuohy catheter is strongly bent on the tip, in order to both avoid excessive contact with the dura mater, and to release the liquid at an angle of approximately 45° respective to the catheter entry point. Epidurals are performed in correspondence of parts of the spinal cord, often in correspondence with L3/L4 vertebrae. Identification of the epidural space is carried out manually by the clinician, relying to perceptive feeling; each time the catheter penetrates one layer of derma, enters the ligament tissue, and passes through the ligament flavum, the different consistency of the tissue changes the forces on the catheter, allowing the clinician to identify the epidural space.

In 2017, the University of Sannio and Sapienza University (Italy) have developed an innovative solution based on a fiber-optic sensor to measure the forces exerted on the Tuohy catheter during epidural insertion [16]. The proposed solution is based on an FBG sensor, integrated with a commercial Tuohy hypodermic needle, which measures the force applied on the needle tip. An FBG, fabricated on an ormoceramic glass compound and coated with 900-μm diameter Hytrel layer, is positioned within a Tuohy needle. The FBG is prestrained within the catheter lumen, in order to make it sensitive to positive and negative pressure variations; the Bragg wavelength shift, detected with a commercial interrogator, is proportional to the pressure (force) exerted on the needle tip. Experiments performed on phantoms and mimicking both correct and incorrect epidural insertions show that in a correct insertion, and two peak pressure values are observed in correspondence to the transitions of tissue from skin to ligament, and from ligament to the epidural space. When the epidural space instead is not correctly reached, the second pressure peak is not observed, resulting in a feedback for the clinician. Repeated experiments on phantoms show that the FBG-based sensor can correctly identify a correct or wrong insertion by analyzing the peak pressure, in all the experiments performed by the research team.

16.5 Conclusions

FOS-based diagnostic devices have the increasing capability to resolve time-space maps, measuring both biophysical quantities and their space distribution even in long-term analysis. This promising capability can provide unprecedented measurement quality, while reducing invasiveness for patients.

The possibility to embed FOS in vivo through minimally invasive catheters can, as a general trend, move diagnostic from qualitative to quantitative layout.

A tangible impact will be achieved when clinicians will be able to accumulate significant amount of data, recorded with FOS-based devices, finding quantitative metrics to discriminate health patients from diagnosed conditions.

References

[1] Arkwright, J. W., et al., "In-Vivo Demonstration of a High Resolution Optical Fiber Manometry Catheter for Diagnosis of Gastrointestinal Motility Disorders," *Opt. Expr.*, Vol. 17, No. 6, 2009, pp. 4500–4508.

[2] Arkwright, J. W., et al., "Design of a High-Sensor Count Fibre Optic Manometry Catheter for In-Vivo Colonic Diagnostics," *Opt. Expr.*, Vol. 17, No. 25, 2009, pp. 22423–22431.

[3] Arkwright, J. W., et al., "Measurement of Muscular Activity Associated with Peristalsis in the Human Gut Using Fiber Bragg Grating Arrays," *IEEE Sensors Journal*, Vol. 12, No. 1, January 2012, pp. 113–117.

[4] Arkwright, J. W., I. Underhill, and N. Blenman, "Optical Device," U.S. Patent 8,724,935 B2, 2014.

[5] Herranz, D., et al., "Novel Catheter Enabling Simultaneous Radiofrequency Ablation and Optical Coherence Reflectometry," *Opt. Expr.*, Vol. 6, No. 9, 2015, pp. 3268–3275.

[6] Nitti, V. W., "Pressure Flow Urodynamic Studies: The Gold Standard for Diagnosing Bladder Outlet Obstruction," *Rev. Urol.*, Vol. 7, Suppl. 6, 2005, p. S14.

[7] Schäfer, W., et al., "Good Urodynamic Practices: Uroflowmetry, Filling Cystometry, and Pressure-Flow Studies," *Neurourol. Urodyn.*, Vol. 21, No. 3, 2002, pp. 261–274.

[8] Poeggel, S., et al., "Fiber-Optic EFPI Pressure Sensors for In-Vivo Urodynamic Analysis," *IEEE Sensor Journal*, Vol. 14, No. 7, 2014, pp. 2335–2340.

[9] Poeggel, S., et al, "Differential *In Vivo* Urodynamic Measurement in a Single Thin Catheter Based on Two Optical Fiber Pressure Sensors," *Journal of Biomedical Optics*, Vol. 20, No. 3, 2015, p. 037005.

[10] Poeggel, S., et al., "Optical Fibre Pressure Sensors in Medical Applications," *Sensors*, Vol. 2015, 2015, pp. 17115–17148.

[11] Raboel, P. H., et al., "Intracranial Pressure Monitoring: Invasive Versus Non-Invasive Methods—A Review," *Critical Care Research and Practice*, Vol. 2012, 2012, p. 950393.

[12] Kawoos, U., et al., "Advances in Intracranial Pressure Monitoring and Its Significance in Managing Traumatic Brain Injury," *Int. J. Mol. Sci.*, Vol. 16, 2015, pp. 28979–28997.

[13] Murtha, L., D. McLeod, and N. Spratt, "Epidural Intracranial Pressure Measurement in Rats Using a Fiber-Optic Pressure Transducer," *J. Vis. Exp.*, Vol. 25, No. 62, 2012.

[14] Roriz, P., et al., "Review of Fiber-Optic Pressure Sensors for Biomedical and Biomechanical Applications," *J. Biomed. Opt.*, Vol. 18, No. 5, 2013, p. 050903.

[15] Gelabert-Gonzalez, M., et al., "The Camino Intracranial Pressure Device in Clinical Practice. Assessment in 1000 Cases," *Acta Neurochir.*, Vol. 148, 2006, pp. 435–441.

[16] Carotenuto, B., et al., "Optical Guidance Systems for Epidural Space Identification," *IEEE J. Select. Topics Quant. Electron.*, Vol. 23, No. 2, 2017, p. 6900309.

CHAPTER 17

Sensing in Thermal Ablation

17.1 Thermal Ablation: Procedures and Rationale

Thermal ablation (TA) is an emerging medical procedure that is rapidly emerging as a minimally invasive treatment for a broad range of applications. According to [1], the TA market was estimated as $8.6 billion in 2013, with an 11.0% annual growth rate. The common point of all TA procedures is the capability to heat or cool human cells over their mortality rate [2], using a minimally invasive (or noninvasive) device. All TA procedures are based on two devices:

1. A source of electromagnetic radiation (radiofrequency, microwave, light, ultrasound), having typical power within 2 to 50W, which is used as an energy source, to be converted into heat.
2. An applicator, consisting of a minimally invasive device that transports the source power to the tissue under treatment. The applicator is positioned by the clinician at the point of treatment, usually percutaneously, and using imaging techniques (magnetic resonance imaging, ultrasounds, or computed tomography) to track the position of the device. When the source is turned on, the electromagnetic power is radiated from the active part of the electrode to the tissue in contact with it and then conveyed in the surrounding tissue by conduction. With respect to other types of electromagnetic irradiances (such as phototherapy), the energy is selectively focused in a specific region of the tissue.

This arrangement allows creating a heat field inside human tissue, with the possibility of accessing deep-seated locations such as brain, heart surface, or inner parts of lungs, with a high spatial selectivity. The goal in TA is to bring cells over their threshold of mortality, well identified in Rhim's outline [2]: temperatures higher than 42°C are cytotoxic for most cells, and within the range of 42°C to 47°C, cellular mortality starts to be significant for long exposure times. Over these values, cellular damage becomes irreversible, and TA treatments start to have a significant impact on cells' survival. The most significant threshold occurs for 60°C, whereas instantaneous protein coagulation occurs: by reaching this threshold, cellular mortality is nearly instantaneous, and TA is extremely effective in killing cells in the portion of tissue exposed to heating. Negative effects instead are often introduced when the temperature of 100°C is reached or exceeded: vaporization of water constituents of the tissues can have a significant impact on the impedance and absorption properties of the tissue, and higher temperature may induce carbonization.

Thus, TA is a procedure that aims to expose a small and spatially confined portion of tissue to 60°C to 100°C heating, with a minimally invasive applicator. TA is opening a plurality of mini-invasive surgical treatments, including correction of cardiac arrhythmia, electrical scalpel, chronic pain management, and other forms of cauterization. In this chapter, we focus on one application that is rapidly emerging and which has the highest impact for biomedical sensors: interventional cancer care. In this application, the tissue exposed to tissue is a tumor, previously diagnosed and localized: TA is then applied by positioning the applicator at the focal point of the tumor, and heating the surrounding tissue such that the entire volume of the tissue (with an additional safety margin) is exposed to ≥60°C. The following discussion reviews TA methods, and their sensing applications, applied to cancer treatment; indeed, most of the methodologies and conclusions are valid for the whole spectrum of TA applications.

Several TA approaches have been documented for tumor treatment, using different principles and applicators (reviewed also in Table 17.1, with references).

17.1.1 Radio Frequency Ablation

Radio frequency ablation (RFA) requires a radio frequency generator (RFG) with 5–50-W power, and 15–500-kHz frequency [3]. An active electrode, with a stem, umbrella, bundle, or spiral shape, is positioned at the focal point of the tumor, while a passive electrode is positioned in a neutral point, usually on patient's spine. A catheter is used to percutaneously guide the applicator to the tumor, and then the electrode(s) are released. As the tissues have typical impedance of 20–150Ω, heat is propagated by initial resistive dissipation, and to the surrounding tissue by conduction. Major RFA strengths are the minimal invasiveness of the catheter and the possibility of embedding multiple electrodes; the main weakness is the vaporization of water constituents of the tissues at 100°C, which causes the tissue impedance to rise.

Table 17.1 Comparison of Main TA Techniques Applied to Cancer Care

Technique	*Reference*	*Tumors*	*Characteristics*
RFA	[3]	Liver, lung, kidney, prostate	Mini-invasive, midsize, sensitive to vaporization
MWA	[4]	Lung, liver, pancreas	More invasive, large treatments, insensitive to tissue vaporization
LA	[5]	Brain, thyroid, pancreas, liver	Microapplicator, small size, spatially confined heating
HIFU	[8]	Liver, prostate, breast, kidney	Noninvasive, mid-size, hard to focus correctly
Cryo	[6]	Prostate, lung	Cold temperatures, lower mortality of cells, mid-size
Nano-rods	[7]	Preclinical	Increase heat diffusion in surrounding tissue

17.1.2 Microwave Ablation

Microwave ablation (MWA) is an ablation technique that makes use of an antenna applicator [4], which applies an electromagnetic field; this produces frictional energy in contact with the target tissue that converts to heat. MWA systems operate at 900 MHz to 2.4 GHz, delivering power up to 100W. MWA has the key advantage of being very insensitive to tissue properties (perfusion, vaporization), but applicators usually have a larger size, although the last generation of applicators has substantially contributed to reduce invasiveness.

17.1.3 Laser Ablation

Laser ablation (LA) makes use of a continuous wave laser, usually in the near infrared, and emitting up to 10W [5]. Light is coupled into an optical fiber, with frontal, radial, or multipoint firing. As light penetrates into tissue, energy is absorbed and converted to heat. LA is a minimally invasive treatment, with applicators having microscopic size, but an ablation pattern is usually confined in a small volume.

17.1.4 High-Intensity Focused Ultrasound

High-intensity focused ultrasound (HIFU) is based on the generation of an ultrasound wave (0.8–3.5 MHz), produced by a piezoelectric crystal and focused with an arrayed applicator to the target tissue. The high-intensity power, together with the precise focusing, allows reaching power density up to thousands of W/cm^2, which is then converted in heat through this sonication process. HIFU is a noninvasive treatment, with growing performance, but is sensitive to focusing issues due to perfusion or blood vessels.

17.1.5 Other Techniques

Cryoablation is used to freeze tumor cells, below –20°C, with a miniature localized iceball [6]. Very recently, RFA and LA techniques are applied in conjunction with metal nanoparticles (usually gold nano-rods), which control stabilize heat irradiance in the tissue [7]. Similar results are obtained with nano-ferromagnetic materials.

17.2 Sensing in Thermal Ablation

Sensors play an essential role in modern thermal ablation techniques. Temperature detection is the main task, because temperature is the driver for cells mortality: the more capillary and densely spaced is temperature sensing, the more accurate is the estimation of the instantaneous temperature values applied in each portion of the tissue. The measurement of temperature achieves the following outcomes:

1. Estimate, point-by-point, the temperature, and therefore the mortality rate, of cells in each part of the tissue under analysis.
2. Predict the outcome of TA, using modeling techniques such as the one described in the IMPPACT project [9].

3. Improve TA efficiency (and its consequence long-term lifetime expectancy) by verifying that the whole tumor area has been treated: an immediate feedback allows the clinician to repeat, or extend, the procedure. Alternatively, when the target treatment volume has been reached, TA can be interrupted avoiding excessive ablation.
4. Gain knowledge on the physical phenomena involved in TA and establish them as a standard.
5. Estimate temperature in organs surrounding the ablation volume, protecting them.
6. Modulate the parameters of the ablation setup, on the basis of the sensors' readout.

Until 2013, temperature measurement has been mainly performed with thermocouples (TCs), and more recently with micro-TCs. However, this approach is not effective; TCs can detect temperature only in a single point and have heat-sink effect that tends to overestimate the temperature [10]. In contrast, thermal imaging techniques based on infrared camera require a line of sight and cannot be performed in vivo, while computerized tomography methods do not achieve sufficient reliability. Since 2013, advances in OFS have promoted multiplexed, semi-distributed, and distributed sensors as the state of the art of sensing technology for thermal detection in TA procedures, particularly in minimally invasive RFA and LA. Several fiber-optic technologies have been documented in application for RFA, LA, and preliminarily in MWA.

Pressure is also a key parameter, particularly for treatment of encapsulated tumors (ETs) such as hepatocellular carcinoma (HCC). In ET, a tissue sheath encloses the tumor, and during ablation may substantially inflate causing a pressure rise; several studies, including [11], reported that an excess of pressure rise is associated to scattered recurrences, which have a negative impact over a 5-year lifetime. Intraparenchymal pressure measurement is therefore a key asset for ET treatment, and the first OFS-based measurement has been performed in 2014 by Tosi et al. [12] with an EFPI probe.

TA is an extreme sensing environment, characterized by steep thermal gradients, need for miniature size, catheterization and insertion problems, and other stringent requirements. Table 17.2 reviews the requirements for temperature and pressure sensors for an efficient detection in mini-invasive TA [13]. The most critical properties of TA-compliant sensors is the microscopic size, in order to avoid altering the ablation pattern, biocompatibility, and tensile strength sufficient to guarantee percutaneous insertion and removal; in addition, thermal sensors are required to sample temperature with millimeter-level resolution, while pressure sensors require to be thermally insensitive to minimize cross-sensitivity. Remarkably, OFSs are the only technology that can operate in this scenario and the only one capable of distributed measurement.

Table 17.2 Main Requirements for Sensors Applied to Temperature Distribution and Pressure Monitoring in Mini-Invasive TA

Requirement	*Temperature Distribution Sensor*	*Pressure Sensor*
Size (diameter)	<0.25 mm	<0.25 mm
Active size	<1 mm	<1 mm
Accuracy	0.5°C	0.1 kPa
Spatial resolution	0.1–1 mm	—
Tensile strength	10N	5N
Biocompatibility	ISO 10993	ISO 10993 (epoxy-free)
Cross-sensitivity	Strain-insulation	< 0.1 kPa/°C
Response time	1 second	1 second

17.3 FBG Arrays

FBG arrays have been applied to RFA [13] and LA [10]; Figures 17.1 and 17.2 describe the setup and main results of FBG sensors applied to RFA experiments ex vivo on porcine liver phantom. In this setup as in Figure 17.1(a), an RF generator (RFG, TAG 100W, Invatec), with 480-kHz sinusoidal carrier was used as source, with emitted power within the 9–20-W range. The device was connected to an applicator, consisting of a stem-shape hollow needle with a 3–4-mm inner-outer diameter, with a brass active electrode with a 1-cm length mounted at 1 cm from the tip. The ablation device was inserted through a sample of porcine liver, serving as phantom, positioned over a metal plate connected to the passive electrode of the RFG. This way a difference of potential is applied between the electrodes, and the dissipated power on the tissue is conveyed as heat in the surrounding volume. The RFG mounts an impedance-meter that disconnects the power when the tissue impedance falls outside of the 20–300Ω: this condition typically occurs when the maximum temperature overcomes 100°C, and tissue starts vaporizing.

On the electrode, an array of 5 FBGs was mounted, and fixed with tape. All FBGs (DTG-1550, FBGS International) were fabricated on the same ormoceramic fiber, and have 1,546–1,553-nm peak wavelengths. Each FBG has a 5-mm length and a 1-cm length between the center of each FBG, thus sampling a length of 4.5 cm with a 1-cm resolution. The FBG array was positioned on the sensor tip, such that FBG4 laid in correspondence of the ablation peak, FBG3 and FBG5 were adjacent to this position, and FBG1-2 were at the margin of the ablation area. Figure 17.1(b) shows the effect of a temperature measurement, whereas each FBG starts to shift, and different temperatures are recorded by each sensing element. By estimating the FBG peak, a space-time analysis can be performed.

Figure 17.1(c) shows the result of temperature measurement in an RFA experiment. As the RFG is turned on, we initially observe a steep temperature rise that affects FBG3-5, with almost a quadratic trend. As the 60°C threshold is overcome by these sensors, the temperature rising slope reduces to an almost linear rise, and this phase has a duration of approximately 140 seconds. At this point, vaporization takes place, and tissue impedance quickly rises well over 300°C, causing the RFG to disconnect. Thus, after reaching the peak, temperature starts to fall, and the

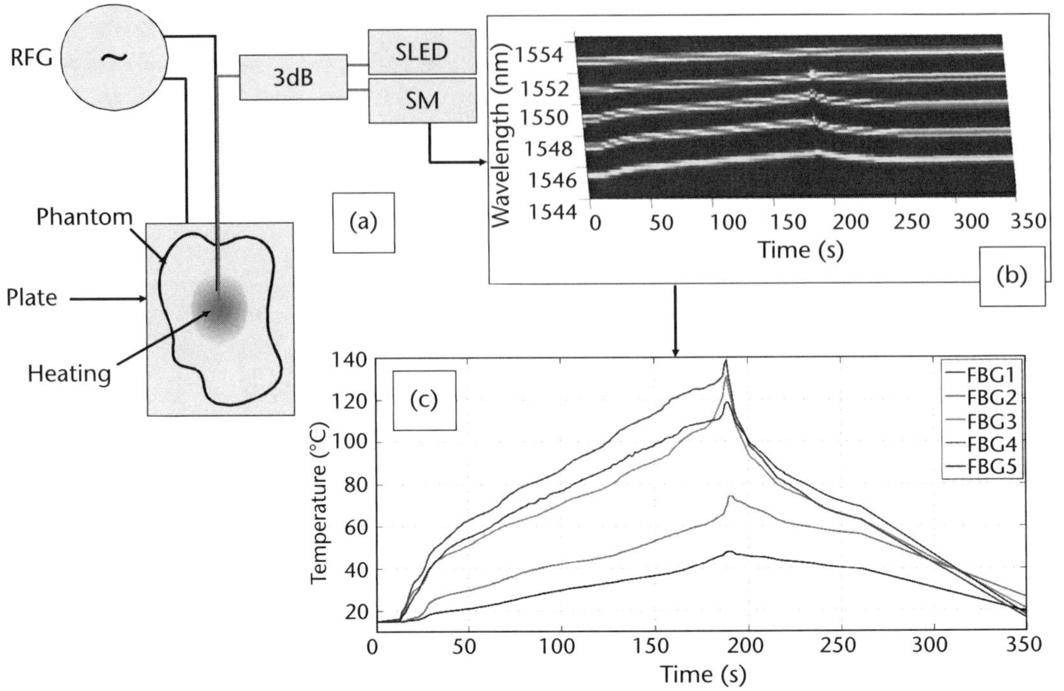

Figure 17.1 Measurement of thermal field in RFA with an FBG array: (a) experimental setup; (b) spectrum of the FBG array, measured by the spectrometer; and (c) temperature estimated by each of the 5 FBGs.

steepest descent occurs near the ablation peak, falling back to ambient temperature after a total of 350-second elapsed time. We notice that FBG4, as it is positioned in proximity of the ablation peak, records the highest temperature, while FBG3 and FBG5 still record peak temperatures well over 100°C. Alternatively, FBG2 also records a peak temperature over the 60°C threshold and is therefore inside the ablation zone; FBG1 is the furthest from the active electrode, and its peak temperature does not exceed the cytotoxicity level.

The measurement performed with FBGs has the advantage of straightforwardness and cost-effectiveness, but a main limitation is the low spatial resolution, limited to 1 cm for the DTG used in experiments. The main value of the FBG-based technique is that it allows investigating the thermal properties of the tissues and heat propagation, with a reliable method, and this can be used to obtain predictive features. In Figure 17.2, three experiments of RFA on porcine liver have been reported as in [13], using the same sensing system and FBG position as in Figure 17.1. To compare data on the same chart, both time and temperature have been normalized (0 = procedure start; 1 = ablation peak). In the second ablation, FBG2 records a lower normalized temperature than other experiments, while FBG1, FBG3, FBG4, and FBG5 the normalized temperature is higher, with a significant deviation for FBG3. Data recorded in experiments 1 and 3 is substantially similar, with some deviations for FBG4. This preliminary study demonstrates that when temperature rises with a steeper rate near the RFA applicator, the heat field tends to remain more confined inside the central part of the tissue, and therefore the volume of tissue exposed to cellular mortality is lower. Instead, as in experiments 1 to 3, a

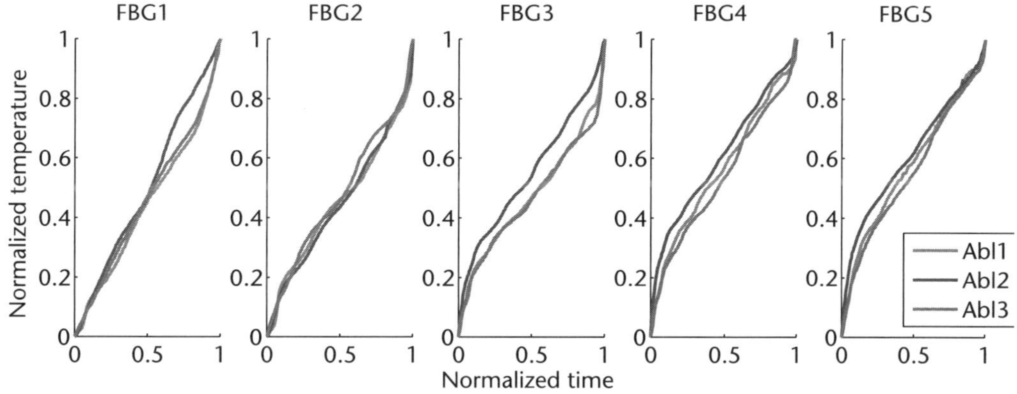

Figure 17.2 Temperature pattern, recorded by each of the 5 FBGs as in Figure 17.1, for 3 different RFA experiments. Both time and temperature have been normalized between 0 (procedure start) and 1 (ablation peak).

slower heating rate near the applicator corresponds to a broader heat expansion, which is essential in treating mid-sized tumors [3]. This information, obtained at the point of treatment, is extremely significant for biomedical engineers to plan the optimal RFA treatment that allows the most controlled heat expansion.

Further work in this direction has been performed by Palumbo et al. from Parthenope University and Federico II University (Naples, Italy) [14], and is depicted in Figure 17.3. In this work, an RF ablation pulsed source (Angiodynamics, Habib 4X Laparoscopic Bipolar Resection Device) having 4 tip electrodes, and controlled by an RF generator, has been modified by inserting 4 FBG sensors, positioned within the center of the electrodes and on their sides and applied for the ablation of liver and kidney porcine phantom. FBGs have been embedded in a metallic needle (Chiba Coloplast, SS7201) with a cyanoacrylate tip; this arrangement is shown in Figure 17.3(a) and allows mimicking the percutaneous penetration of the applicator through the tissue. The arrangement includes 5 FBGs, mounted in 5 different fibers, and controlled by a commercial interrogator (Fibersensing): 4 FBGs are positioned in correspondence to the center of the 4 RF electrodes ($x = 0$ cm), and on its side with 0.5 cm progressive shift (i.e., $x = 0.5$, 1.0, and 1.5 cm), with an additional FBG at $x = -1.0$ cm for temperature referencing. The 4 Chiba needles enclosing the FBGs have been attached to the external side of the RF probe, as in Figure 17.3(b), obtaining a prototype applicator with on-board sensors. Compared to the work proposed in Figure 17.2, in this case sensors are displaced perpendicularly to the ablation electrodes and mounted in different fibers, whereas in the previous work, the sensors were positioned along the fiber axis in a single FBG fiber array. Most notably, Figures 17.3(c, d) show the result of the temperature pattern estimation, performed by extrapolating the thermal data in a three-dimensional thermal map. Each pulse produced by the RF applicator drives the inner temperature, within the electrodes, well over 100°C, while the 60°C threshold for cellular damage extends within 1.0 cm and 1.2 cm in radius from the electrode side. Overall, [14] provides an alternative positioning of FBGs in a multistep RF probe with respect to [13], and shows a useful methodology for extrapolating a thermal map from FBG readouts.

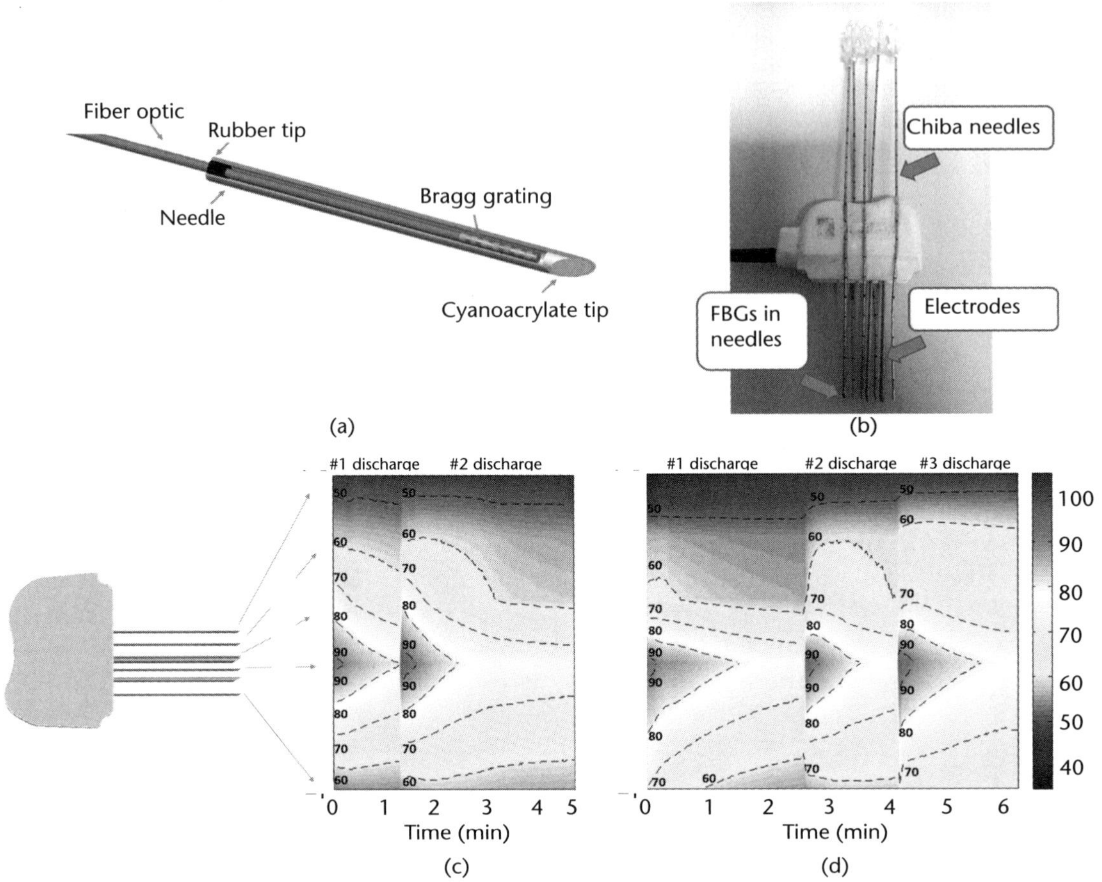

Figure 17.3 Application of FBG sensors in a RF ablation applicator. (After: [14].) (a) FBG package in cyanoacrylate tip. (b) RF ablation probe with FBG sensors mounted on it. (c, d) Temperature profile during two TA procedures in liver phantom.

17.4 High-Density Temperature Sensing

As FBGs have spatial resolution problems, other OFS technologies for distributed and semidistributed sensing are emerging for temperature measurement with narrow, submillimeter, spatial resolution. In particular, two technologies are under the spotlight: chirped FBG [15] and distributed temperature sensors [16].

17.4.1 Chirped FBG

The main investigation on the use of CFBG in thermal ablation has been carried out by Tosi et al. in 2014 [15], using the same setup of the previous section and Figure 17.1 for RFA, but replacing the sensing element with a linearly chirped FBG (LCFBG). The LCFBG used in the experimental activities has a length of 15 mm, chirp coefficient of 2.22 nm/mm, and 94.5% reflectivity. As described in Section 17.5, the FBG is used for semidistributed sensing, detecting variations of the reflected spectrum as a function for the temperature applied in each point of the grating.

The main problem of the LCFBG is the demodulation of the grating spectrum, which is still an open task. In the study, a simple detection method was used, based on computing the ratio between the measured spectrum and the reference spectrum (computed during calibration, prior to RFA start) and further processing to estimate the temperature slope between the high-temperature region and the low-temperature region. This methodology allows an approximate estimation of the temperature variations occurring in each part of the grating, and its nominal spatial resolution (70 μm) is equal to the wavelength resolution of the spectrometer (156 pm) divided by the chirp coefficient. However, this method is effective only for monotonic temperature pattern, as it estimates temperature on the 2 edges of the grating, and then computes the slope between these 2 points. For this reason, it has been applied for longitudinal temperature measurement, whereas the LCFBG tip is positioned adjacently to the tip of the RFA ablator, such that temperature pattern is monotonic; other experiments, reported in [13] were performed along the direction perpendicular to the applicator, maintaining a monotonic temperature pattern.

Figure 17.4 shows a temperature space-time analysis recorded with the LCFBG. The *xy* plane reports the time elapsed after RFA start, and longitudinal distance from the ablation peak (where 0 mm is the hottest point), reported in the 0 to 12-mm range. The high spatial resolution allows performing measurements with a much higher density than FBG arrays, and this allows an estimation of the ablation threshold with great accuracy. Similar to Figure 17.2, temperature quickly rises with a fast quadratic rate in the initial phase and then settles to a linear growth until vaporization is reached. This corresponds to a quick temperature rise until the RFG is disconnected, and then tissue starts cooling. This methodology allows an accurate estimation of the heat propagation into the tissue depth and is an excellent tool to perform an early stop of RFA when small tumors, or small cauterizations, are treated.

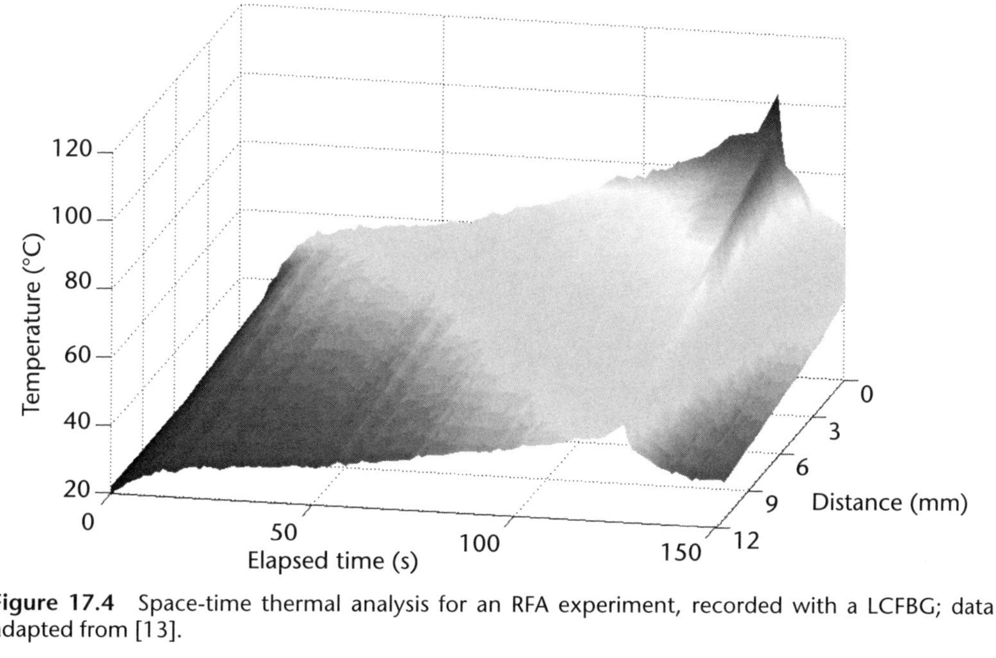

Figure 17.4 Space-time thermal analysis for an RFA experiment, recorded with a LCFBG; data adapted from [13].

Following this investigation, a very recent and early-stage study has been carried out [17], with the intent of demodulating the LCFBG also for nonmonotonic thermal patterns. This task is essential in order to study heat propagation, particularly in LA where the ablation peak tends to slightly move in space during the procedure. The experimental LA setup was based on a solid-state Nd:YAG laser (Echolaser X4, Elesta s.r.l.) emitting 5-W power at 1,064 nm, and operating in continuous wave. The laser beam was coupled into a large-core fiber having 0.3-mm core diameter, frontal-firing. Next to the LA fiber applicator, an LCFBG was positioned; the sensor has 5-cm active length, 50-nm bandwidth (1 nm/mm chirp coefficient), and 92% reflectivity. The optical interrogation setup was the same of the previous experiments, using a spectrometer (Ibsen Photonics, I-MON-512-USB) for detection. In this case, the LCFBG is positioned such that the ablation peak point is positioned close to the LCFBG center, so that temperature gradients can be estimated on both sides of the grating. Figure 17.5 shows a preliminary result, whereas the LCFBG spectrum (as recorded by the spectrometer) is reported at different times during LA, which has a duration of 85 seconds. The chart shows that a noticeable deformation of the LCFBG is obtained, during the heating and cooling phases: The inset shows that the spectrum shifts towards higher wavelengths due to heating, and a spectral peak appears in correspondence to the ablation peak position. Spectral demodulation is an ongoing research task, using an iterative optimization algorithm: preliminary results appear promising, but decoding length is too long (10 to 30 seconds), not compatible with real-time processing.

17.4.2 Distributed Sensing

Only one study has been proposed for DTS application in RFA [16] and expanded in [13]. In this study, a commercial DTS system (Luna OBR4600) has been applied to distributed temperature detection. The system, based on a complex white light interferometry setup, was configured in order to have 0.1–1-mm spatial resolution, and 1-Hz sampling time; a single-mode fiber SMF-28 was used as sensing element.

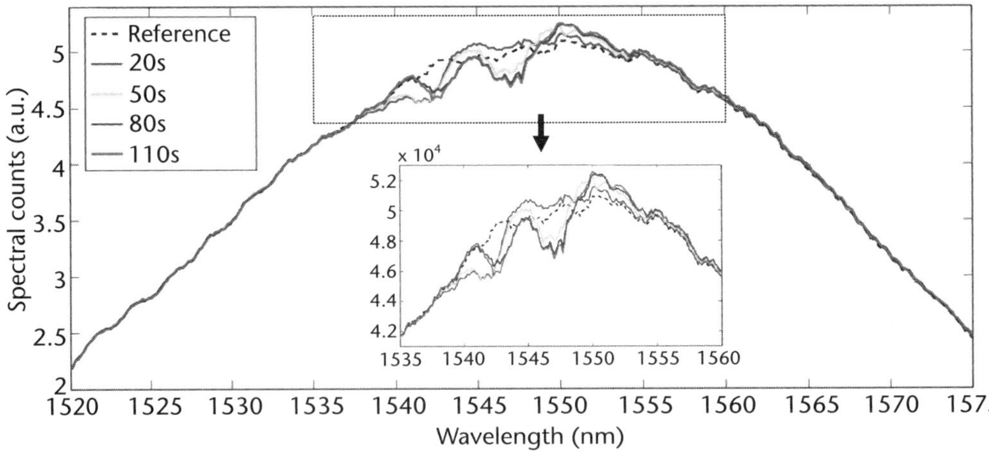

Figure 17.5 Experimental results of an LCFBG sensor applied to LA; the chart reports the LCFBG spectrum (with a zoom on the inner region) recorded by the spectrometer at different times during the ablation process, having a 85-second duration.

Two configurations have been studied: in the first configuration, a chamber containing a liver phantom was prepared, and the sensing fiber was prepared to perform measurement along 8 different directions from the ablation center: this study, reported in [16], aimed at exploring the methodology for controlling the ablation pattern and verify its symmetry on the planar plane, as well as recording thermal gradients in different directions. The study reported in [13] instead resembles the insertion of a fiber probe in vivo, as the sensing fiber is positioned parallel to the RFA applicator.

Figure 17.6 shows the output of the DTS instrument, which is capable of capturing data with narrow spatial resolution, and fast response. In Figure 17.6(a), the analysis has been performed both in the space domain: setting the ablation threshold at 60°C, it is possible to notice that the radius of tissue exposed to ablation is 1.2 cm. This sensing technique, like LCFBG, is very effective in estimating in real time the ablation margin, as it allows tracking the amount of tissue exposed to the 60°C threshold in real time. In Figure 17.6(b), the analysis is carried out in the time domain (normalized), in order to understand how thermal heating evolves at various distances from the applicator. It is possible to show that moving further from the applicator, heating has an almost linear trend, with a rapid heating at the end; instead, closer to the applicator, heating has an initial fast-rising phase that is well highlighted in the chart.

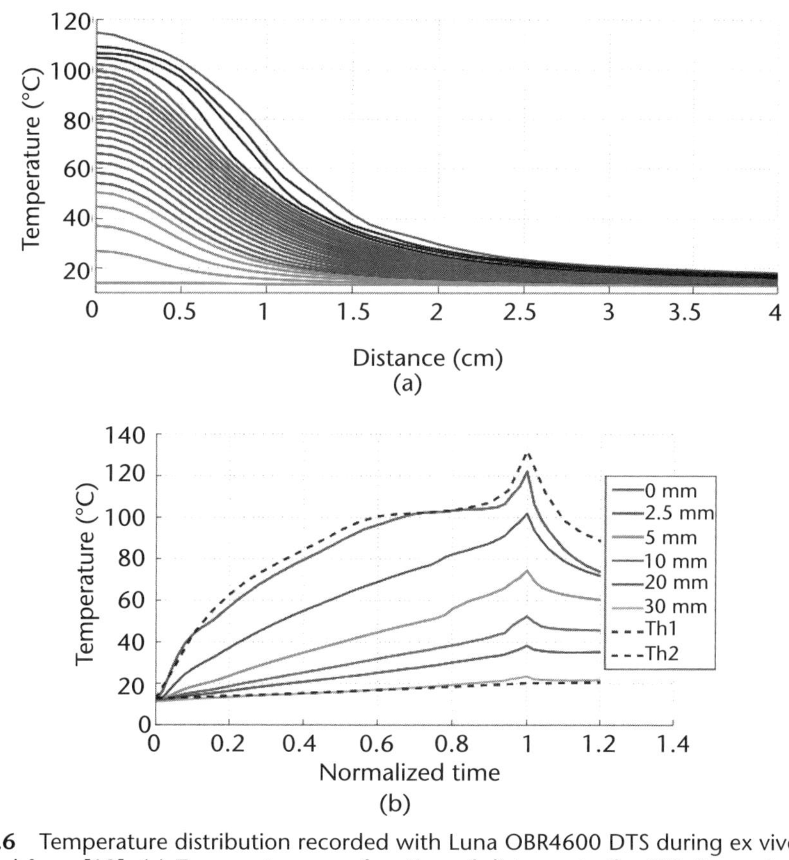

Figure 17.6 Temperature distribution recorded with Luna OBR4600 DTS during ex vivo RFA; data are adapted from [13]. (a) Temperature as a function of distance to the RFA tip; each curve is reported every 10 seconds during heating. (b) Temperature as a function of time, reported at different distances from the tip, and compared to two thermistors (Th1-2).

The study is conclusive in measuring thermal heating in space-time, and is the base for analysis in all TA procedures, obtaining comparative results. Despite the effectiveness, two problems appear in this analysis: (1) as the DTS works with Rayleigh backscattering, it requires a perfectly cleaved fiber end, or a fiber termination; otherwise, the strong backrefletion overlaps the backscattering in the final section of the fiber and this problem is not affecting ex vivo analysis, but may affect in vivo insertions in percutaneous catheters; and (2) a tight trade-off among accuracy, sampling time, and spatial resolution exists, and accuracy may be low (1°C or worse) when the spatial resolution and sampling time are jointly minimized. However, the methodology is an excellent fit for the application, and we envision further work in exploring present and emerging TA techniques with DTS. Our study also demonstrates that, despite the steep spatial gradients, the bandwidth of the DTS detector is not limiting the measurement of temperature gradients in RFA.

17.5 Pressure Sensing

The first pressure measurement was performed ex vivo and in vivo by Kotoh et al., using a MEMS sensor [18]. Due to the thermal cross-sensitivity of the sensor, the measurement was not performed at the point of treatment, but at a 3-cm distance, and therefore it has poor relevance with respect to intraparenchymal pressure. Recently, this measurement has evolved thanks to OFS sensors based on EFPI: most notably, the study reported in [12] reported the intraparenchymal pressure measurement in RFA, while in [19] the pressure measurement has been reported for LA.

Both studies have been performed using an EFPI sensor, similar to the one reported in Section 16.2, developed by the University of Limerick. Besides the standard advantages in terms of size, biocompatibility, and accuracy, this sensing architecture has a tremendous advantage in minimizing the thermal cross-sensitivity, which is the main issue for pressure measurement in TA. The MEMS sensors used in [18] has a thermal error of 0.5 kPa/°C, while most of the all-glass EFPI structures proposed until 2013 had a thermal detuning of 1.7 kPa/°C or higher. Temperature compensation is not an effective strategy in mini-invasive TA: due to the steep spatial gradients, both temperature and pressure values change abruptly in a small space, and temperature sensors (used to compensate pressure, in a EFPI/FBG structure) are positioned in a different location and therefore record a different temperature than the one acting on the pressure probe. The only viable solution is adopting an EFPI probe with a very small thermal cross-sensitivity, and this is an essential requisite. Both proposed studies have therefore been performed with an EFPI probe having sensitivity of 1.6 nm/kPa, and thermal sensitivity of –0.05 nm/°C, therefore with a detuning coefficient of 0.03 kPa/°C, acceptable for TA requirements.

The results of this study is shown in Figure 17.7, reporting the comparison results highlighted in [19]: using both RFA and LA setups previously discussed, an EFPI probe has been positioned to record pressure in different experiments, at different tip-to-tip distance (i.e., the geometrical distance between the EFPI fiber tip and the applicator tip). The results show that in LA procedures, which have a short duration, pressure has an immediate rise and tends to stabilize to a plateau after 50 to 70 seconds, with absolute maximum values of 71 kPa and 48 kPa. Conversely, in RFA pressure remains low for the majority of the procedure and quickly rises when

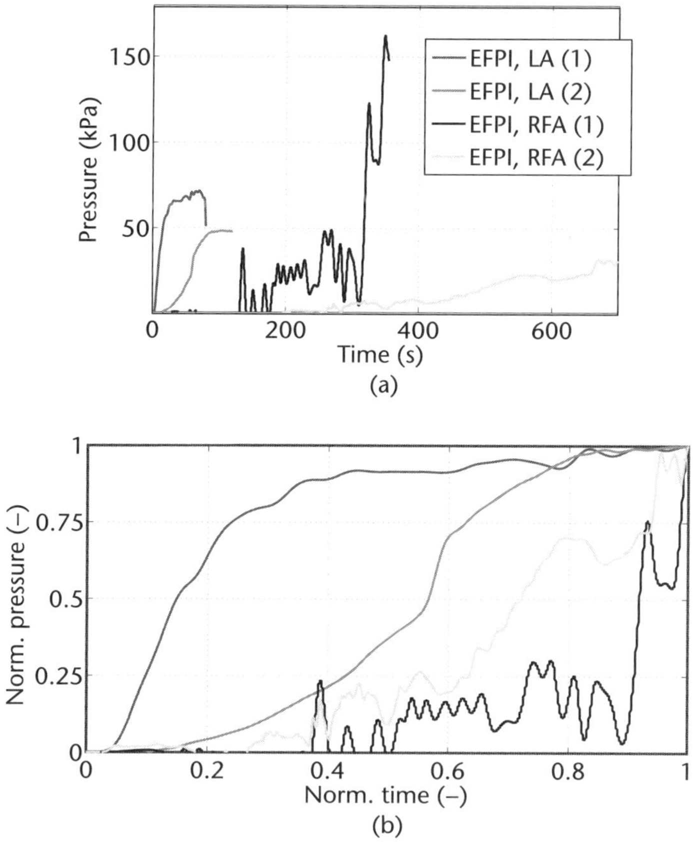

Figure 17.7 Result of the pressure measurement in LA, performed at tip-to-tip distance of 1 mm (1) and 5 mm (2) [19], and in RFA measurement with tip-to-tip distance of 0.5 mm (1) and 1.2 mm (2) [12]. (a) Pressure values versus elapsed time; and (b) normalized values for both pressure and time (0 = ablation start, 1 = ablation peak). (Data adapted from [19].)

the tissue starts vaporizing, reaching the peak value of 162 kPa in proximity to the applicator tip and 32 kPa at a 1.2-mm distance. The preliminary conclusions of the comparative studies are that EFPI probes are a reliable technology for pressure detection in mini-invasive RFA and LA, and that in LA the maximum pressure (that is underlying the scattering recurrences problem) appears to be more predictable, while in RFA pressure rises abruptly only when the tissue starts the vaporization process.

17.6 Concluding Remarks and Future Perspectives

OFS are rapidly emerging for measurements applied to TA, which is itself a fast growing market in medical devices. Until 2013, the application of OFS to TA was a marginal task, but current OFS technologies have really advanced the field, and have established as the state of the art, at least at precommercial level. It is therefore expected that disruptive innovation can enable TA to substantially improve performances, in terms of medical performances, related to 5 to 10 years of lifetime expectancy.

In terms of sensing technologies, we observe that a plurality of technologies (FBG, LCFBG, DTS, EFPI) has been demonstrated and benchmarked, as a measurement tool. Alternatively, no clear winner is emerging from the groundwork. FBGs represent the most overall cost-effective solution, with both sensor and interrogator costing a fraction of the TA equipment, but the spatial limitation is a key issue. DTS has very low-cost applicator, but the interrogation setup is 4 to 12 times more expensive than the TA setup, and limitations to the accuracy/sampling rate and the need for a surface-cleaved fiber appear as key issues. LCFBG have probably the main potential, with the same FBG interrogation setup, and with the possibility to abate the sensor cost by high-volume manufacturing, but a reliable and fast detection technique is still an unsolved task, although preliminary works start to be disseminated [20]. Similarly, no TA technology has strongly overcome the others, for cancer care applications, although RFA and LA start to establish as the main mini-invasive procedure while MWA appears as the main alternative to resection for treatment of larger tumors.

The main task is now to consolidate and commercialize the technologies, achieving a tight integration of TA applicators and OFS in medical catheters that allow also flexible positioning, radio marking, and a tight referencing of the applicator, and its sensors, to the tumor. The integration between OFS developers, medical device companies, and clinicians is hereby essential, because of the extreme characteristics of the working environment of sensors (thermal and pressure gradients, sudden changes of tissue biology, heat propagation).

In contrast, the best rationale for OFS is when they are used not only to monitor TA, but also to control its fundamental parameters, adjusting them over time. This concept, namely smart TA, was first introduced in [13] and is depicted in Figure 17.8. The applicator is a catheter that contains the TA applicator, such as a multitip RFA ablator as in the figure, and a plurality of OFS that allow a capillary rendering of temperature distribution, pressure, and eventually biological parameters (that can discriminate healthy tissue from tumors). Additional sensors can be externally inserted with a microcatheter, to ensure regions external to the tumor are protected from ablation. During TA, a multichannel interrogator can record sensors spectra and estimate measured parameters. Data should be then processed by a controller that in a closed-loop can adjust: (1) the main parameters of the source,

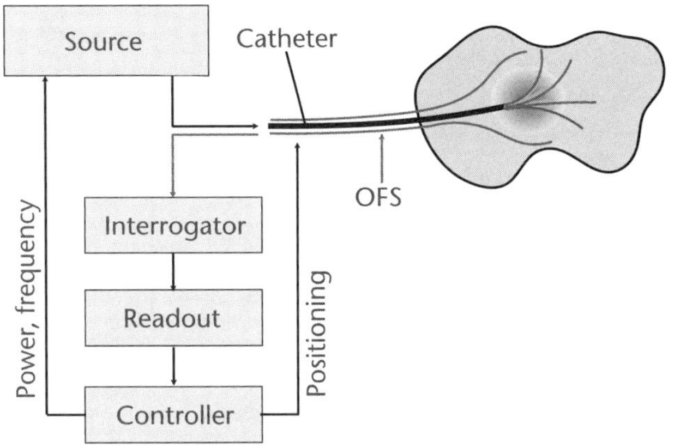

Figure 17.8 Outline of smart-TA concept.

such as frequency and power; and (2) micropositioning of the ablation device, whereas a stereotactic applicator is used. The first feature is currently possible, as many research-grade RFA/MWA applicators run LabVIEW software that controls power and frequency; the second task may be enabled by robotic microsurgery applications, described in Chapter 19, which already mount OFS onboard. The overarching goal is to be able to match with the best possible accuracy the heat field shape with the tumor profile, but also avoid undesired effects such as early tissue vaporization, over-inflation of encapsulated tumors, excessive heat propagation. OFS, particularly the lab-in-a-fiber concept, is a key enabling technology for smart TA.

References

[1] BCC Research, "Ablation Devices: Technologies and Global Markets," HLC163A, 2014.

[2] Rhim, H., et al., "Essential Techniques for Successful Radio-Frequency Thermal Ablation of Malignant Hepatic Tumors 1," *Radiographics*, Vol. 21, Suppl. 1, 2001, pp. S17–S35.

[3] Goldberg, S. N., "Radiofrequency Tumor Ablation: Principles and Techniques," *Eur. J. Ultrasound*, Vol. 13, No. 2, 2001, pp. 129–147.

[4] Simon, C., D. Dupuy, and W. Mayo-Smith, "Microwave Ablation: Principles and Applications 1," *Radiographics*, 2005.

[5] Saccomandi, P., et al., "Theoretical Analysis and Experimental Evaluation of Laser-Induced Interstitial Thermotherapy in Ex Vivo Porcine Pancreas," *IEEE Trans. on Biomed. Eng.*, Vol. 59, No. 10, 2012, pp. 2958–2964.

[6] Tatli, S., et al., "Percutaneous Cryoablation Techniques and Clinical Applications," *Diagnostic Interv. Radiol.*, Vol. 16, 2010, pp. 90–95.

[7] Letfullin, R. R., et al., "Laser-Induced Explosion of Gold Nanoparticles: Potential Role for Nanophotothermolysis of Cancer," *Nanomedicine*, Vol. 1, 2006, pp. 473–480.

[8] Kennedy, J. E., "High-Intensity Focused Ultrasound in the Treatment of Solid Tumours," *Nat. Rev. Cancer*, Vol. 5, 2005, pp. 321–327.

[9] European Union, "FP7-STREP IMPPACT, 2008-2011," http://www.imppact.eu/.

[10] Saccomandi, P., E. Schena, and S. Silvestri, "Techniques for Temperature Monitoring During Laser-Induced Thermotherapy: An Overview," *Int. J. Hyperthermia*, Vol. 29, No. 7, 2013, pp. 609–619.

[11] Inokuchi, R., et al., "Percutaneous Microwave Coagulation Therapy for Hepatocellular Carcinoma: Increased Coagulation Diameter Using a New Electrode and Microwave Generator," *Oncol. Rep.*, Vol. 24, 2010, pp. 621–627.

[12] Tosi, D., et al., "Fiber Optic Combined FPI/FBG Sensors for Monitoring of Radiofrequency Thermal Ablation of Liver Tumors: Ex-Vivo Experiments," *Applied Optics*, Vol. 53, No. 10, 2014, pp. 2136–2144.

[13] Tosi, D., E. G. Macchi, and A. Cigada, "Fiber-Optic Temperature and Pressure Sensors Applied to Radiofrequency Thermal Ablation in Liver Phantom: Methodology and Experimental Measurements," *Journal of Sensors*, Vol. 2015, 2015, p. 909012.

[14] Palumbo, G., et al., "Temperature Profile of Ex-Vivo Organs During Radio Frequency Thermal Ablation by Fiber Bragg Gratings," *J. Biomed. Opt.*, Vol. 21, No. 11, 2016, p. 117003.

[15] Tosi, D., et al., "Fiber-Optic Chirped FBG for Distributed Thermal Monitoring of Ex-Vivo Radiofrequency Ablation of Liver," *Biomedical Optics Express*, Vol. 5, No. 6, 2014, pp. 1799–1811.

[16] Macchi, E. G., et al., "Optical Fiber Sensors-Based Temperature Distribution Measurement in Ex Vivo Radiofrequency Ablation with Submillimeter Resolution," *Journal of Biomedical Optics*, Vol. 19, No. 11, 2014, p. 117004.

[17] Tosi, D., et al., "Towards Inline Spatially Resolved Temperature Sensing in Thermal Ablation with Chirped Fiber Bragg Grating," *IEEE International Symposium on Medical Measurements and Applications (MeMeA)*, Benevento, Italy, May 15–18, 2016.

[18] Kotoh, K., et al., "Evaluation of Liver Parenchymal Pressure and Portal Endothelium Damage During Radio Frequency Ablation in an In Vivo Porcine Model," *Liver Int.*, Vol. 25, 2005, pp. 1217–1223.

[19] Tosi, D., et al., "Intra-Tissue Pressure Measurement in *Ex Vivo* Liver Undergoing Laser Ablation with Fiber-Optic Fabry-Perot Probe," *Sensors*, Vol. 16, No. 4, 2016, p. 544.

[20] Gassino, R., et al., "Toward the Development of a Distributed All-Fiber Temperature Sensor for Biomedical Applications," *IEEE International Instrumentation and Measurement Technology Conference (I2MTC)*, Taipei, 2016.

CHAPTER 18

Application and Detection of SPR Sensors

18.1 Genesis and Present Status

Our interest is focused on biosensor technologies, that are capable of detecting biological concentrations or analytes in real time, by means of optical fiber sensors technologies, much like pressure or temperature measurements are performed. In this framework, the maturity of this research topic is developing, but the strong technological premises and the advent of SPR and TFBG is providing traction to this potentially disruptive field. In terms of applicability of the FOS technologies, a determinant factor is how the sensing system is functionalized to the detection of a specific analyte. Refractive-index sensors are effective, despite exhibiting a certain cross-sensitivity and a thermal response not negligible, but the real breakthrough occurs when the sensing structure is functionalized to a specific biorecognition, through a binding process.

In this context, the true status of this application field with a huge potential was described in the feature article by Albert in 2014 [1], which reports the current status and trend regarding the lab-on-fiber concept. This concept appears as the promising continuation of the optofluidics-based lab-on-chip solutions that make use of miniature and well-sized designs of fluidic channels that collect samples of liquid drops (such as blood) and drive them through a chemical reagent that binds with some of the liquid constituents, such as glucose, antibodies, or other biomarkers, converting then this chemical reaction into a current or voltage drop, detectable on the chip. As much as this process is an acquaintance in the analysis of chemical substances, and can be performed in several contexts, when looking at biomedical applications, the procedure of biodetection is inherently tied to the collection of samples. The analysis of blood and organic substances is performed by collecting the sample with an invasive method, and subsequently analyzing it, using methods such as the lab-on-chip. Tissues are instead obtained through a biopsy procedure, more invasive, that requires the collection of samples and their subsequent analysis. The whole procedure of collection and analysis requires a time within 20 minutes and several days.

FOS, compared to other optical, electrical, or electrochemical solutions, can provide advantages on two different levels. At the first conservative level, corresponding to the solutions reported so far, FOS technologies have been proposed as an alternative to other technologies and require the collection of the sample (liquid or tissue) from a patient. Biosensors then perform the measurement of analyte in a subsequent time. The advantages of FOS with respect to alternatives is mainly performance-based: FOSs typically are more sensitive to variations of analyte constitutive parameters than other technologies, and due to the possibility to enable WDM, it is possible to have multipoint and/or compensated structures [2] that operate at any temperature. Compared to electrical sensors, they also have fewer problems in operating with fluids. The improvement of resolution in terms of refractive index detection, and the chase for the smallest variations of RIU, is essential for the most advanced biosensor applications, including the search for cancer biomarkers that is becoming a priority [3]. In the past 5 years, solutions based on SPR, either directly induced in the fiber or supported by fiber tapers or TFBG, are dominant in terms of performance and ease of functionalization [4].

At a second breakthrough level, FOS technologies can perform real-time continuous measurements, a feature unmatched by other technologies. This characteristic, when combined to the other precious advantages of the technology, can be the game-changer. At first, we can detect not only biological concentrations, but also their change over time: this opens the door for imaginative applications in personalized drug delivery, where the chemical effect of drugs and effectiveness of the delivery process can be measured in real time. Furthermore, cutting-edge applications, such as the detection of cancer stem cells, require the detection of a very small amount of cells, which are not always detectable at the sensor location, and only a long-term monitoring in vivo can be effective. The applicability of FOS technologies to long-term biosensors is a research field still in the early stage, despite the latest investigations show promising results in terms of working principles. A weakness of this approach is the catheterization, which is required to position the sensors in situ and to allow both fiber protection and binding to the functionalized fiber; as discussed in Chapter 14, to date, a FOS catheter for long-term, in vivo analysis has not been documented yet, most measurements based on FOS are more similar to a biopsy then to a real-time operated catheter described in Chapters 15 to 17. In this chapter, we will discuss the latest applications of SPR-based FOS to detection of cancer biomarkers, thrombine, and glutaminase. All the works reviewed in this chapter make use of the SPR technology, either fabricated in the fiber or in conjunction with a TFBG structure, which appears to outperform alternative approaches in terms of sensitivity.

18.2 Lung Cancer Biomarker Detection

The structure based on TFBG-SPR has been consolidated within the last 3 years and is now appearing in the first applicative contexts (particularly by the research groups at University of Mons, Belgium, and Carlton University, Canada). The work reported by the first group [5] provides a key insight on the attempt for early-detection of lung cancer through protein-based biomarkers; in this research effort, fiber-optic specialists have interacted with biologists for the surface chemistry and biosensor functionalization.

The principle of operation of the sensor is the SPR assisted with a TFBG. The TFBG having 1-cm length and tilt angle between 7° and 9° has been fabricated using a tilted phase mask and an argon-ion 244-nm laser. Subsequently, the fiber surrounding the TFBG has been sputtered, using 99.99% gold sputtered with a vacuum sputter coater (Leica EM SCD500), followed by a high-temperature annealing process. The use of a TFBG+SPR principle results in a spectrum different from standard TFBG: this is because the SPR dip intercepts a small subset of TFBG cladding modes, which appear to be the most sensitive ones. This arrangement improves the sensitivity, because while SPR is characterized by a broad dip, the bandwidth of each TFBG cladding mode is much smaller, and then the SPR spectrum itself acts as a modulator on the narrow peak. In addition, it requires the use of a detector with a narrow resolution: the authors used a Luna Optical Vector Analyzer, which interrogates modulus and phase of the sensor spectrum with 1.6-pm resolution, but at the same time has cost well over one order of magnitude higher than FBG interrogators and spectrometers described in Chapter 10. The sensor has been designed for the detection of cytokeratin 7 (CK7) full protein and CK7 peptide. The functionalization, described in depth in the article, is based on a standard procedure for biosensors, and the authors have implemented a well-established bioreception methodology based on antigen/antibody. ELISA (enzyme-linked immunosorbent assay [6]) has been used to study the CK7 binding process. The CK7 antigen detection by the bioreceptor is based on the specific chemical reaction with its corresponding antibody (AbCK7).

Measurements have been performed at different concentration of CK7 in phosphate buffer saline solution, in order to have different variations of refractive index due to the CK7/AbCK7 interactions. The functionalized fiber has been immersed in the solution, detecting the spectrum every 30 seconds; the analysis is entirely focused on the most sensitive SPR mode, described in [4], measuring the amplitude at the dip wavelength. The authors observe a decrease of optical intensity of 1.0 dB when CK7 concentration increases from 1 ng/mL to 10 mg/mL for low-concentration detection, while the sensor appears to saturate over a 100-mg/mL concentration, mainly due to the low variation of refractive index over these values. The work proposed by University of Mons is a milestone in the application of SPR biosensors for cancer care: CK7 proteins are used in the early-stage detection of lung cancer. CK7 peptides, having the size of a few nanometers, have been detected through the binding process, despite their small size, confirming that once the functionalization has been achieved, the SPR platform is a sound and robust method for detection. In terms of concentration sensitivity, the obtain results are close to 1 ppm (part per million), which is a good accuracy for a SPR-based sensor, with the potential to be increased improving the detection software.

18.3 Thrombin Detection

The case study of thrombin detection has been investigated by University of Porto [7], through an aptamer-based SPR biosensor, and by Carleton University, through a TFBG-SPR biosensor [8]. The case-study of thrombin is a benchmark for the application of biosensors in the diagnostic of cardiovascular diseases, since ischemia and cerebral aneurism tend to release significant concentrations of thrombin. The challenge for biosensors in this area is the early-stage evaluations of cardiovascular risk factors: thrombin is directly implicated in vasoconstriction and is a possible biomarker for the atherosclerosis and ischemia diseases, besides being involved in blood clots.

The first system, reported by Coelho et al., is shown in Figure 18.1 and is based on a thrombine-binding aptamer (TBA); the process of protein purification and aptamer identification is reported in the work and shown in Figure 18.1(a). After identification of the thrombin-binding aptamer (TBA), the sensor has been fabricated with a process similar to [8]. A single-mode fiber has been treated with etching along a 1-cm longitudinal section in order to remove the cladding; then, a thin film (16 nm) of gold was deposited surrounding the core, and followed by a 100-nm layer of titanium dioxide. After the SPR structure has been fixed, the sensor has been incubated to activate the TBA. The characterization of the sensor shows that the SPR wavelength shift is 3.5 nm when analyte concentration goes from unbound to 10 nM (equal to 10 mol/dm^3), and a further 2-nm shift is observed when the concentration reaches 50 nM, after which the sensor saturates. As a very interesting feature, the authors also developed, with the similar binding principle, a sensor based on long-period grating (LPG) for the same application, comparing the performance with the SPR probe: the sensitivity of the SPR, evaluated in terms of wavelength shift per variation of analyte concentration, is much superior, and the behavior of the sensor is more accurate. The estimated resolution of the SPR sensor is 0.54 mM.

A similar sensor was proposed by Shevchenko et al. [9], but using the TFBG-SPR platform with a tilt angle of 10°C and a thin film of gold with 50-nm thickness. Thrombin aptamer has been identified and then immobilized using a standard procedure. The work carried out by the authors is based on human alpha-thrombin and protein buffer, whose adhesion to the SPR layer has been verified under confocal microscopy. As described in Section 18.2, the SPR acts as a modulator for one cladding mode of the TFBG, which is then monitored in amplitude: the authors

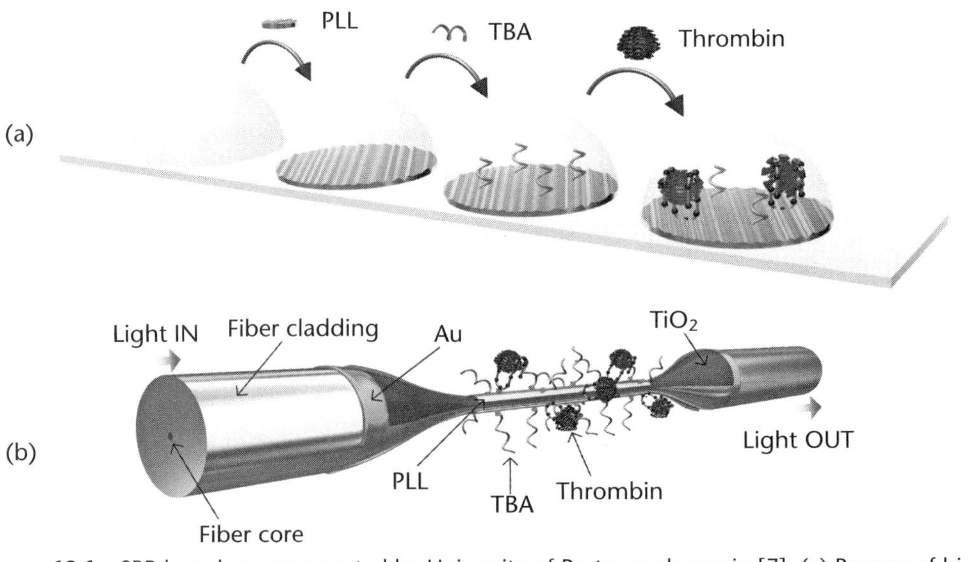

Figure 18.1 SPR-based sensor reported by University of Porto, as shown in [7]. (a) Process of biosensor binding, in 4 steps: starting from a clean surface, the surface is functionalized with poly-L-lysine (PLL), then incubated with thrombin-binding aptamer, which is capable of binding to thrombin. (b) Schematic of the SPR sensor functionalized: etched single-mode fiber coated with Au and TiO2.

thus monitor the signal change as a function of thrombin concentration. The sensitivity has an inferior value to the previous work, as the signal variation is approximately 0.02 dB for a variation of protein level from 100 nM to 5,000 nM.

18.4 DNA Detection

The work proposed by Pollet et al. [10] shows a first development of an immunosensor based on fiber-optic SPR, and functionalized for DNA hybridization and DNA-protein interactions detection. The sensing system is based on a standard single-mode fiber, subsequently sputtered with a 50-nm gold layer, which was then covered with a protein repulsive self-assembled monolayer of mixed polyethylene glycol. In the first part of the work, the authors described how ssDNA (single-strand DNA) is immobilized on streptavidin coated SPR probes to detect complementary ssDNA in the concentration range from 500 nM to 5,000 nM. Noncomplementary DNA induced no significant SPR resonance shift, hence demonstrating the low cross-sensitivity of the sensor. In this part of the work, the authors obtained an almost linear calibration curve, with a sensitivity of 140 pm/nM, in terms of SPR wavelength shift per variation of analyte concentration. In the second part of the work, pointing in the direction of immunosensing the fiber-optic SPR biosensor was functionalized with ssDNA aptamers against human immunoglobulin E (hIgE) [11]. In this case, the SPR exhibits a nonlinear calibration curve in the range 0 to 500 nM, with a maximum wavelength shift of 3 nm.

18.5 Celiac Disease Diagnostic

The work presented by National Research Council, Naples, Italy [12] showed the development of an SPR sensor on a POF fiber, and its application to the diagnostic of celiac disease. Also, this sensor is developed according to the antigen/antibody principle, but its implementation is performed on a larger fiber. As the POF is a large-core, strongly multimode fiber, approaches based on gratings are not effective; hence, the authors focused on direct SPR on a layer of fiber; the POF used in experiments is a standard 980-μm core PMMA fiber, which is embedded in a resin layer to facilitate the layer deposition. In order to obtain a good wavelength selectivity, the optical sensor system was realized by removing the cladding of a plastic optical fiber along half the circumference, spin coating on the exposed core a buffer of photoresist (Microposit S1813), and finally sputtering a thin gold film using a sputtering machine. This process was first described in [13].

The POF sensor proposed by CNR targets the diagnostic of celiac disease, performed through the analysis of trans-glutaminase (tTG) present in concentration ranging from 0.4 nM to 3 mM. The binding process is performed through the anti-tTG antibody. The SPR functionalization has been performed using guinea pig tTG, through two-step deposition of droplets of a protein solution of 1 mg/mL. The obtained result is that the SPR wavelength dip has an almost linear dependence on the antibody concentration within 3 nM to 3,000 nM, when expressed in logarithmic units: the SPR wavelength changes is equal to 4.3 nm/log(Ab concentration) in this setup, which is a relatively large sensitivity. The system does not exhibit a very high linearity (0.97%). The sensitivity is slightly inferior to the previous works, but it is consistent with the fact that the fiber used is a large-core POF, operating in strong multimode regime. In terms of applications, due to the easier light coupling, the favorable Young's modulus, and the lower constraints in terms of fiber protection, the work proposed by CNR opens important perspectives in terms of fiber catheterization and real-time sensing.

18.6 Perspectives

In agreement with [1], FOS biosensors appear as a key emerging technology. SPR is the key principle of operation, either as a stand-alone method or coupled with TFBG or tapered fiber, the first solution being more attractive for applications as tapered fibers are extremely fragile and hard to handle. SPR and TFBG-SPR can be considered as the golden standard and the platform on which biologists and surface chemistry specialists can operate the functionalization and the bioreception. In these approaches, a key advantage, particularly compared to in situ methods such as BiaCORE machines (www.biacore.com/lifesciences/index.html), is the easiness to couple light from the source and to the spectrometric detector; the complexity of light interaction on the boundaries of the analyte is managed by the fiber-optic probe. The antigen/antibody approach to biosensors, using the ELISA method first established in the 1970s [6, 11], can be combined with SPR, thanks to the fact that gold and silver layers, which are the most used thin film constituents in SPR, are also suitable for functionalization. Also, studies as [2, 14] show how it is possible to use WDM to obtain compensated or multisensitive probes, overcoming several performance barriers. To date, notably, the SPR-based sensors described in this chapter are well competitive with other SPR methods, such as the Biacore instruments class, in terms of performance and cost.

Application-wise, biosensors are several steps behind physical sensors, particularly because of the complications in catheterizations. Despite several authors [5, 9] having performed in situ measurements, the sensors yet do not perform as real-time systems, and biopsy is implicitly needed to obtain samples of measurands. Whereas in previous chapters we documented sensors that effectively operate in vivo inserted in patients via anatomic channels or percutaneously (or in endoscopic/laparoscopic way) and up to several hours or days, using catheters that allow fiber insertion, positioning, operation, and removal with minimum invasiveness and negligible or compensated artifacts, for biosensors this achievement is only starting to appear on the radar. In a near-future perspective, SPR biosensors can become a tool for early-stage cancer detection with the possibility to outperform other competing technologies, on the pure basis of sensitivity. However, the hardest task that requires searching for low concentration of recurrent cells would require a long-term monitoring platform, which can be enabled only by catheters. Designing catheters for fiber-optic SPR probes, which are extremely fragile and tend to suffer from damages to the thin coating, require a correct positioning and contact with the measurand, is a priority for elevating SPR FOS as a biosensing standard. These considerations are in line also with the latest studies from Gupta et al. [15] and Sharma et al. [16], which show a comprehensive outline of the fabrication of SPR probes, particularly in terms of sensitivity to analyte as a function of the metallic layer thickness, and SPR materials.

References

[1] Albert, J., "How We're Shrinking Chemical Labs onto Optical Fibers," *IEEE Spectrum*, 2014.

[2] Fallauto, C., et al., "Compensated Surface Plasmon Resonance Sensor for Long-Term Monitoring Applications," *IEEE Transactions on Instrumentation and Measurement*, Vol. 63, No. 5, May 2014, pp. 1287–1292.

[3] Sawyers, C. L., "The Cancer Biomarker Problem," *Nature*, Vol. 452, No. 7187, April 3, 2008, pp. 548–552.

[4] Caucheteur, C., V. Voisin, and J. Albert, "Near-Infrared Grating-Assisted SPR Optical Fiber Sensors: Design Rules for Ultimate Refractometric Sensitivity," *Optics Express*, Vol. 23, No. 3, February 9, 2015, pp. 2918–2932.

[5] Ribaut, C., et al., "Small Biomolecule Immunosensing with Plasmonic Optical Fiber Grating Sensor," *Biosensors and Bioelectronics*, Vol. 77, March 15, 2016, pp. 315–322.

[6] Engvall, E., and P. Perlmann, "Enzyme-Linked Immunosorbent Assay, ELISA III. Quantitation of Specific Antibodies by Enzyme-Labeled Anti-Immunoglobulin in Antigen-Coated Tubes," *The Journal of Immunology*, Vol. 109, No. 1, July 1, 1972, pp. 129–135.

[7] Coelho, L., et al., "Aptamer-Based Fiber Sensor for Thrombin Detection," *Journal of Biomedical Optics*, Vol. 21, No. 8, August 1, 2016, p. 087005.

[8] Coelho, L., et al., "Sensing Structure Based on Surface Plasmon Resonance in Chemically Etched Single Mode Optical Fibres," *Plasmonics*, Vol. 10, No. 2, 2014, pp. 319–327.

[9] Shevchenko, Y., et al., "In Situ Biosensing with a Surface Plasmon Resonance Fiber Grating Aptasensor," *Analytical Chemistry*, Vol. 83, No. 18, August 22, 2011, pp. 7027–7034.

[10] Pollet, J., et al., "Fiber Optic SPR Biosensing of DNA Hybridization and DNA-Protein Interactions," *Biosensors and Bioelectronics*, Vol. 25, No. 4, 2009, pp. 864–869.

[11] Engvall, E., and P. Perlmann, "Enzyme-Linked Immunosorbent Assay (ELISA) Quantitative Assay of Immunoglobulin G," *Immunochemistry*, Vol. 8, No. 9, September 1, 1971, pp. 871–874.

[12] Cennamo, N., et al., "An Innovative Plastic Optical Fiber-Based Biosensor for New Bio/Applications. The Case of Celiac Disease," *Sensors and Actuators B: Chemical*, Vol. 176, January 31, 2013, pp. 1008–1014.

[13] Cennamo, N., et al., "Low Cost Sensors Based on SPR in a Plastic Optical Fiber for Biosensor Implementation," *Sensors*, Vol. 11, 2011, pp. 11752–11760.

[14] Sciacca, B., and T. M. Monro, "Dip Biosensor Based on Localized Surface Plasmon Resonance at the Tip of an Optical Fiber," *Langmuir*, Vol. 30, No. 3, January 15, 2014, pp. 946–954.

[15] Gupta, B. D., and R. K. Verma, "Surface Plasmon Resonance-Based Fiber Optic Sensors: Principle, Probe Designs, and Some Applications," *Journal of Sensors*, August 11, 2009.

[16] Sharma, A. K., R. Jha, and B. D. Gupta, "Fiber-Optic Sensors Based on Surface Plasmon Resonance: A Comprehensive Review," *IEEE Sensors Journal*, Vol. 7, No. 8, August 2007, pp. 1118–1129.

CHAPTER 19

Minimally Invasive Robotic Surgery

19.1 Robotic Microsurgery and Its Feedback

Starting from the 1980s, the scientific community posed great interest in robotic surgery, with the intent to partially or totally replace, or complement, manipulation of tissues performed by a manual operator with a robotic system. Progressing up to date, robotic surgery has evolved towards the microsurgery frontier, whereas a set of minimally invasive surgical tools are used for tissue manipulation at the microscopic scale [1]. In minimally invasive robotic surgery (MIRS), a robot with high precision is used to manipulate the tissue, using tools that mimic the operations (tissue grabbing, manipulation, removal, cauterization) of a surgeon. The precision of robotic tools is superior to a surgeon, who is limited by hand tremor, in terms of positioning and force control. For this reason, MIRS provides advantages over traditional surgery in terms of minimization of incision size and bleeding, better aesthetic outputs, and an overall slight reduction of hospitalization stay. Currently, many clinics around the world are starting to integrate their surgical tools with robots, and it is expected that the worldwide market for robotic surgery will be $20.8 billion by 2024 (Grand View Research, Inc. Report ID: 978-1-68038-231-0).

A MIRS approach is a master-slave system: the surgeon operates a master system, located remotely with respect to the patient, which controls a slave manipulator that combines one or more tools. The connection between slave and master is typically performed in an electrical way, separating the actuators. This allows an almost complete separation between the tissue manipulator, and the front control panel operated by the surgeon. Indeed, a key advantage of MIRS is the fact that the surgeon operates remotely. In a very first and imaginative demonstration dating to 2002, Marescaux et al. [2] demonstrated the tele-operated surgery concept, by successfully performing a cholecystectomy (i.e., a surgical removal of the gallbladder). In this work, surgeons based in New York performed an operation on a patient in Strasbourg, at a 7,000-km distance. The surgeon operated with a dedicated console, featuring two calibrated arms and an imaging system; surgeon's action were converted into a data stream, transmitted with small delay (155 ms) through a high-speed fiber-optic telecommunication system to the receiver side. Each datum was translated into a control for two robotic arms, featuring microactuators for positioning and tissue manipulation, and a third arm for an endoscopic camera.

More recently, tele-operated remote surgery came to media attention, and an article by BBC [3] showed the clinical prospects for tele-operated surgery in Canada. As MIRS enables tele-operated surgery, with a master-slave approach, several organizations have invested towards its development, not without conflicts of ethics. Public (NASA) and private (SpaceX, Virgin Galactic) spaceflights carriers have scouted MIRS as a key enabling technology for long-haul human space travels to outer space. It is known also that the U.S. Army has largely invested in remote MIRS for military healthcare. In addition, several countries have run feasibility tests for MIRS with the aim of extending hospital-level healthcare to remote villages.

The robot described and used in both [2, 3] is the ZEUS Robotic Surgical System, manufactured by the U.S.-based Computer Motion with a price close to $1 million and approved by the FDA in 1994. Later, the ZEUS was progressively replaced by the da Vinci robot, manufactured by Intuitive Surgical for a cost close to $2 million, which is currently almost unanimously considered the state of the art. Figure 19.1 shows pictures of the da Vinci robot, both the console (surgeon's side) and the manipulator (patient's side).

Aside from the cost and tele-health considerations, most surgical operations carried out by commercial robots can be also performed by trained surgeons, using manual instruments for minimally invasive treatments. However, the overarching goal is to be able to exploit MIRS to access surgical interventions that are not accessible, or can cause significant risks, due to physical limitations of surgeon's performance (hand tremor, fatigue, tactile perception) and technical limitations of surgical instruments. In order to really advance surgery, we would need to be able to provide a feedback to the surgeon that exceeds the standard imaging control. This is the scenario where FOSs play an important role: estimating contact forces, on the tip of a surgical instruments and on its sides; estimating contact forces between the instrument and the tissue; and estimating the bending effects of the surgical tool into position. These tasks can be barely performed with any non-FOS sensing technology, due to the extreme limitations of the sensing environment and the requirements for small diameter and lightweight onboard of a MIRS instrument. Provided that advances in microactuators for positioning and manipulation are effectively able to outperform manual operations, MIRS is still limited by the lack of a haptic feedback: the surgeon cannot sense, with an unambiguous and physically

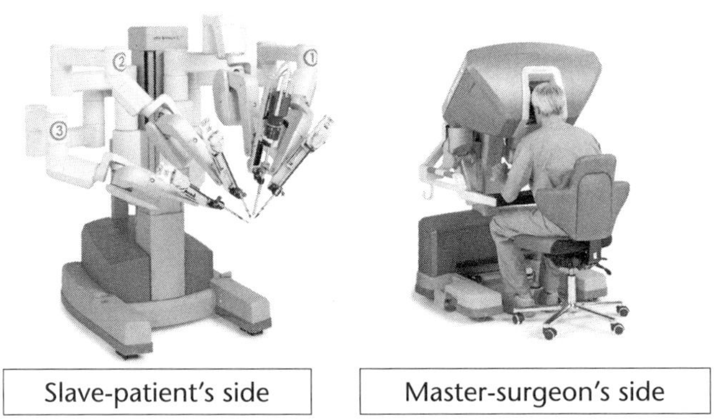

Figure 19.1 Da Vinci robotic system (Intuitive Surgical). Left: manipulator, based on up to 4 robotic arms with microactuators and cameras. Right: console for surgeon's operation and control.

interpretable way, the effect of the tool on the tissue and the contacts between tools and tissue. All this information is intuitive for well-trained surgeons and cannot be replaced by imaging. In this chapter, we will focus on the FOS systems that enable advanced robotic microsurgery. We will largely focus our effort on few case studies, under investigation and very active in terms of research: vitreoretinal surgery (VRS) and IBS sensors.

19.2 Vitreoretinal Surgery: Case Study

Vitreoretinal surgery (VRS) is one of the main case studies for MIRS, based on FOS. MIRS-VRS involves manipulation of eye tissues, usually involving the retina or the macula, using robotic surgical tools that estimate the contact forces on the tip and on the side of the tool shaft, usually below the threshold of human perception. The major constraint in VRS is the size and weight of sensing system, which is mounted on-board of the surgical tool inserted in the eye: the overall dimension of the tool should not exceed 0.9 to 1.0 mm in diameter, including the shaft, the tool functionality, and sensors. Also, accurate force detection is needed: Gupta et al. [4] estimated the human sensory feedback as ~7.5 mN, while MIRS system is required to operate below the millinewton (mN) threshold [5]. VRS is a key case study for MIRS, as the requirement for minimally invasive instruments is the most demanding, contact forces are below sensory threshold, and the bending of the surgical instruments needs to be tracked. VRS is also a risky procedure, often performed in critical cases, whereas retinal detachment or malfunctioning cannot be treated with alternative procedures. The largest advances in FOS for robotic surgery have been implemented in the area of VRS by Johns Hopkins University (JHU), Laboratory for Computational Sensing and Robotics, mainly working on an eye phantom specifically developed for the experiments in membrane peeling and sclerotomy [6]. Systems operated at JHU have been mounted on a Steady-Hand Eye Robot (SHER).

19.2.1 Tool-Tip FBG Force Detection

Figure 19.2 shows the principle and positioning of the tool developed at JHU for tool-tip contact force estimation with SHER [7]. The system is integrated on the tool shaft, and is operated inside the eye. The tool is a 2-degree of freedom (2-DOF) operator, which embeds a micropick on the tip that allows grasping small amount of tissue. Force sensing is performed with 3 FBG sensors, mounted around the shaft. Here each FBG acts as a cantilever beam: when a transversal force is applied in one direction, it induces a strain on the correspondent FBG. Having three FBGs allows, as explained in [6], to extract the forces on x and y direction. In first place, the common-mode strain (and the temperature effect) is eliminated by removing the average wavelength shift from each FBG reading. The remaining wavelength shift for each of the 3 FBGs, hereby called $\underline{\Delta\lambda} = [\Delta\lambda_1, \Delta\lambda_2, \Delta\lambda_3]^T$ is linearly related to the force array $\underline{F} = [F_x, F_y]^T$ by means of a transfer matrix $\underline{K}$: $\underline{F} = \underline{K} \times \underline{\Delta\lambda}$. Since all FBGs are fabricated with the same technology, they have the same strain-optic and thermo-optic coefficients; furthermore, the colocation of the FBG into the eye ensures that they are held at the same temperature. In this arrangement, the transfer

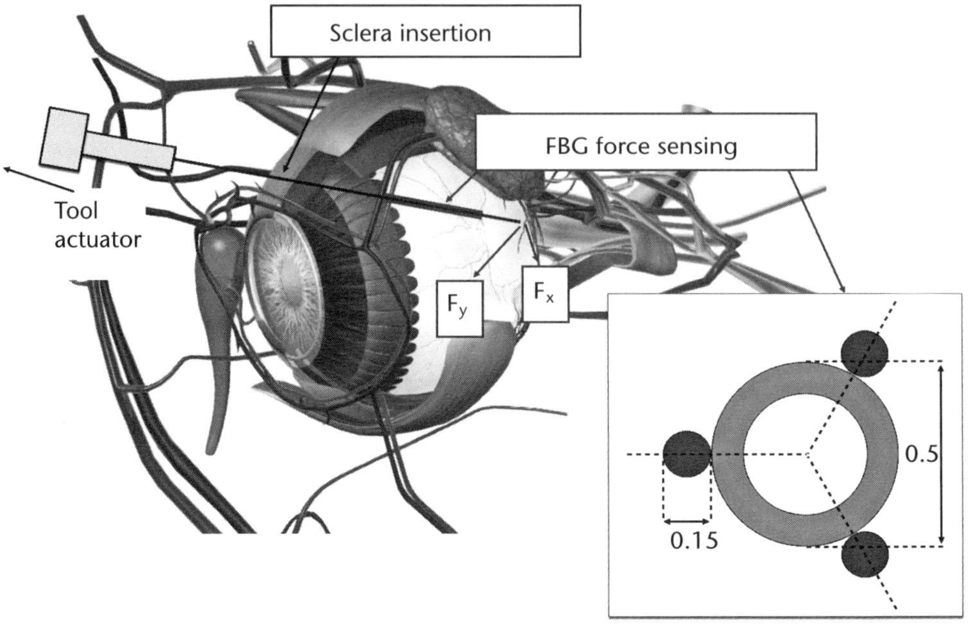

Figure 19.2 2-DOF force-sensing tool, mounted on a micropick. The tool is designed to detect two-dimensional forces with 3 FBGs, detecting forces as a cantilever beam. The pick is inserted through the sclera, in a neutral position, and penetrates through the eye and reaching the retina. FBGs are positioned inside the eye and detect forces acting on the plane perpendicular to the tool axis. The inset shows the tool layout (cross-sections), with dimensions in millimeters.

matrix can be calibrated in a phantom, applying different transversal forces in both directions and therefore estimating the 6 coefficients of the transfer matrix.

The positioning of the FBGs with respect to the tool is also essential. The pick shaft has 0.5-mm tubular radius, compatible with a dimension of <0.9 mm as required for eye surgery. Penetration is performed through the sclera, in a neutral position. The FBGs have a 1-cm length, which allows all sensing elements to be fully positioned within the eye (which has approximately 2.4 cm vertical length). FBGs have 0.15-mm diameter, inclusive of the outer protective jacket, and they are located in proximity of the distal tip in order to maximize sensitivity. The force detection method has a sub-mN estimated accuracy.

19.2.2 Multi-FPI Force Detection

A step forward in terms of force sensing capability for VRS has been documented in [5]. In this case the sensing system is based on Fabry-Perot multiple cavities fabricated directly at the distal tip of 3 fibers, positioned at 120° symmetry as in the previous tool of Figure 19.2. Apart from the FPI interrogation, which is performed with optical coherence tomography (OCT) rather than with spectrometric technique, the two aspects of interest for this design are: the possibility to perform a three-dimensional force estimation, introducing the force detection also on the z direction, parallel to the tool shaft axis and perpendicular to the xy plane of the previous 2-DOF tool; and the embodiment of the three sensors around the tool shaft, in a Nitinol flexure to which 3 fibers are connected.

The principle of estimation of forces is similar to the previous methodology, extended to lateral forces, which are estimated through the analysis of common-mode strain. In this case, the paper still demonstrates a linear dependence, through a 3 × 3 transfer matrix, between a linear combination of each sensor readout (in terms of FPI cavity displacement variation) and the forces on each axis. The sensitivity of this approach is higher than for FBG estimates, as the cavity displacement variation is in the order of 6 nm/mN for axial calibration and up to 27 nm/mN for lateral forces. In performance term, it would be significant to study the response of the FPI interferometric system using spectrometric detection and particularly in the cepstrum domain as described in Chapter 11.

19.2.3 Tool-Shaft and Transverse Force Detection

In [8], a system for dual force sensing with FBG sensors is presented. The force-sensing tool is based on 3 arrays of 3 FBGs each, each having 3-mm length; the whole tool has 45-mm length. The tool has a helicoidal shape, and the fibers (0.2-mm diameter) are packed within the tool, reducing the tool diameter to 0.5 mm, well below the prior state of the art. Figure 19.3 shows the tool positioning within the eye. All FBG arrays have identical layout and dimensions and are positioned at 120° symmetry as in the previous tool. Each array consists of a distal FBG, positioned near the fiber tip inside the eye, and two FBGs positioned at 22-mm distance from the distal FBG and with a 4.5-mm distance center to center. In this arrangement, the distal FBG is used to detect transverse forces, with an approach similar to the previous arrangement, and including the common-mode compensation. The key innovation consists in the tool-shaft force estimation, which allows the estimation of sclerotomy location (i.e., the access point on the sclera and its lateral contact force); both FBGs are located outside the eye, and this approach allows separating the transverse forces and the forces acting on the sclera. With this approach, the authors substantially split the analysis between the 3 distal FBGs, and the 6 FBGs proximal to the tools used to estimate the sclerotomy forces and location. As in previous works, mN-level forces are detectable with the FBG arrays.

The work in [8] also introduced also the variable admittance robot control, enabled by the FBG sensors on-board of the tool shaft and with the SHER robotic

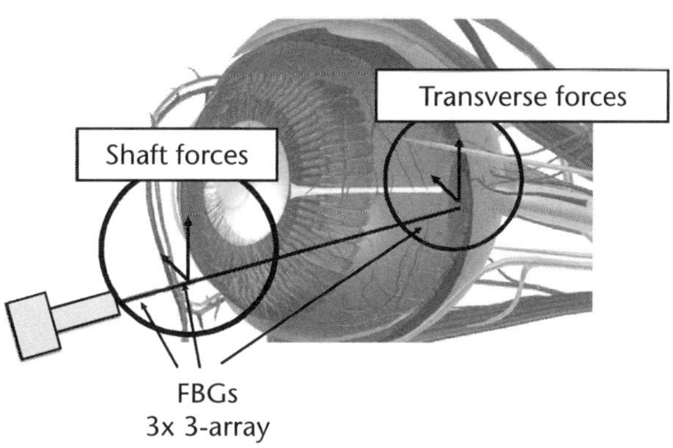

Figure 19.3 Graphical representation of the VRS force-sensing system presented in [8], allowing dual detection of transverse and tool-shaft forces with 3 arrays of 3 FBGs.

system. The system is based on a hands-on cooperative control, where the surgeon and the robot control the surgical tool. The user input force applied on the master handle controls the velocity of the robot performing the same operation. In this loop, the FBG sensors are detected by an FBG interrogator operating at 2 kHz and further postprocessed returning the forces' estimate, which is returned by the SHER to the surgeon. The system has been applied for epiretinal membrane peeling, a task that involves the removal of small portions of retinal tissues to restore a patient's sight and remove visual distortions. The multiforce sensing feature, onboard a SHER robot, is progressing in terms of technology readiness level and is converging into a standard design of the FBG sensor arrays.

19.3 Intensity-Based Sensors

Whereas FBG and FPI are mainly used in MIRS structures for multipoint sensing, there has been also a substantial research in inexpensive sensors, based on bare optical fibers with IBS approach, still having the capability to detect forces in robotic systems. Recent works using an intensity-based approach, based on a plurality of sensing point, have been proposed by Ahmadi et al. from Concordia University, Canada [9, 10]. Figure 19.4 shows the outline of the recent work proposed in [9], which is based on a solid fiber-optic sensor, based on three fibers, for the detection of force and position of a lump within a lump or a blood vessel. The sensor is based on the IBS principle: six fibers are laid inside three v-grooves etched within a semiflexible beam microfabricated from a silicon wafer. Within each v-groove, two fibers are faced leaving an inner gap, each having a different gap length. Gaps are arranged in different position to obtain a different response depending on the localization of force acting on the sensor beam. A SLED source (843 nm) is coupled in each input side, and three photodetectors are used to record the output. The sensor beam is used to detect solid obstructions, such as a lamp, acting on a soft tissue in contact with the sensor: the contact force between the obstruction and the beam induces a variation of all gap lengths, which depend on both the intensity and the position of the lump. Thus, an intensity variation of each photodiode output is related to the gap length variation. The sensing system proposes a localization algorithm, which helps estimating the location of the lump. The simplified algorithm for force detection is based on three sequential steps. In first place, the location estimation is performed, discriminating whether the force is applied on the left or right side of the sensor. In the second step, a disambiguation is performed between the inner part or the edge of the beam, thus achieving a detection within 4 regions (left edge, left center, right center, right edge). Finally, depending on the sensor

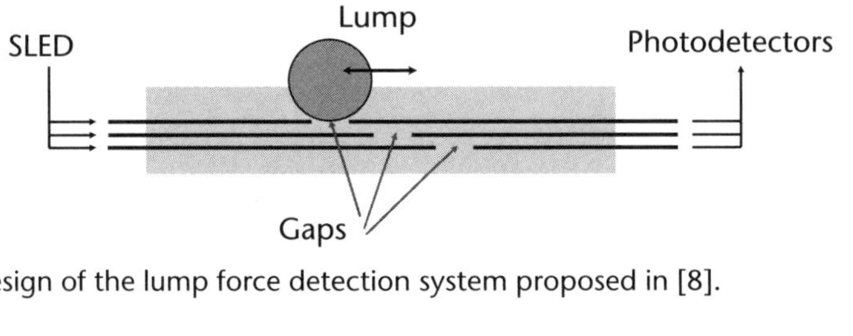

Figure 19.4 Design of the lump force detection system proposed in [8].

position, the algorithm selects two sensor outputs, computes the differential, and from it estimates the force. In order to operate with unbiased estimates, it is necessary to have the same transimpedance gain and same fiber losses on each channel or to compensate eventual mismatches.

The research on IBS systems applied in MIRS proposed by Ahmadi et al. aims at reducing the cost of sensors on-board of surgical tools, hence making it compatible with disposable systems, and operating with endoscopic instruments returning a fingertip feeling to the surgeon. In contrast to FBG or FPI sensors, which clearly select an active area, IBS sensors require a plurality of fibers in order to correctly reference the sensor to the location of force events. The possibility to integrate the manufacturing of the sensing beam with the fibers can further improve miniaturization, providing a low-cost alternative to other FOS solutions.

19.4 Distributed Sensing: An Industrial Prospect

One of the main industrial outputs for recent FOS system has been obtained in MIRS application and refers to the acquisition of Luna Inc.'s distributed shape sensing technology by Intuitive Surgical, the manufacturer of the da Vinci robot, in January 2014. Although several details of the commercial deals are undisclosed, they allow the confirmation that distributed shape sensing is a promising technology to be installed on-board of future robotic surgery systems for detecting the directional bending of tools and also, from this information, extracting the contact force on the lateral side of the shaft. As shown by Liu et al. [11], shape sensing is also a key asset for dexterous continuum manipulators (DCM), which are used to perform complex surgical operations in remote and hardly accessible locations.

The Luna distributed shape sensing system is largely based upon two patents: [12, 13]. The work from Froggatt et al. is applied in an OFDR-based interferometer, which detects strain through a Rayleigh scattering detection system. The system described in [13] instead is based on a weak-FBG array, and an OFDR detection system. FBGs are fabricated on a multicore fiber having the structure as in Figure 19.5. A central core is located at the center of the fiber, along the neutral axis; three other cores are helically wound, with a 120° geometry that keeps twisting around the fiber. This way, the cross-section having tri-axial symmetry is maintained in each section of the fiber. For this system, a set of continuous FBG is inscribed in the fiber; this allows having a higher reflectance than the Rayleigh backscattering for each fiber section, allowing reducing the spatial resolution. Luna

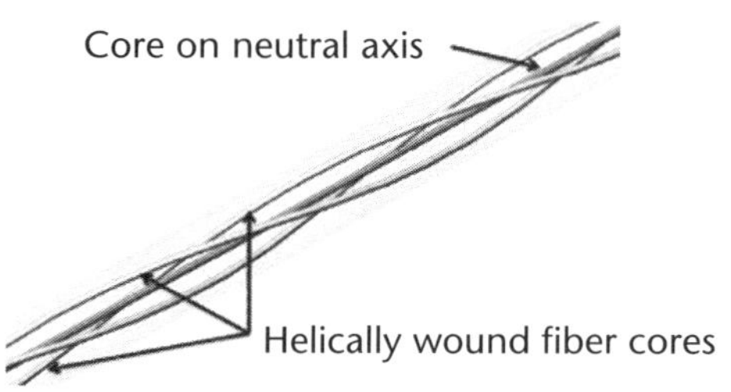

Figure 19.5 Layout of Luna distributed shape sensing fiber sensor (adapted from manufacturer's white paper "Fiber Optic Shape Sensing").

shape systems based on OFDR operate on maximum lengths ranging from 1.5m to 30m. In MIRS, we are interested on the shortest length, which is achieved by the continuous-FBG system.

Current characteristics of Luna systems that have been disclosed through datasheets and white papers are not yet compliant with the state of the art of MIRS sensors in terms of technical specifications. In first place, the fiber has a diameter of 250 μm, which is larger than the 125 to 150 μm currently in use. Furthermore, the typical accuracy for shape sensing, in terms of localization of bending changes, is estimated as 1% of the total length of the sensing tract, and results as a trade-off with detection precision and scanning speed. Assuming a sensing length of ~3 cm, which in VRS includes the entry point detection on the sclera, lateral contact forces within the eye, and forces acting on the tip, the detection error is >0.3 mm which is still relatively high. However, significant progress has been performed since 2010, both in terms of the fiber transducing element and for OFDR interrogation; and VRS is a critical MIRS application, whereas bend detection has lower requirement in many endoscopic instruments. Alternatively, shape sensing is a present task in dexterous continuum surgical tool, particularly for endoscopic surgery; to date, this is limited to manual or semiautomated instruments, without a full robotic control, operated on a sensing region with longer length. Hereby, shape sensors are used to locate obstructions and find the least invasive path to the location of intervention.

19.5 Future Evolution

The future evolution of robotic surgery is tied to improvements of the sensing technologies. Whereas the development of precision robotic system is adequate to satisfy the demand for several microsurgery scenarios, and endoscopic as well as MRI imaging is an acquaintance, it is still not completely possible to access several applications, as a clinical practice, due to the lack of feedback, perception of the tissue under manipulation. The hand of a surgeon is an extremely sensitive and intuitive tool, whereas a robotic tool can detect with greater precision, but limitedly to few points. The future work will involve an expansion of the number of sensing points, and an overall complexity of the sensing scheme: moving from the 3 to 4 sensing points of a FBG- or FPI-based architecture, to a chirped grating semidistributed sensing scheme can expand a lot the functionalities of robotic tools. The measurement of contact forces can be replaced by the localization and measurement of contact forces and thus add unprecedented features.

Also, fiber size needs to be reduced, moving from the 125 to 150 μm used in current manipulators (inclusive of protective jacket) to 40 to 80 μm. Current manipulators are limited to ~1 mm in diameter size, but in order to perform clinical trials, it is envisaged to reduce the thickness below 0.5 mm, preserving the strength of the device and of each fiber sensor. In VRS, tests on humans are severely protected by ethical issues; thus, clinicians perform in vivo tests on rabbits (although new European Union [EU] regulations are limiting such tests), which require a smaller tool size than humans due to a lower size of the iris and vitreous body. In this framework, it is essential to pursue the fabrication of FBGs with techniques (draw tower, femtosecond laser) that allow maintaining the fiber jacket to avoid the use of a naked FBG, too fragile, or of a recoating that would increase the thickness. We envision the fabrication of CFBG with draw-tower approach, in order to obtain robust and repeatable semidistributed sensors for transverse and axial force measurement and localization. The success case of Luna Technologies is a standpoint for the FOS community and a benchmark for the applicability of FOS so medical devices; the latest advances in multicore fibers and OFDR systems will push performance further, and the aim of integrating force detection with other parameters (pressure, temperature, biological quantities) is in clear view.

References

[1] Camarillo, D. B., T. M. Krummel, and J. K. Salisbury, "Robotic Technology In Surgery: Past, Present, and Future," *Amer. J. Surg.*, Vol. 188, No. 4A, 2004, pp. 2–15.

[2] Marescaux, J., et al., "Transcontinental Robot-Assisted Remote Telesurgery: Feasibility and Potential Applications," *Annals of Surgery*, Vol. 235, No. 4, April 1, 2002, pp. 487–492.

[3] BBC, "The Surgeon Who Operates from 400 km Away," http://www.bbc.com/future/story/20140516-i-operate-on-people-400km-away.

[4] Gupta, P. K., P. S. Jensen, and E. de Juan, "Surgical Forces and Tactile Perception During Retinal Microsurgery," *Proceedings of the Medical Image Computing and Computer-Assisted Intervention—MICCAI'99*, Vol. 1679/1999 of Lecture Notes in Computer Science, Springer, 1999, pp. 1218–1225.

[5] Liu, X., et al., "Miniature Fiber-Optic Force Sensor Based on Low-Coherence Fabry-Pérot Interferometry for Vitreoretinal Microsurgery," *Biomed. Opt. Express*, Vol. 3, No. 5, May 2012, pp. 1062–1076.

[6] Gupta, A., et al., "Human Eye Phantom for Developing Computer and Robot-Assisted Epiretinal Membrane Peeling," *2014 36th Annual International Conference of the IEEE Engineering in Medicine and Biology Society*, Chicago, IL, 2014, pp. 6864–6867.

[7] Iordachita, I., et al., "A Sub-Millimetric, 0.25 mN Resolution Fully Integrated Fiber-Optic Force-Sensing Tool for Retinal Microsurgery," *International Journal of Computer Assisted Radiology and Surgery*, Vol. 4, No. 4, June 2009, pp. 383–390.

[8] He, X., et al., "A Multi-Function Force Sensing Instrument for Variable Admittance Robotic Control in Retinal Microsurgery," *IEEE International Conference on Robotics and Automation (ICRA)*, June 2014, pp. 1411–1418.

[9] Ahmadi, R., et al., "Micro-Optical Force Distribution Sensing Suitable for Lump/Artery Detection," *Biomedical Microdevices*, Vol. 17, No. 1, February 1, 2015, pp. 1–2.

[10] Ahmadi, R., et al., "Discretely Loaded Beam-Type Optical Fiber Tactile Sensor for Tissue Manipulation and Palpation in Minimally Invasive Robotic Surgery," *IEEE Sensors Journal*, Vol. 12, No. 1, January 2012, pp. 22–32.

[11] Liu, H., et al., "Shape Tracking of a Dexterous Continuum Manipulator Utilizing Two Large Deflection Shape Sensors," *IEEE Sensors Journal,* Vol. 15, No. 10, 2015, pp. 5494–5503.

[12] Froggatt, M. E., and R. G. Duncan, "Fiber Optic Position and/or Shape Sensing Based on Rayleigh Scatter," U.S. Patent 7.772,541 B2, 2010.

[13] Childers, B. A., et al., "Fiber Optic Position and Shape Sensing Device and Method Relating Thereto," U.S. patent 7,781,724 B2, 2010.

CHAPTER 20

Smart Textiles and Wearable Sensors

20.1 Introduction

The growing interest in smart textiles and wearable sensors, including those based on optical fibers, has three main drivers:

1. *Continuous monitoring:* There is a generally growing availability of low-cost miniature technologies, together with sensor network devices such as Arduino boards, dataloggers, memories for data storage, and wireless transceivers based on modern networking protocols such as Wi-Fi, Zigbee, or Dash7, and data mining software. All these conditions lead to the main assessment that the wider the availability of biophysical (and, in the near future, biological) parameters recorded for an individual, the higher the effectiveness of healthcare in terms of prevention of diseases and treatment of current conditions. When continuous monitoring is performed with no or minimal invasiveness, healthcare benefits have little or no impact on human lifestyle. Continuous monitoring is extremely significant in infants, for preventing sudden infant death syndrome (SIDS), and in elderly people, mainly to prevent heart-related diseases.
2. *Personalized medicine:* This includes personalized drug delivery (PDD). The effectiveness of therapies may differ from each individual, having significant impact particularly on critical drugs; different responses from each organism to therapies can be monitored with sensors, which can lead to the identification of the optimal therapy for each patient. Although PDD is mainly a genomic study, monitoring biophysical parameters is essential: an example is the possibility to identify drugs effective in reducing the overall blood pressure, but also the understanding of the effect of altitude on respiratory apparatus from each individual.
3. *Remote healthcare:* The high cost of hospital-based diagnostics, through the whole hospitalization cycle, suggests transferring as more as possible the effort for simple diagnostic and data-logging operations outside of hospitals, and digitalize as much as possible the diagnostics. Remote healthcare has been developed and successfully implemented in several cases: examples are local communities with no hospital access, first responders for

human operators in harsh environments (e.g., detection of deadly H_2S in oil fields), and healthcare in space missions.

A study of IDTechX [1] shows that revenues for wearable technologies are on the trajectory to grow from $14 billion market in 2014 to a $70 billion market by 2025, overpowering most of the external technologies. Wearable technologies are mainly addressed by MEMS technologies, particularly when operated with organic electronics. However, the contribution of FOS is not to be underestimated, and successful research has been conducted in setting the premises for the applications of optical fibers in wearable, or similar, technologies.

In this sphere, we can divide FOS systems in 2 main applicative areas: (1) smart textiles, clothing functionalized with sensors; and (2) on-board sensors, which are mounted on board of pocket devices such as smartphones or watches. In the first case, the whole sensing network (including data acquisition and recording) is embedded into clothing, or garments, making the design more complex, but also opening the door to a deep integration between sensors and textiles and to a continuous monitoring of parameters. In the second context, sensors rely on the support of an existing device, exploiting its internal resources (camera, memory, third-party applications) in conjunction with an external fiber-optic sensing architecture.

20.2 Smart Medical Textiles

Smart textiles, or e-textiles, are clothing or fabrics that enable digital operations such as sensing and monitoring, energy harvesting, and computing; sensing is the main area of research, particularly for monitoring of biophysical and biological parameters. The genesis of modern research on FOS for smart textiles is within the framework of the European OFSETH project [2]: in this project, only optical sensing technologies are used to monitor vital parameters such as respiratory rate, cardiac rate, and pulse oxymetry, mainly during MRI. The project mainly aimed at the development of devices for respiratory detection, having fibers embedded within the yarns in two specific points. The key rationale of the project is that, under MRI, standard sensing technologies such as transistors, accelerometers, and other electrical devices are not allowed. The main scientific results of this research have been reported by Witt et al. [3].

Figure 20.1 shows photographs of the OFSETH project. The team developed two sensors, both for strain detection: (1) an FBG sensor, fabricated on a silica standard fiber, and embedded into an elastic textile; (2) a macrobending sensor developed by Multitel (Belgium) and interrogated via OTDR. The two sensors have been combined into a harness, MRI-compatible, that allows for measurement of both thoracic (FBG) and abdominal (macrobending) respiration. Similar results have been obtained by Carmo et al. [5], who investigated FBG sensors for respiratory detection, with a single sensing point.

As from Massaroni's review [4], the vast majority of FOS for smart textile research reported to date (86%) addresses monitoring of respiratory function, using either FBG or IBS (typically macrobending). The principle of operation of respiratory sensors is simply to convert breathing movements into strain, easily detectable. Most of the complication comes from the embodiment of the whole electronics in a

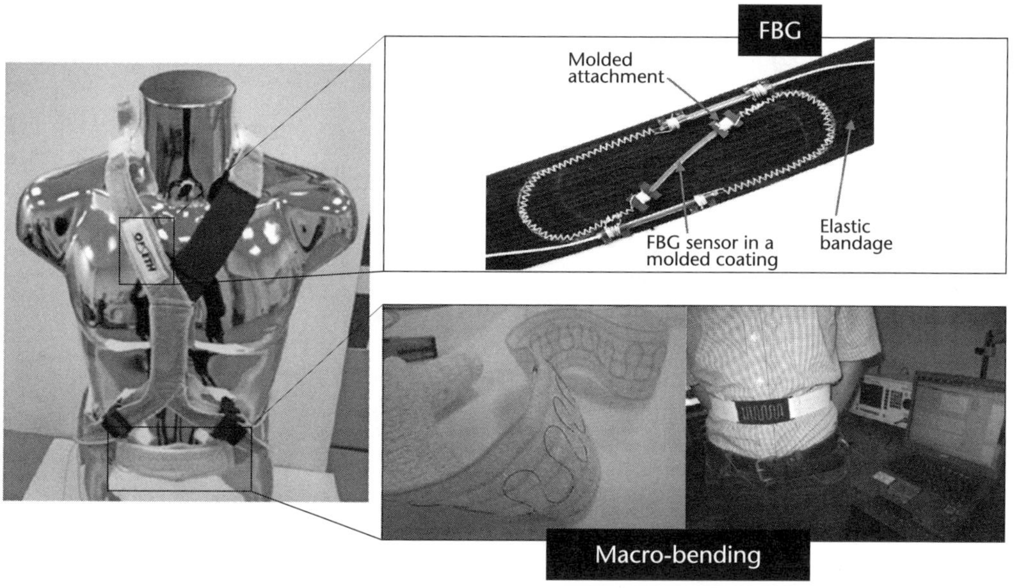

Figure 20.1 Layout of the OFSETH project [4]. Left: design of the MRI-compatible sensing harness with on-board sensors. Right, top: design of the FBG sensor; right, down picture of macrobending sensor.

wearable and battery-powered device: therefore, typical architectures are based on single-wavelength detection with a laser source, or portable spectrometry. Often, the system acts as a first-responder unit. Even when sensing is performed alongside an extended active length, the unit integrates all the spatial information, usually displaying a respiratory chart as in [3]. In the works presented so far, strain has always been detected over an extended length, and then processed as an individual data. More advanced features may in the future be enabled by spatially resolved sensors, which however are not miniaturized to a wearable format, and to date can only be used as external units limiting patient's mobility.

Alternatively, some emerging applications are starting to appear. The work from Li et al. [6] is concerned with temperature detection and is presented as intelligent clothes for early detection of high body temperature in fever. In this application, accuracy is the dominant requirement, and the authors have expanded the accuracy of a standard silica FBG by using a polymer coating: with the proposed packaging, the sensitivity is 150 pm/°C is achieved, 15 times higher than standard FBG, leading then to the hundredth of degrees Celsius accuracy. Alternatively, the polymer package is not atoxic, due to the use of naphthenate and methyl ethyl in the compound; however, packaging is low-cost and suitable for integration with the clothing, as in Figure 20.2. Li et al. provided a methodology for embedment of FBG into woven fabrics; a small half-cylindrical roll with the FBG is covered by woven fabrics, creating a structure that can propagate heat from the body to the sensor. In this work, according to mathematical modeling, 8% to 37% of the body heat is transferred to the sensor, according to its position.

In 2015, the first smart textile for biochemical detection was proposed by Esmaeilzadeh et al. [7], achieving a true milestone in this research. Figure 20.3 shows the sensor design and its deep integration into wool fabrics, which is one of the

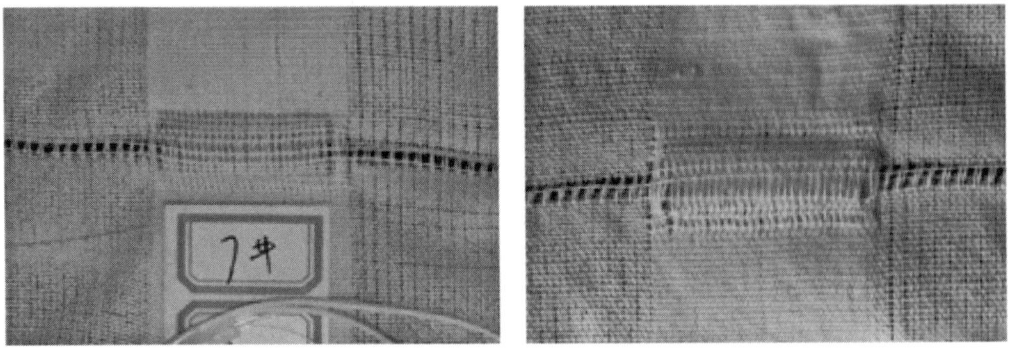

(a) (b)

Figure 20.2 Smart textile embedding FBG sensors for continuous body temperature monitoring; the picture shows the embodiment of a single FBG in fabric. (From: [6].)

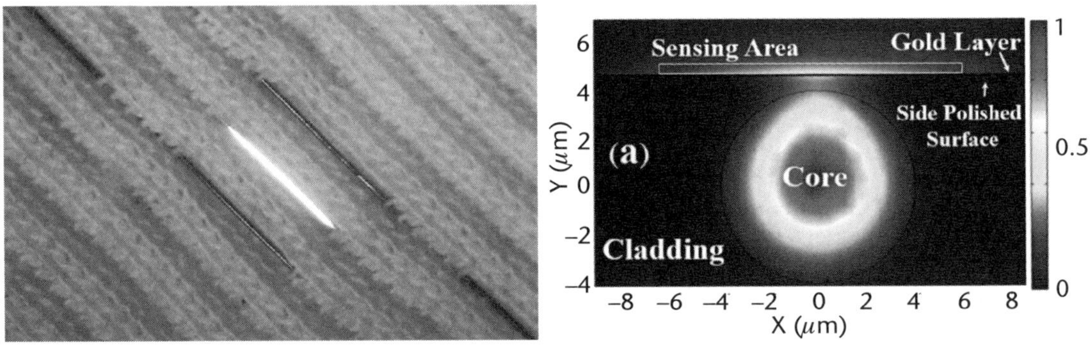

Figure 20.3 The first SPR-based FOS for smart textiles, for dew detection. (From: [7].) Left: SPR sensor (illuminated in the middle), with other fibers (in dark gray) embedded in a textile. Right: Cross- section of the SPR sensor, showing power distribution.

most significant designs achieved so far. The sensor detects humidity and is mainly addressed to detection of condensation and water evaporation. The figure shows the power distribution within the fiber and in the thin gold film where plasmonic mode is excited.

The most attractive part of this research is the possibility to use a fiber-optic SPR detector within a textile. Although humidity is not a primary method for diagnostic, sweat is an extremely relevant fluid for genomics and biomarkers detection, and it comes in contact with clothes. In 2014, the Skin Project, as reported in [8], started exploring the possibility to replace sweat analysis with the traditional blood analysis, and the possibility to combine it with wearable sensors can completely disrupt biosensing. The Skin Project, which has not disclosed partial results, has U.S. military funding and is mainly concerned with vital parameters. However, recently, analytical chemistry and genomics research have already demonstrated the possibility to use sweat analysis [9] for biomarkers for bladder, prostate, and lung cancer, diabetes, metal excretion, and drugs, among others.

20.3 Wearable Sensors

There is very recent, and early-stage, research in FOS for smartphones, which can potentially change the philosophy of cabled sensors. Normally, interrogation of FOS is based on ad hoc devices and software, but within the last 5 years, the explosion of smartphones has given us devices that are worn for the vast majority of the time, which already mount hardware that can be also used for sensing purposes. All smartphones embed a CCD-based camera, which can be turned into a spectrometer, a flash light that can be used as LED source, a battery for power supply, and allows developing third party apps that can control all the hardware. Thus, virtually, it is possible to combine the smartphone technology with FOS. There is a close affinity between smartphone-based sensors and smart-textile, as smartphones are pocket devices that are carried by the user almost like clothes. Alternatively, protocols for sensing networks are very different: smart textiles are usually meant to perform continuous monitoring and to perform data processing within the interrogator; conversely, smartphones are limited in terms of power consumption, and most of the internal hardware that can potentially be used for signal processing is dedicated to the phone, and cannot be accessed: for example, it is typically not allowed to mount a photodiode inside the phone and use its signal conditioner to detect an optical power.

The emergence of smartphones, however, is pushing the IEEE Sensors Council to call for rigorous sensor standards, and by intewith smartphone manufacturers, it is likely that in the near future the possibility of having customizable hardware and advanced programming interfaces will become a reality, expanding sensing capabilities of our pocket devices.

We can trace back the first pioneer work to 2015 [10]: with the evocative lab-in-a-phone name, a research team of the University of Sydney has developed an optical (but not fiber-optic) sensor for the detection of pH of water. The system is based on a three-dimensional printed holder, with an internal structure that allows minimal beam steering, and makes use of the flash LED of the phone covered with a colored filter, and the CMOS camera of the phone as spectrometer. The system operates with an Android application.

A similar approach has been taken to the next level by Bremer [11], from Hanover Centre for Optical Technologies, that sets the first-ever work on a SPR sensor for smartphones. This work has been spread out by several media [12], pointing out how Bremer's SPR technology on board a smartphone can potentially replace pregnancy tests. The system developed by Bremer minimizes the use of components external to the phone hardware and software, as in Figure 20.4: only three external components are used. A diffraction grating is used to substantially turn the smartphone camera into a spectrometer, which is then used by just capturing an image or a video. The white flash is used as light source. The whole sensing system is based on a 400-μm fiber, which is polished at 45° to allow efficient coupling from the source and to the detector. The SPR sensor has been achieved by silver coating, after a chemical process that results in the refractive index dependence as in Figure 20.4 (right); the active length was mounted on a sample holder, and the rest of the fiber was glued to the phone. A sensitivity of $5.96 \cdot 10^{-4}$ RIU/pixel has been documented, whereas performances are limited mainly by the camera resolu-

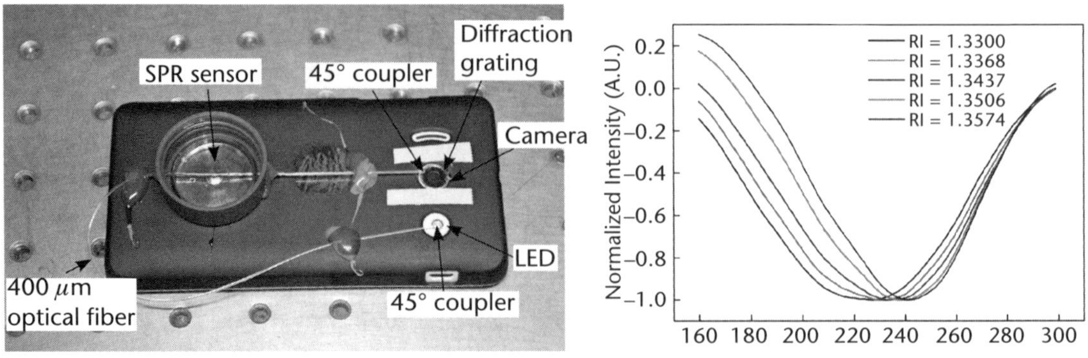

Figure 20.4 Layout of an SPR sensor for smartphone. (From: [11].) Left: Photograph of the sensor, realized on a commercial Android smartphone. Right: SPR spectrum recorded by the phone camera, for different values of the refractive index of the solution detected in the sample holder.

tion. Although very preliminary, and yet not functionalized to a specific analyte, the work of Bremer has opened the door for smartphone-based biological sensing.

20.4 Future Prospects

FOS for smart textiles and wearable systems is a research line that has appeared recurrently in the scientific community only since 2015, and thus we are in the early stages of an enormous potential. The real impact of FOS in this biomedical context will depend on how the interaction between optical fibers and fabrics will be strengthened, on the advances in spatially resolved sensing (and their miniaturization), and on the biosensing capability, all aspects that differentiate FOS from other sensing technologies.

A first game changer is woven optical fibers. Reference [13] reported several emerging innovations in the field of smart textiles. To our interest, the possibility of interlacing standard fabrics with optical fibers is a breakthrough: this allows one to position either standard fibers (for IBS), FBG, or SPR sensors directly inside the tissue during fabrication, changing the process of integrating optical fibers with garments. In the same direction goes the recent trend on fibers made of hydrogel, which is increasingly applied in textiles. Such technologies abate the barrier between the fabrication of clothing and embodiment of sensors. Photovoltaic textiles are another breakthrough with immediate applications in smart textiles. A major weakness of FOS is the avid need for power supply, particularly for the light sources. Energy harvesting can come in support, and photovoltaic textiles as developed in Dephotex project [14] can be used for such scope. Research has demonstrated novel fibers that can be integrated into flexible textiles and, in the future, possibly used as light sources.

References

[1] Wearable Technology 2015-2025: Technologies, Markets, Forecasts, 2014, http://www.idtechex.com/research/reports/wearable-technology-2015-2025-technologies-markets-forecasts-000427.asp.

[2] OFSETH project, FP6-IST-2005-027869: http://www.ofseth.org/.

[3] Witt, J. F., et al., "Medical Textiles with Embedded Fiber Optic Sensors for Monitoring of Respiratory Movement," *IEEE Sensors Journal*, Vol. 12, No. 1, 2012, pp. 246–254.

[4] Massaroni, C., P. Saccomandi, and E. Schena, "Medical Smart Textiles Based on Fiber Optic Technology: An Overview," *J. Funct. Biomater.*, Vol. 6, 2015, pp. 2014–2021.

[5] Carmo, J. P., et al., "Application of Fiber Bragg Grating to Wearable Garments," *IEEE Sensors Journal*, Vol. 12, No. 1, 2012, pp. 261–266.

[6] Li, H., et al., "Wearable Sensors in Intelligent Clothing for Measuring Human Body Temperature Based on Optical Fiber Bragg Grating," *Opt. Expr.*, Vol. 20, No. 11, 2012, pp. 11740–11752.

[7] Esmaeilzadeh, H., et al., "Smart Textile Plasmonic Fiber Dew Sensors," *Opt. Expr.*, Vol. 23, No. 11, 2015, pp. 14981–14992.

[8] Live Science, "U.S. Military's Skin Sensors Could Reveal Science of Sweat," http://www.livescience.com/46072-military-skin-sensors.html.

[9] Jadoon, S., et al., "Recent Developments in Sweat Analysis and Its Applications," *Int. J. Anal. Chem.*, 2015, p. 164974.

[10] Hossain, A., et al., "Lab-in-a-Phone: Smartphone-Based Portable Fluorometer for pH Measurements of Environmental Water," *IEEE Sensors Journal,* September, 2015, Vol. 15, No. 9, pp. 5095-5102.

[11] Bremer, R., "Fibre Optic Surface Plasmon Resonance Sensor System Designed for Smartphones," *Opt. Expr.,* Vol. 23, No. 13, 2015, pp. 17179–17184.

[12] "Could Your Smartphone One Day Tell You You're Pregnant?" http://www.sciencedaily.com/releases/2015/07/150701152319.htm.

[13] Cherenack, K., and L. van Pieterson, "Smart Textiles: Challenges and Opportunities," *J. Appl. Phys.*, Vol. 112, 2012, p. 091301.

[14] Dephotex FP7 project, http://www.dephotex.com.

CHAPTER 21

Fiber-Optic Spectrometric Sensors

Fiber-optic spectrometry, briefly recapped in Chapter 9, is not a core topic of this book; however, the analysis of recent spectrometric sensors provides a cue for future applications for FOS systems, particularly in cancer care. In this chapter, we will describe a few applications for cancer diagnostic, addressed with spectrometric detection, which can be potentially expanded through FOS biosensors. Our aim is providing the readers a framework for complementing the biophysical sensing in interventional cancer therapies (Chapter 17) and emerging biological sensors for cancer diagnostic (Chapter 18), with the possible upgrades of spectroscopic sensors in this field.

21.1 Detection of Liver Tumors

Evers et al., with Philips (Netherlands) [1], have provided a comprehensive study of different spectroscopic techniques, based on optical fibers, for the comparison of metastatic liver versus healthy tissue, with the aim of identifying biomarkers that can discriminate the presence of a tumor. In a first-level analysis, absorption spectroscopy results show that metastatic liver tumors have a higher absorption in the visible and NIR windows, particularly around 600 to 950 nm. The work reports a substantial discrimination between metastatic and healthy liver, but suggests that steatosis or cirrhosis are more hardly separable from a metastatic liver. Then, a second-level analysis based diffused reflectance spectroscopy (DRS) has been reported, collecting DRS spectra within 500 nm and 1,600 nm; in this investigation DRS spectra have been used to discriminate different biomarkers, between healthy liver and tumors. A selection of biomarker was analyzed, and the most promising results appear for level of total hemoglobin, fat concentration, and bile percentage, whereas instead the amount of DRS or Mie scattering is a poorer discriminator. In this study, 94% of 828 patients were correctly diagnosed. The study from Evers et al. paves the road for cancer diagnostic with FOS systems, replacing the spectroscopic approach with labeled biosensors capable of quantitative detection of the tissue properties.

21.2 Detection of Lung Tumors

The study was continued in [2], using DRS and fluorescence spectroscopy (FS). The aim is to diagnose peripheral lung tumors, using a similar approach based on measurement of total absorption, and single biomarkers. In this study, the main discrimination factors are the level of total hemoglobin, collagen concentration, and NADH coenzyme. Both studies with DRS and FS are based on a single optical fiber, inserted at the tumor point, and with real-time reading of reflected and fluorescence spectra, with postprocessing.

21.3 Portable Tissue Scanner

A different approach has been reported in [3] and involves a tissue scanner. The same principles of DRS and FS have been implemented in a scanning system photographically and schematically shown in Figure 21.1, based on a *xy* scanner with a motion controller capable of moving the front end of a probe. On the probe, two fiber bundles have been mounted, one for each spectroscope. The system is used to rapidly scan the tissue with spectroscopy and obtain reflected spectra that map the tissue under investigation. The goal of the study is to detect in real time the contour (two-dimensional) of a tumor during a surgical intervention. Thermal ablation research can possibly extend the result of this study, incorporating FOS with fiber-optic imaging for determining tumor peripheral contours.

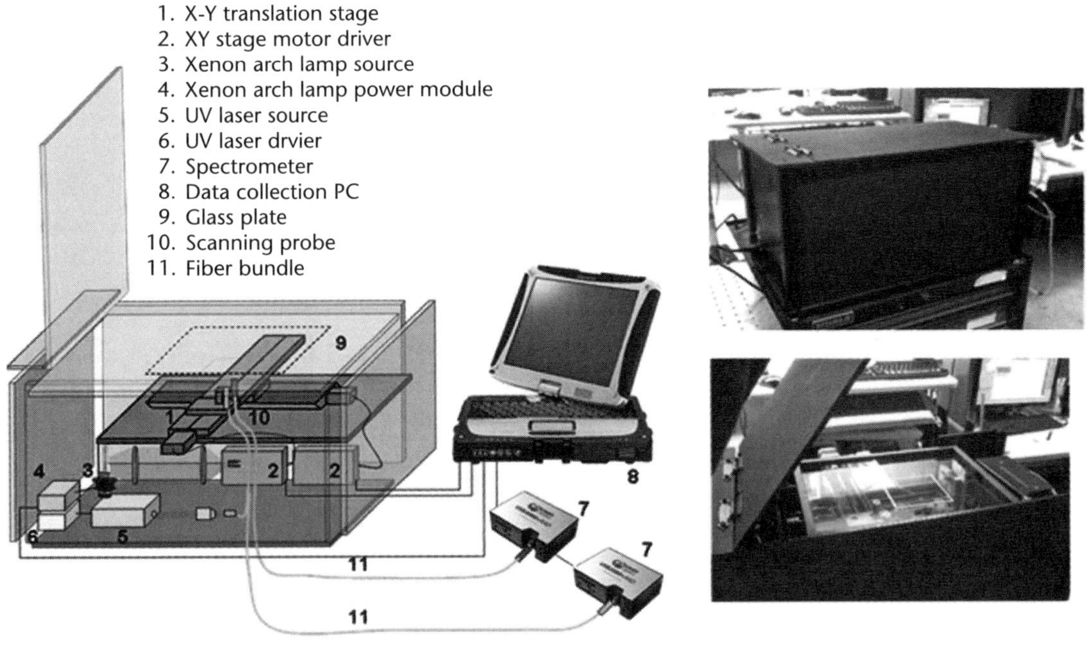

Figure 21.1 Tissue scanner based on fiber optic DRS approach, reported by Lue et al. [3]. The system consists of a two-axis scanner, capable of moving a probe formed by two fiber bundles capable of recording DRS and FS. (From: [3].)

21.4 Application Perspective

From the FOS application point of view, spectroscopy has several weaknesses in terms of sensitivity, discrimination, and cross-sensitivities. The optical fiber sensing principle requires the fiber to be inserted at the tumor point in order to sense the tissue properties; such a task can be performed only with a prior localization and diagnosis of the tumor. Alternatively, the online measurement of tissue properties during interventional procedures is hampered by the large amount of cross-sensitivities (temperature, pressure, strain, and changes of cells status and coagulation). However, the future trend of medical devices foresees the integration of sensors and spectroscopic systems, particularly operating with a WDM approach, in order to functionalize a single fiber for several types of detection tasks. It is noteworthy that a combination of sensing and imaging is also an approach that can provide significant upgrades, covering the mutual weaknesses.

References

[1] Evers, D. J., et al., "Optical Sensing for Tumor Detection in the Liver," *European Journal of Surgical Oncology (EJSO)*, Vol. 39, No. 1, January 31, 2013, pp. 68–75.

[2] Spliethoff, J. W., et al., "Improved Identification of Peripheral Lung Tumors by Using Diffuse Reflectance and Fluorescence Spectroscopy," *Lung Cancer*, Vol. 80, No. 2, May 31, 2013, pp. 165–171.

[3] Lue, N., et al., "Portable Optical Fiber Probe-Based Spectroscopic Scanner for Rapid Cancer Diagnosis: A New Tool for Intraoperative Margin Assessment," *PLoS ONE*, Vol. 7, No. 1, 2012, p. e30887.

CHAPTER 22

Conclusions and Future Perspectives

In this chapter, we provide an all-encompassing look on the recent and emergent field of fiber-optic sensors for biomedical applications. Our analysis has focused on three parts: understanding the main principles of FOS and configurations suited for microscopic sensing; building effective sensing systems and understanding the protocols and requirements for their biomedical use; and learning from recent works and applying FOS to specific biomedical applications. We identified a major gap between sensors, particularly fiber-optic sensors, and their biomedical applications, and we hope that future efforts will be devised to application-driven development to meet the growing demand for medical probes and measurement instruments.

The future role of FOS for biomedical engineering will be decided by how scientists can convert the hundreds of configurations proposed every year (and most of them destined to remain an academic exercise) into few, effective platforms that can effectively perform disruptive innovations. In order to connect this enormous potential of performance, miniaturization, sensing capacity with applications that can really become marketplaces, we provide 10 takeaways from the book that we hope will be useful for the scientists operating in this field.

1. *Make measurable:* In the famous quote of Galileo Galilei, "Measure what is measurable, and make measurable what is not so." FOS scientists should focus on the second part. The major task of FOS is to enable the measurement of key parameters, in places and conditions where the traditional technologies fail. The human body is indeed a widely unexplored sensing environment as the lack of access points, the need for narrow spatial resolution, the combination of physical and biological parameters, and the strict biocompatibility requirements, on top of limitations for interrogation systems, have caused technologies to perform only a small fraction of the available monitoring challenges. FOS technologies are expensive, for both interrogators and disposable probes, and their deployment is meaningful only when they unlock new and unexplored applications, inaccessible for other MEMS, electrochemical, spectroscopic, or optofluidic system: make measurable what it not so. Measuring blood pressure with FOS is no more than an academic exercise, but enabling multiple FFR pressure measurements is a potential breakthrough.
2. *Cost-benefit paradigm:* Despite their intrinsic expensiveness, which can be only partially abated through volume manufacturing (as it happened for FBG), FOS commercialization is highly dependent upon the whole

paradigm of cost and benefits. Costs refer to both the disposable probe and the interrogator equipment as a whole; also, for this reason, distributed systems are attractive, as they use standard inexpensive fibers rather than inscribed or extrinsic sensors. The quantification of benefits is much harder to undertake, in part because it involves subjective and not quantitative entities (such as invasiveness or patient's discomfort or embarrassment), in part because some sensing information can be obtained in other ways (e.g., through laparoscopic surgery, biopsy, or endoscopic imaging). In this uncertainty, one benefit that is highly quantifiable is hospitalization cost, which is extremely high in the whole Western world. According to *Kaiser State Health Facts*, the average cost for inpatient treatment in United States is $1,878/day, while the same figure is 832 Euros ($950) in Italy according to the Italian Ministry of Health, and hospital care has an average cost higher than $60/day for nonsanitary services, and this is why interest for tele-health systems is high. For this reason, FOS-based systems that are capable of moving hospital-based to portable-based diagnostics, as well as performing conclusive diagnostics (that negates the requirement for additional analysis), have an immediate and exploitable market opportunity.

3. *Focus on extreme applications:* As a consequence of the previous points, FOS developers should focus on extreme applications, whereas the sensing environment is the harshest and requirements are the strictest. FOSs are the only sensing technology that easily allows embedding several sensing point on a single device and multiparametric sensing, but also have key advantages in every application that is devised, and these advantages have to be exploited by the traction of advanced applications. Thermal ablation is a perfect example of this point: the sensing environment is extremely hard to access and can be easily altered, and thermal and pressure spatial gradients are among the steepest to be observed in medicine: but this is where key innovations are made, and in less than 3 years FOS became the golden standard for benchmarking TA techniques.
4. *Integration and distribution:* In our vision, the key assets for next-generation biomedical FOS are maximizing the number of parameters that can be measured on a single fiber and their spatial distribution. Distributed sensors based on white light interferometry are now an acquaintance, but they usually can read only one parameter (typically strain or temperature). FBGs are suitable for multiplexing, but their resolution is coarse. Semidistributed sensors such as LCFBG are a recent trend, as well as tilted FBG. In contrast, EFPI sensors are the golden standard for pressure sensing, and most diagnostics in medicine are based on pressure readings. Thus, in order to fully exploit FOS systems, we need to break down the current barriers and be able to combine sensors having narrow and wide spectra, as well as adding dimensionality to the XDM sensing networks, where X refers to time (TDM), wavelength (WDM), cepstrum (CDM), spatial (SDM), or other quantities. The research on innovative encoding and decoding technique is still strong, and the transition from single-point to distributed sensing is happening in less than a decade. In particular, spatial division multiplexing is a very recent development, but the possibility to fabricate

FBGs and other sensors on each core can disrupt the potential of fiber-optic sensing.

5. *Cost reduction and standardization:* Through the past two decades, the development of FOS was strictly tied to optical communications: the devices made available in large volumes and relatively low cost were used in FOS, while additional devices were mainly unaffordable. As FOSs are reaching maturity, and thanks to development of spectroscopic, fluorescence, and endoscopic imaging devices, we have a broader availability of components at a lower cost. Fiber-coupled spectrometers, in particular, have reduced price by more than one order of magnitude, and they are now the building block of FOS networks. However, costs are often high for disposable medical devices. In order to further approach commercialization, a further cost reduction is needed, for both interrogation systems and disposable device. While traditional optical fibers have been standardized by the IEEE, custom-made devices do not obey specific standards and therefore are often hard to be integrated in a nutshell. The evolution of sensing devices imposes a need for standards that do not account only for the optical fibers, but also for the integration into medical devices. The IEEE Sensors Council is involved into standardization, which is mainly addressed to pocket or smartphone-based devices; a broader standardization for endoscopic devices and for catheterizations is needed currently.
6. *Multifunctionality over sensing:* Medical devices are intended to perform several operations, and sensing is one of them, often the only one. Endoscopic devices typically embed multiple functionalities including manipulation and grabbing of tissue samples, cauterizations, electrical incisions, thermal ablation, scattering spectroscopy, and fiber-based imaging. The growing miniaturization of medical devices into laparoscopic or arthroscopic form factors provides the opportunity to incorporate FOS into medical devices to add sensing features. In this type of medical devices, catheters sustain the penetration into tissue and the positioning at the point of treatment, as well as removal after treatment; this can simplify the development of the sensing system.
7. *Overcoming biopsy:* Targeted SPR-based biosensors are a very recent acquaintance, and as a sensing principle, are not as established as FBG, EFPI, or interferometric sensing. However, the potential is immense, as they allow identifying the presence of specific biomarkers and analytes, localized in specific control points. The structure based on light diffraction, through a tilted FBG or a tapered fiber, combined with a nano-thick gold layer functionalized with bioreceptors and inducing SPR, has demonstrated its operation. The evolution of this architecture will strongly depend on which applications will be addressed, and on the capability to minimize cross-sensitivity and to bind biomolecules for a reasonably long time without deteriorations. However, the potential is disruptive: replacing biopsy, a minimally invasive procedure that requires obtaining and processing a tissue sample, and requires both manual intervention and a long time, with an almost real-time detection that can be extended over a long time to maximize the chance to catch specific substances. In this direction is, for

example, the research aimed at identifying cancer stem cells, present in extremely small concentrations. Once again, SPR-based sensors need to address the same value proposition in terms of cost-benefit and unlocking unprecedented applications. In order to achieve this overarching goal, it is necessary to understand how SPR sensors operate in vivo in a biological environment and so far no data have been reported, to the best of our knowledge. The next 5 years will be the key in understanding if fiber-optic SPR sensors will really complete the trajectory of overcoming biopsy, or will remain confined in an isolated niche.

8. *End-user expertise:* The application-driven paradigm requires making research and development tasks in larger and multidisciplinary teams. A complete medical device requires optics and electronics developers, microfabrication expertise, biomedical engineers, clinicians, and typically a medical device company that integrates the sensor in a complete system. The role of the clinician (i.e., the end-user of the sensing device) is essential and should be incorporated as early as possible in the development tasks because he or she can identify flaws and performance gaps of the device under test, but also can recognize other applications for the same device. Connecting clinicians and engineers in the same development team is not a common practice in nearly any research group; instead, highly influential clinicians are often called in as consultants closer to commercialization, at a price that is usually unaffordable for any research grant. In our experience, the interaction between engineers and clinicians is at first bizarre: we were often asked to perform tasks outside of the sensing world or physically impossible, and in contrast, simple and well-established devices of the previous generation were regarded as disruptive innovation. However, after taking over the initial adjustments, this method has been extremely effective in connecting technologies with applications, and this is how, for example, we established our work in the fields of urology and thermal ablation for cancer care.
9. *Biomarkers and metrics:* Measuring parameters is the end-point for sensors specialists, but is the starting point for diagnostic. For example, FFR or ICP pressure measurements are useless without the knowledge of the acceptable, low-risk, and risk values, and these are found and established after extensive clinical tests, which are often several years long as patient lifetime is recorded. As we unlock new applications, we may be successful in measuring, but the lack of biomarkers and useful metrics is frustrating because no significance is found for the sensing system. However, this is an unavoidable problem, and the strategy of working only in fields where biomarkers are already established would limit sensing systems to do what has already been done. Instead, scientists should find support from phantoms and other physical models to be able to obtain at least a qualitative statement over expected measurement values and their significance in diagnostic. Again, clinicians have the key role in supporting sensor technologies with their diagnostic meaning.
10. *Robotic technologies:* Robotic systems are disrupting several technological fields, and medicine is no different. Robotic microsurgery aims at enabling

several surgical tasks, where hand tremor and the lack of access points is a limitation, as for vitreoretinal surgery. In robotic surgery, even more than in manual treatments, the role of sensors is a key to detect forces and shape. Sensors working in robotic surgery are required to have an extremely miniature size and great accuracy, with a plurality of sensors required on a single submillimeter surgical tool. The robotic surgery industry is expanding at a double-digit growth rate, and recently Intuitive Surgical, the manufacturer of the da Vinci robot, has acquired a branch of Luna Technologies, related to distributed shape sensing. In a broader sense, even multifunctional endoscopic devices are starting to embed a limited number of robotic features, with microactuators allowing for stereotactic movements at the microscale. In all closed-loop systems, sensors have a key role as they allow the measurements required in the feedback loop: an example is the smart thermal ablation concept.

Appendix A
FBG and EFPI Generators

In the following, 2 MATLAB functions are provided for generation of FBG and EFPI structures (Chapters 5 and 7), and used in 3 scripts to generate FBG arrays, chirped FBG, and FBG/EFPI sensors. Readers are free to take and modify the code to simulate all structures based on FBG and FP cavity, in the spectrum domain. We also include, in our analysis, the SNR (signal-to-noise ratio), defined as the ratio between the maximum of reflectivity of the sensor and the noise variance.

FBG Generator: FBGgeneration.m

```
function [ Reflectivity ] = FBGgeneration ( lambdaB,
lambda, kL, L, dneff, neff, v )

% Creates a FBG (Reflectivity) starting from:
% - lambdaB = Bragg wavelength (in m)
% - lambda = wavelength axis (in m)
% - kL = grating strength
% - L = grating length (in m)
% - dneff = r.i. modulation amplitude
% - neff = effective r.i.
% - v = fringe visibility

k = (kL/L) / v; % Computes k coefficient

% Computes sigma, delta, and sigma^hat
sigma = pi./lambda*dneff;
delta = 2*pi*neff*(1./lambda - 1./lambdaB);
sigmahat = delta + sigma;

% Finally, computes reflectivity
Reflectivity = ( sinh(L*sqrt(k.^2 - sigmahat.^2)).^2 ) ./ (
( cosh(L*sqrt(k.^2 - sigmahat.^2)).^2 ) - (sigmahat.^2)./
(k.^2));

end
```

EFPI Generator (2-Mirror Extrinsic Structure): EFPIgeneration.m

```
function [ Reflectivity ] = EFPIgeneration ( lambda, LFP,
d, ni, nFP, nd, no )

% EFPI generator: computes the reflectivity of a 2-cavity
(3-mirror) EFPI
% Inputs:
% lambda: wavelength axis
% LFP: Fabry-Perot cavity length
% d: diaphragm length
% ni = ref. index input medium
% nFP = ref. index Fabry-Perot cavity
% nd = ref. index diaphragm
% no = ref. index outer medium

% First step: computes the reflectance and transmittance of
each mirror
r1 = abs((ni-nFP)/(ni+nFP)); t1 = sqrt(1 - r1.^2);
r2 = abs((nFP-nd)/(nFP+nd)); t2 = sqrt(1 - r2.^2);
r3 = abs((nd-no)/(nd+no)); t3 = sqrt(1 - r3.^2);


% Second step: creates the transmission matrix for each
mirror
T_Mirror_1 = ((i*t1).^-1) * [-1 r1; -r1 1];
T_Mirror_2 = ((i*t2).^-1) * [-1 r2; -r2 1];
T_Mirror_3 = ((i*t3).^-1) * [-1 r3; -r3 1];

% Third step: for each wavelength, computes the
transmission matrix for
% propagation

for ii = 1:length(lambda)

    T_prop_L = [exp(i*2*pi*LFP*nFP./lambda(ii)) 0; 0
exp(-i*2*pi*LFP*nFP./lambda(ii))];
    T_prop_d = [exp(i*2*pi*d*nd./lambda(ii)) 0; 0
exp(-i*2*pi*d*nd./lambda(ii))];

    % Fourth step: obtain the total transmission matrix
    Tcavity = T_Mirror_1 * T_prop_L * T_Mirror_2 * T_prop_d
* T_Mirror_3;

    % Final step: obtain the S11 parameter as T(1,2)/T(2,2)
    S11 = Tcavity(1,2) / Tcavity(2,2);
    Reflectivity(ii) = abs(S11)^2;

end
```

FBG Array Analyzer: FBGarray.m

```
% Script: creates an array of FBG having different
wavelengths and
% properties, in reference and heating conditions

clear all; close all; clc

% Initialize FBG array
lambdaB = [1540 1545 1550 1555 1575]*1e-9; % Bragg
wavelengths
kL = [0.8 1 1 1 4]; % kL coefficients
L = [1e-2 1e-2 1e-2 1e-2 1e-2]; % grating length
neff = 1.5;
dneff = 1e-5;
v = 1;

% Wavelength grid
lambda = 1535e-9 : 1e-12 : 1585e-9; % in m

% Thermal coefficient and temperature
kT = 10.2e-12; % in meters/degC
DeltaT = [10 -10 35 45 -20]; % Temperature variation, in
degC

% Noise parameters
SNR_dB = 45; % SNR, in dB
ActivateNoise = 1; % Set to 0 or 1 to turn off/on noise

% Partial transmission
Transmission_partial_ref = ones(1, length(lambda));
Transmission_partial_sens = ones(1, length(lambda));

for ii = 1:length(lambdaB)

    % Without temperature variation
    Reflectivity_ref = FBGgeneration(lambdaB(ii), lambda,
kL(ii), L(ii), dneff, neff, v);

    % With temperature variation
    Reflectivity_sens = FBGgeneration(lambdaB(ii) +
kT*DeltaT(ii), lambda, kL(ii), L(ii), dneff, neff, v);

    % Update transmission
    Transmission_partial_ref = Transmission_partial_ref .*
(1-Reflectivity_ref);
    Transmission_partial_sens = Transmission_partial_sens
.* (1-Reflectivity_sens);

end
```

```
Reflectivity_Ref = 1-Transmission_partial_ref;
Reflectivity_Sens = 1-Transmission_partial_sens;

% Corrupt with noise
Noise_Variance = max(Reflectivity_Ref) ./
(10.^(SNR_dB/10));
Reflectivity_Ref = Reflectivity_Ref + ActivateNoise *
Noise_Variance * randn(1,length(lambda));
Reflectivity_Sens = Reflectivity_Sens + ActivateNoise *
Noise_Variance * randn(1,length(lambda));

figure
plot(lambda, Reflectivity_Ref,'b');
hold on
plot(lambda, Reflectivity_Sens,'r');
xlabel('Wavelength (nm)');
ylabel('Reflectivity');
```

LCFBG Analyzer: LCFBG.m

```
% Script: simulates a LCFBG as a discrete cascade of FBGs

clear all; close all; clc

% Initialize LCFBG
N = 200; % Number of sections for discretization
lambda_start = 1520e-9; % Lower limit wavelength
lambda_stop = 1570e-9 ; % Upper-limit wavelength
Grating_length = 5e-2; % Total length of the grating
dneff = 1e-5;
neff = 1.5;
v = 1;
L = Grating_length / N; % Length of each step
kL = ones(1, N) * 0.4; % Assumes constant kL strength

% Wavelength axis
lambda = 1510e-9 : 10e-12: 1580e-9;

% Bragg wavelengths
lambdaB = linspace(lambda_start, lambda_stop, N);

% Temperature pattern
kT = 10.2e-12; % Temperature coefficient, in m/degC
DeltaT = 120*exp(-((1:N)-100).^(2)/200); % Arbitrary
temperature function

% Noise parameters
SNR_dB = 50; % SNR, in dB
ActivateNoise = 1; % Set to 0 or 1 to turn off/on noise
```

```
% Layer-by-layer, compute the reflectivity
Transmission_partial_ref = ones(1, length(lambda));
Transmission_partial_sens = ones(1, length(lambda));

for ii = 1:N

    % Reference grating
    Reflectivity_ref = FBGgeneration(lambdaB(ii), lambda,
kL(ii), L, dneff, neff, v);

    % Temperature-shifted grating
    Reflectivity_sens = FBGgeneration(lambdaB(ii) +
DeltaT(ii)*kT, lambda, kL(ii), L, dneff, neff, v);

    % Update transmission
    Transmission_partial_ref = Transmission_partial_ref .*
(1-Reflectivity_ref);
    Transmission_partial_sens = Transmission_partial_sens
.* (1-Reflectivity_sens);

end

Reflectivity_Ref = 1-Transmission_partial_ref;
Reflectivity_Sens = 1-Transmission_partial_sens;

% Corrupt with noise
Noise_Variance = max(Reflectivity_Ref) ./
(10.^(SNR_dB/10));
Reflectivity_Ref = Reflectivity_Ref + ActivateNoise *
Noise_Variance * randn(1,length(lambda));
Reflectivity_Sens = Reflectivity_Sens + ActivateNoise *
Noise_Variance * randn(1,length(lambda));

figure
plot(lambda,Reflectivity_Ref,'r');
hold on
plot(lambda,Reflectivity_Sens,'b');
xlabel('Wavelength (m)');
ylabel('Reflectivity');
```

Dual FBG/EFPI Sensor Analyzer: FBGandEFPI.m

```
% Script: creates the spectrum of an FBG/EFPI probe and
tests its spectrum
% with pressure and temperature variations

clear all; close all; clc

%%%%%%%%
% Part 1: EFPI/FBG generation
```

```
% Initialize FBG
lambdaB = 1550e-9;
kL = 1.5;
L = 1e-2;
neff = 1.5;
dneff = 1e-5;
v = 1;

% Initialize EFPI
LFP = 230e-6; % Cavity length
d = 1.5e-6; % diaphragm length
ni = 1.5; nFP = 1; nd = 1.5; no = 1;

% Wavelength axis
lambda = 1520e-9 : 50e-12 : 1580e-9;

% Noise parameters
SNR_dB = 40; % SNR, in dB
ActivateNoise = 1; % Set to 0 or 1 to turn off/on noise

% Reflectivity FBG
ReflectivityFBG = FBGgeneration (lambdaB, lambda, kL, L,
dneff, neff, v);

% Reflectivity EFPI
ReflectivityEFPI = EFPIgeneration (lambda, LFP, d, ni, nFP,
nd, no);

% Total reflectivity
ReflectivityTotal = 1 - (1-ReflectivityEFPI).*(1-
ReflectivityFBG);

% Noise
Noise_Variance = max(ReflectivityTotal) ./
(10.^(SNR_dB/10));
ReflectivityTotal = ReflectivityTotal + ActivateNoise *
Noise_Variance * randn(1,length(lambda));

figure
plot(lambda,ReflectivityTotal);
xlabel('Wavelength (m)');
ylabel('Reflectivity');


%%%%%%%%
% Part 2: P/T-dependent spectrum

DeltaT = -20 : 2 : 100; % All temperature values (degC)
DeltaP = 0 : 0.2 : 5; % All pressure values (kPa)
```

```
% Sensitivity data from Bremer et al., IEEE Sens. J. 12(1),
2012
Coeff_EFPI_P = 4.4e-9; % Sensitivity EFPI to pressure (m/
kPa)
Coeff_EFPI_T = 1.48e-9; % Sensitivity EFPI to temperature
(m/degC)
Coeff_FBG_P = 2e-15; % Sensitivity FBG to pressure (m/kPa)
Coeff_FBG_T = 10.3e-12; % Sensitivity FBG to temperature
(m/degC)


% Temperature-dependence chart
figure; hold on
for ii = 1:length(DeltaT);

    % Computes spectrum
    ReflectivityFBG = FBGgeneration (lambdaB + DeltaT(ii) *
Coeff_FBG_T, lambda, kL, L, dneff, neff, v);
    ReflectivityEFPI = EFPIgeneration (lambda, LFP -
DeltaT(ii)*Coeff_EFPI_T, d, ni, nFP, nd, no);
    ReflectivityTotal = 1
- (1-ReflectivityEFPI).*(1-ReflectivityFBG);
    ReflectivityTotal = ReflectivityTotal + ActivateNoise *
Noise_Variance * randn(1,length(lambda));

    % Plot spectrum
    Temp_value = ones(1,length(lambda)) * DeltaT(ii);
    plot3(lambda, Temp_value, ReflectivityTotal);


end

view(3)
xlabel('Wavelength(m)');
ylabel('Temperature (^oC)');
zlabel('Reflectivity');

% Pressure-dependence chart
figure; hold on
for ii = 1:length(DeltaP);

    % Computes spectrum
    ReflectivityFBG = FBGgeneration (lambdaB + DeltaP(ii) *
Coeff_FBG_P, lambda, kL, L, dneff, neff, v);
    ReflectivityEFPI = EFPIgeneration (lambda, LFP -
DeltaP(ii)*Coeff_EFPI_P, d, ni, nFP, nd, no);
    ReflectivityTotal = 1
- (1-ReflectivityEFPI).*(1-ReflectivityFBG);
    ReflectivityTotal = ReflectivityTotal + ActivateNoise *
Noise_Variance * randn(1,length(lambda));

    % Plot spectrum
```

```
    Press_value = ones(1,length(lambda)) * DeltaP(ii);
    plot3(lambda, Press_value, ReflectivityTotal);


end

view(3)
xlabel('Wavelength(m)');
ylabel('Pressure (kPa)');
zlabel('Reflectivity');
```

Appendix B
FBG and EFPI Interrogation

In this appendix, we provide the MATLAB code used for FBG and EFPI interrogation, mainly described in Chapter 11.

FBG Wavelength Shift Tracking: FBGtracking.m

```
function [FBGwavelength] = FBGtracking (lambda, Spectrum,
Selection)

% FBG Tracking Function
% lambda = wavelength of the spectral frame
% Spectrum = FBG spectrum
% Selection = 1 for centroid; 2 for 3dB bandwidth; 3 for
polynomial fit

if (Selection == 1)
    % Apply centroid
    FBGwavelength = sum(lambda.*Spectrum) ./ sum(Spectrum);

else
    % Define P/Q resampling rate; use arbitrary
coefficients
    P = 131; Q = 5;

    % Resample spectrum
    SpectrumR = resample(Spectrum,P,Q);

    % Rewrite lambda
    lambdaR = linspace(min(lambda), max(lambda),
length(SpectrumR));

    % Define threshold and perform thresholding
    Threshold = 0.5 * max(SpectrumR);
    [IndexT] = (SpectrumR > Threshold);
    lambdaT = lambdaR(IndexT);
    SpectrumT = SpectrumR(IndexT);
```

```
    if (Selection == 2)

        FBGwavelength = min(lambdaT) +
0.5*(max(lambdaT)-min(lambdaT));

    end

    if (Selection == 3)

        % Quadratic fit
        QFit = polyfit(lambdaT, SpectrumT, 2);
        FBGwavelength = -QFit(2) / (2 * QFit(1));

    end
end
```

Short-Cavity EFPI Tracking: EFPIshort.m

```
function [gamma] = EFPIshort (Spectrum, LeftIndex,
RightIndex)

% Estimates gamma function for a short-cavity EFPI
% Spectrum = EFPI spectrum values
% LeftIndex = Set of indexes for left part computation of
gamma
% RightIndex = Set of indexes for right part computation of
gamma
% gamma = output

gamma = sum(Spectrum(LeftIndex)) ./
sum(Spectrum(RightIndex));
```

Long-Cavity EFPI Tracking: EFPIlong.m

```
function [Cepstrum] = EFPIlong (Spectrum)

% Estimates Cepstrum with FFT

Cepstrum = abs(fft(Spectrum - mean(Spectrum)));
```

KLT for FBG and EFPI Tracking: KLT.m

```
function [xi_N, xi] = KLT (Spectrum, Selection)

% Performs KLT on the Spectrum, and returns the top
eigenvalue xi_N of the
% eigenvalue array xi.
```

```
% Selection = 1 for FBG tracking (short spectrum); 2 for
EFPI tracking
% (downsampling required)

if (Selection == 1)

    SpectrumP = Spectrum;
else
    if (Selection == 2)

        % Set downsampling rate
        DownsamplingRate = 5;
        SpectrumP = Spectrum(1 : DownsamplingRate : end);
    end
end

% Performs FFT
F = (fft(SpectrumP - mean(SpectrumP)));

% Extends to Toeplitz matrix
M = toeplitz(F);

% Performs SVD
[V,D] = eig(M);

% Returns xi and xi_N
xi = diag(D);
xi_N = xi(end);
```

Capon Estimator: CaponPSD.m

```
function [PSD] = CaponPSD (Spectrum, Taps, Fs, Frequency)

% Capon power spectral density estimator
% The function returns the PSD estimated over the selected
frequency axis.
% The PSD is normalized and expressed in dB
%
% Input parameters:
% Spectrum => optical spectrum signal
% Taps => number of taps of the FIR filter (set to
500-2500)
% Fs => sampling frequency (set to 1 for digital frequency)
% Frequency => vector containing the frequency axis
%
% Output:
% Gx => normalized PSD in dB
%
% Note: it is advisable to have length(x) > 2.5*Taps
```

```
% Note2: if the Rxx matrix is close to singular, add some
Gaussian noise to
% x, de-commenting the related line

% Upsample spectrum
P = 131; Q = 5;
x = resample(Spectrum,P,Q);
N = length(x);

% Noise addition (remove if spectrum is already noisy)
x = x .* randn(1,N)/10000;

figure
plot(x)
size(x)

% Build matrix X
X = ones(Taps, N - Taps);
for p = 1 : N-Taps
    X(:,p) = x(Taps-1+p : -1 : p);
end

% Build matrix Rxx
Rxx = X * X' / Taps;

% Inversion of Rxx
% (replace for complexity reduction)
RxxInv = inv(Rxx);

%%%%%%%%%%%%%%%%%%
% PSD estimation %
%%%%%%%%%%%%%%%%%%

n = 0 : Taps-1;

for d = 1 : length(Frequency)

    % Normalized frequency
    fD = Frequency(d) / Fs;

    % Build steering vector
    a = (exp(-j*2*pi*fD*n)).';

    % PSD estimation
    G(d) = real(1 / (a' * RxxInv * a));

end


% Normalization
PSD = 10*log10(G/max(G));
```

About the Authors

Daniele Tosi is an assistant professor at Nazarbayev University School of Engineering in Astana, Kazakhstan, and head of the Laboratory of Biosensors and Bioinstruments at National Laboratory Astana. From 2012 to 2014 he was a Marie Curie Intra-European Fellow at the University of Limerick in Ireland. He obtained a B.Sc. and M.Sc. in telecommunication engineering and a Ph.D. in electronics and communications engineering at Politecnico di Torino, Italy, in 2004, 2006, and 2010, respectively. His research interests are in fiber-optic sensors and biosensors, biomedical optical sensors, smartphone-based optical systems, and signal processing. He is an associate editor of the *IEEE Sensors Journal* since 2016.

Guido Perrone holds a Ph.D. in electromagnetics/photonics from Politecnico di Torino (Technical University of Turin, Italy) where he is currently a professor in charge of courses on microwaves and photonic devices in the Department of Electronics and Telecommunications. He has almost 30 years of experience in the development of optical components, with a particular focus in the last 15 years on optical fiber sensing systems and high-power lasers for industrial and biomedical applications.

Throughout his academic career, Professor Perrone lectured in more than 100 courses and supervised more than 100 graduate-level thesis activities and about 15 Ph.D. students. The outcomes of his research have led to over 200 papers in journals and conferences, several patents on fiber sensors and lasers, and the creation of a company that manufactures devices for laser beam management in industrial and biomedical applications. He has also been the promoter of a public/private joint research facility on innovative laser devices.

Professor Perrone is a member or senior member of several organizations, including the Italian Professional Engineers Association, the Optical Society of America (OSA), the American Institute of Physics (AIP), the Photonics Society, the Microwave Theory and Technique Society of the Institute of Electrical and Electronic Engineers (IEEE), the Italian Optical Society (SIOF), and the Italian Society of Electromagnetism (SIEM).

Index

A

All-grating fibers (AGF), 44
All-optical pressure sensors, 55
Anatomical model platforms, 179
Apodized FBG, 68–69
Application-integration design, 46–47
Attenuated total reflectance (ATR), 129

B

Bandwidth tracking, 152
Bends, 117
Biochemical detection, 251–52
Biomarkers and metrics, 264
Biomedical sensors
  biological measurements, 43
  biopsy and, 263–64
  OFDR and, 96
  optical fibers, 11–12
  physical parameters, 43
  standards for, 163–69
  tools, 177–79
  working principles, 44
Biopsy, 263–64
Biosensor catheterization, 188
Blood pressure measurement, 192–93
Blood pressure sensor, 198–99
Boundary conditions, at interface of two dielectrics, 6

C

Calibration
  function, 44
  OFDR system, 94–95
  process, 45
Capon estimator, 277–78
CaponPSD.m, 277–78
Cardiovascular (CV) sensors
blood pressure measurement, 192–93
  blood temperature sensor, 198–99
  conclusions, 199
  defined, 191
  fractional flow reserve, 194–95
  heart-assistive devices, 196–98
  pressure measurement, 191–92
Cardiovascular simulators, 178
Celiac disease diagnostic, 235–36
CE mark, 168–69
Cepstrum-division multiplexing (CDM), 140, 262
Chirped FBG (CFBG), 44, 69–72
  defined, 69
  FBG comparison, 72
  interrogation software, 160–61
  linearly (LCFBG), 69–72
  phase mask inscription, 79
  short active length, 87
  thermal ablation (TA), 222–24
  See also Fiber Bragg gratings (FBGs)
Circulators, 29–30
CMT-modeled FBG, 62
Coherent OFDR, 95–96
Colonic analysis, 204
Continuous monitoring, 249
Cost-benefit paradigm, 261–62
Coupled-mode theory (CMT)
  defined, 59
  Erdogan's, 61
  spectrum evaluation, 63
Couplers
  defined, 30
  fabrication, 30–31
  wavelength-insensitive, 31–32
  wavelength-sensitive, 33
Coupling loss, sensors based on, 53–56
Cranial intraventricular catheter, 187
Cross-sensitivities, 131
Cryoablation, 217

Curvature sensors
POF-based, 53
sensitivity and, 52
TFBG and, 74

D

Decibel (dB), 11
Dexterous continuum manipulators (DCM), 245
Diagnostics
in gastroscopy, 201–6
in neurology, 210–12
in urology, 206–10
Diaphragm etching, 108
Dielectric waveguides
defined, 7
examples of, 7
mode profiles, 10
representation of, 8
Differential pressure system, 208–9
Direct KLT, 156–58
Displacement sensors
displacement, 55
schematic representation, 54, 55
transmission-based, working principle, 55
Distributed Bragg reflector (DBR), 27
Distributed feedback (DFB) lasers, 25
Distributed sensing
instrument, 225
minimally invasive robotic surgery (MIRS), 245–46
system, 224
TA techniques with, 226
thermal ablation (TA), 224–26
units, 143–45
Distributed sensors
defined, 41
introduction to, 87–88
microwave photonics (MWP), 91–92
optical frequency-domain reflectometry (OFDR), 88–91
perspectives, 96–97
practical considerations, 94–96
sensing elements, 92–94
DNA detection, 235
Draw-tower method
defined, 80
schematic, 81
weak grating production, 80–81
Dual FBG/EFPI sensor analyzer, 271–74
Dual sensing, 204
Dual VAD (BiVAD), 198

E

EFPI/FBG probe, 142
EFPIgeneration.m, 101, 268
EFPI generator, 268
EFPIlong.m, 276
EFPI probes, 146
blood pressure measurement, 192–93
for CV applications, 193
heart-assistive device, 197–98
pressure monitoring, 192
sensitivity, 199
thermal ablation (TA), 226–27
EFPIshort.m, 276
EFPI tracking
KLT, 158, 276–77
long-cavity EFPI, 155–56, 276
short-cavity EFPI, 154–55, 276
Elective interrogation systems, 119
Electromagnetic interference (EMI), 191
Electromagnetic spectrum, 2
Electromagnetic waves, 1
Emerging lab-on-filter (LoF), 44
End-user expertise, 264
Epidural
defined, 212
force exertion measurement, 213
force-sensing device for, 186
EPPI/FBG all-glass probe, 105–6
Esophageal detection, 203
Ethical issues, 174–75
Evanescent field absorption, 115–18
Extreme applications, 262
Extrinsic FPI (EFPI), 99
Extrinsic FPI (EFPI) sensors, 104
all-glass configurations, 106, 109
all-glass probe views, 109
broadband spectrum, 158
cavity length variation, 142
combining, 105

commercial devices, 145–48
defined, 102
design, 145
diaphragm-based, 106, 108
dual, 105–6, 107
fabrication of, 106–10
gastroscopy, 202
overview, 103
polishing, 108
pressure-sensing architectures, 110
pressure sensitivity, 143
pressure variation, 105
for refractive element modulation, 104
response of, 104
spectrum, 105, 108
tracking, 154–56
See also FBG arrays; FBG package
Ex vivo, 172

F

Fabrication
couplers, 30–31
EFPI sensors, 106–10
FBG, 76–82
Fabry-Perot cavity, 103
Fabry-Perot interferometers (FPIs)
in biomedical field, 99
extrinsic, 99
in fiber-optic sensor networks, 134–36
formation, 99
free spectral range (FSR) and, 101
implementation, 100
principle, 99–102
self-mixing interferometry (SMI), 110–12
in sensing system, 101
single-cavity structure, 101, 102
transmission matrix analysis, 99
FBGandEFPI.m, 271–74
FBGarray.m, 269–70
FBG arrays
analyzer, 269–70
behavior, 67
defined, 66
design, 67
form factor, 67
KLT applied to, 157
in in-line temperature or strain compensation, 67
operation, 66–67
thermal ablation (TA), 219–22
See also Fiber Bragg gratings (FBGs)
FBGgeneration.m, 62, 267
FBG generator, 267
FBG interrogators, 146, 147
FBG package
in capillary, 83
micropackages, 83
overview, 82–83
for thermal detection, 82
FBG tracking
bandwidth tracking, 152
centroid, 151–52
goal of, 151
methods, 151–54
polynomial fitting, 153–54
wavelength shift, 275–76
FBGtracking.m, 275–76
Femtosecond laser (FSL) systems, 81–82
Fiber Bragg gratings (FBGs)
apodized, 68–69
bandwidth, 62
chirped, 44, 68–72
CMT-modeled, 62
description of, 59
draw-tower method, 80–81
fabrication of, 76–82
fabrication techniques, 77
in fiber-optic sensor networks, 134–36
gastroscopy, 202, 204
interfering beams, 79–80
optical, 60
phase mask inscription, 77–79
photosensitivity and, 76
point-by-point technique, 81–82
principle of operation, 60
recent advances, 83
reflection spectrum, 61
strain and temperature evaluation, 65
strain sensitivity, 64
technology, 43
temperature spectral response, 65
thermal and mechanical response, 63–66
thermal compensation, 66

Fiber Bragg gratings (continued)
tilted, 44, 73–76
tracking, 151–54
uniform, 59–63
Fiber-optic biosensors
evanescent field absorption and, 115–18
introduction to, 115
principles of, 115–24
surface plasmon resonance (SPR), 118–24
Fiber-optic guidewire, 188
Fiber optic links, 20
Fiber-optic manometer, 186
Fiber-optic sensing
application-integration design, 46–47
classification, 40–41
couplers, 30–33
definitions, 39–40
isolators and circulators, 29–30
light sources, 19–27
optical spectrum analyzers (OSAs) and spectrometers, 35–36
photodiodes, 27–29
polarization optics, 33–35
principles of, 39–47
schemes, 40
sensor performance analysis, 44–46
systems, 19
in thermal ablation, 217–19
working principles, 41–44
Fiber-optic sensor networks
amplitude detection methods, 134–37
commercial devices, 145–48
commercial list, 148
conclusions, 148
distributed sensing units, 143–45
experimental research, 133
FBG and, 134–36
FPIs and, 134–36
introduction to, 133–34
multiparametric sensors, 142–43
plastic fiber-based, 136–37
scanning-source interrogators, 140–42
white light-based setup, 137–40
Fiber-optic sensors (FOSs)
advantages over MEMS/piezoelectric sensors, xiii–xiv
applications for, xv
biomedical categories, xv–xvi
cost reduction and standardization, 263
defined, xiii, 39
in extreme applications, 262
future role in biomedical engineering, 261–65
integration and distribution, 262–63
materials, xiv
multifunctionality, 263
places in vivo, 164
sensing principles, xviii
See also Sensors
Fiber-optic spectrometric sensors
application perspective, 259
defined, 257
liver tumor detection, 257
lung tumor detection, 258
portable tissue scanner, 258
Fiber tapering, 117
Fiber under test (FUT), 144
Field distributions, 8
FISO signal conditioning systems, 146
Force-sensing device for epidural anesthesia, 186
Fourier transform (FT), 2
Fourier transform infrared reflectometer (FTIR), 129
FPI/FBG dual sensors, 105–6
FPI probes, 105
Fractional flow reserve (FFR)
defined, 194
measurement, 194–95
measurement illustration, 194
Free spectral range (FSR), 101
Full-width half maximum (FWHM) bandwidth, 62

G

Gastroscopy
colonic analysis, 204
defined, 201
dual sensing, 204
esophageal detection, 203
multitool, 205–6
present and future perspectives, 204–6
sensing element, 202–3

shape sensing, 205
spatial resolution, 205
transducer, 205
Glass reinforced polymer (GRP), 118
Graded index (GRIN), 12

H

Heart-assistive devices
defined, 196
EFPI pressure probes, 197–98
intra-aortic balloon (IAB), 196–97
ventricular assist device (VAD), 198
High-density temperature sensing
chirped FBG (CFBG), 222–24
distributed sensing, 224–26
Higher-order modes (HOMs), 9
High-intensity focused ultrasound (HIFU), 217
Holey fibers. *See* Photonic crystal fibers (PCFs)

I

IEC 60601, 165–67
Insertion, catheter, 183
Institutional review board (IRB), 176
Intensiometric sensors. *See* Intensity-based sensors (IBSs)
Intensity-based sensors (IBSs)
based on coupling loss, 53–56
defined, 41, 49
displacement, 54
exploiting the modulation of losses, 51–53
interrogator, 136
introduction to, 49–51
limitations of, 51
minimally invasive robotic surgery (MIRS), 244–45
overview, 41–42
PR and PT power ratio, 54, 55
sensing for operation with FBG or FPI, 135
simplicity of, 50
spatial multiplexing for, 51
surface-modifying approaches, 52
system schematic, 50
Interfering beams, in FBG fabrication, 79–80
Interrogation software
chirped FBG (CFBG), 160–61
conclusions, 161
defined, 151
direct Karhunen-Loeve transform (KLT), 156–58
EFPI direct tracking, 154–56
FBG tracking direct methods, 151–54
LabVIEW implementations, 161
multi-EFPI systems, 158–59
tilted FBG (TFBG), 159–60
Interrogators
FBG, 146, 147
IBS, 136
POF, 136, 137
scanning-source, 140–42
Intra-aortic balloon (IAB), 196–97
Intracranial pressure (ICP) measurement
defined, 210
as FOS application example, 211
performance of, 210–11
probe positioning, 211
Intracranial pressure (ICP) trace, 212
Inverse Fourier transform (IFT), 89, 90
Investigational device exemption, 176–77
in vitro, 172–73
in vivo, 172, 176
ISO 10993, 163–65
ISO 13485, 165
Isolators, 29–30

K

Karhunen-Loeve transform (KLT)
applied to EFPI tracking, 158, 276–77
applied to FBG tracking, 157
direct, 156–58
performance of, 156
KLT.m, 276–77
Kretschmann configuration, 118, 119

L

LabVIEW software, 78, 80
implementation, 161
research-grade RFA/MWA applicators, 229
Laser ablation (LA)
defined, 217
FBG arrays and, 219

Laser ablation phantoms, 179
Laser diodes (LDs)
  amplifier, 23
  astigmatic far-field output beam, 26
  butterfly package, 25
  LED comparison, 22
  light generation, 24
  as light source, 19
  oscillations, 23
  pigtailed packages, 26
  slope efficiency, 24
  structure illustration, 23
  tunable, 27
LCFBG.m, 72, 270–71
Left VAD (LVAD), 198
Light-emitting diodes (LEDs)
  characteristics, 22
  defined, 19
  driving, 21
  emitted power, 21
  LD comparison, 22
  light output, 20
  mode excitation, 117
Light sources, 19–27
Linearly chirped FBG (LCFBG)
  analyzer, 270–71
  behavior, 72
  in biomedical engineering, 70
  defined, 69
  evolution of spectrum, 72
  flat-top, 69
  MATLAB script, 72
  nonuniform temperature gradients and, 71
  sensing principle, 70
  spectra examples, 70
  thermal ablation (TA), 223–24
  thermal pattern identification during TA, 71
Linear polarizers, 33–34
Liver tumor detection, 257
Localized SPR (LSPR), 120
Lock-in amplifier (LIA), 137
Long-cavity EFPI tracking, 155–56, 276
Long-period gratings (LPGs), 44, 117
Lump force detection system, 244
Lung cancer biomarker detection, 232–33
Lung tumor detection, 258

M

Mach-Zender interferometer (MZI), 112
Magnetic resonance imaging (MRI), 191
Maxwell equations, 2, 6
Measurement instruments, 39
Medical catheters
  characteristics of, 181–84
  disposable format and sterilization, 184
  form factor and rigidity, 182–83
  FOS protection and functionality, 183
  insertion, 183
  invasiveness and size, 181–82
  metrologic issues, 184
  positioning, 183
Medical sensor standards
  additional, 168
  CE mark, 168–69
  IEC 60601, 165–67
  ISO 10993, 163–65
  ISO 13485, 165
  main, 163
MEMS switches, 139
Microstructured fibers. *See* Photonic crystal fibers (PCFs)
Microwave ablation (MWA)
  defined, 217
  FBG arrays and, 219
Microwave photonic filters (MWPF), 44
Microwave photonics (MWP), 88, 91–92
  coherency time, 91–92
  configurations, 91
  detection, 92
  incoherent detection, 91
  SDM and, 93–94
  sensing system, 93
  setup schematic, 145
  tap amplitude, 93
  in wavelength shift detection, 93
  working principle, 91
Minimally invasive robotic surgery (MIRS)
  defined, 239
  distributed sensing, 245–46
  future evolution, 246–47
  haptic feedback limitation, 240–41
  intensity-based sensors (IBSs), 244–45

as master-slave system, 239
overview, 239–41
tele-operated surgery, 240
vitreoretinal surgery (VRS), 241–44
Molecule selective behavior, 116
Moral norms, 171
Multi-EFPI systems, 158–59
Multi-FPI force detection, 242–43
Multifunctionality, 263
Multimode fiber (MMF), 10–11
core diameter, 15
number of modes in, 14
See also Optical fibers
Multiparametric sensors, 40, 142–43
Multiplexed sensors, 40
Multipressure catheter, 185–86
Multitool gastroscopy, 205–6

N

Needle thermotherapy probe, 187–88
Neurology
fiber-optic sensors in, 210
ICP measurement, 210–12

O

OFSETH project, 250, 251
Opsens system, 146
Optical backscattering meter (OBR), 43
Optical coherence tomography (OCT), 242
Optical connectors, 15, 16
Optical fibers
biomedical sensing, 11–12
defined, 10
definition of numerical apertures, 12
illustrated, 7, 15
meridional and skew rays, 13
practical aspects of using, 14–17
propagation of, 11
slices, 16
SMF, 11
SPR sensors in, 121–23
structure cross-section, 11
See also Multimode fiber (MMF); Single-mode fiber (SMF)
Optical fiber sensors (OFS), 39
Optical fiber spectroscopy
cross-sensitivities and, 131
for medical detection, 130
overview, 129–31
Optical frequency-domain reflectometry (OFDR), 88–91
in biomedical sensing, 96
calibration, 94–95
coherent, 95–96
demonstration, 88–90
detection, 90–91
forward propagation component, 89
implementation, 90
practical considerations, 94–96
Rayleigh-based, 93, 94
scattering phenomena, 89
setup schematic, 144
spatial resolution, 95
as temperature sensor, 95, 96
theory, 88–91
Optical spectrum analyzers (OSAs), 35–36
Optical time-domain reflectometry (OTDR), 88
Organization, this book, xvi–xviii

P

Personalized medicine, 249
Phantoms
defined, 173
effective, 175
laser ablation, 179
sensor testing on, 175–76
thermal ablation, 178–79
three-dimensional (3-D) printed, 179
Phase mask inscription
chirped FBG (CFBG), 79
defined, 77
evolution of FBG parameters during, 79
process, 77–78
refractive index modulation and, 78–79
setup for FBG, 78
Phase shift, 9
Photodiodes (PDs)
defined, 27
performance in biosensing applications, 28
P-N junction, 27, 28
semiconductor, 28

Photonic bandgap (PBG), 124
Photonic crystal fibers (PCFs)
in biomedical field, 124
defined, 123
hollow-core, 124
light confinement properties, 124
solid-core, 123
SPR sensors in, 123–24
Photosensitivity
defined, 76
FBGs and, 76–77
PhysioBank database, 178
Plane waves
reflection of, 5–7
refraction of, 5–7
TE, 6
TM, 6
Plastic optical fibers (POFs), 12, 16, 49
in fiber-optic sensor networks, 136–37
interrogator, 136, 137
P-N junction, 27, 28
Point-by-point technique, 81–82
Polarization
in encoding light variations, 43
optics, 33–36
in *x*-*y* plane, 3–4
Polarizers
defined, 33
linear, 33–34
schematic representation, 34
Polynomial fitting, 153–54
Portable tissue scanner, 258
Position, catheter, 183
Power splitters, 31
Premarket approval (PMA), 176
Pressure sensing
architectures, 105
measurement, 191–92
thermal ablation (TA), 226–27
Principle of 3Rs, 173–74

R

Radiation modes, 8
Radio frequency ablation (RFA)
application of FBG sensors in, 222
defined, 216
FBG arrays and, 219
measurement of thermal field in, 220
Raman spectroscopy, 131
Rayleigh backscattering, 92
Rayleigh-based OFDR, 93, 94
Reduction, 173
Refinement, 173
Reflection, of plane waves, 5–7
Reflection coefficient, 100
Refraction, of plane waves, 5–7
Refractive index, 3
Remote healthcare, 249–50
Replacement, 173
Research and test methodologies, 175–76
Respiration actuators, 179
Retarders, 34
Right VAD (RVAD), 198
RJC, 146
Robotic technologies, 264–65
Round trip time (RTT), 88

S

Scanning-source interrogators
illustrated, 141
principle of, 140
setup, 140–41
white light-based setup and, 141–42
Self-mixing interferometry (SMI)
defined, 110
extended-range systems, 112
feedback, 111
output, 110
principle, 110, 111
signal detection, 111
Semiconductor optical amplifier (SOA), 23
Semidistributed sensors, 41
Sensitivity
curvature sensors and, 52
defined, 45
fiber SPR, 121
strain, 64–65, 74
Sensor catheterization
biosensor catheterization, 188
cranial intraventricular catheter, 187
fiber-optic guidewire, 188
fiber-optic manometer, 186

force-sensing device for epidural anesthesia, 186
FOS, 185–88
medical catheter characteristics and, 181–84
needle thermotherapy probe, 187–88
perspectives, 189
urologic pressure and multipressure catheter, 185–86
Sensor performance analysis, 44–46
Sensors
cardiovascular (CV), 191–92
distributed, 41, 87–97
FPI-based, 102
intensity-based, 41–42, 49–56
multiparametric, 40, 142–43
multiplexed, 40
optical, 16
semidistributed, 41
single-point extrinsic, 40
single-point intrinsic, 40
spectrometric, 257–59
wearable, 253–54
Short-cavity EFPI tracking, 154–55, 276
Short-term Fourier transform (STFT)
application to TFBG sensing, 160
window, 159–60
Single-cavity FPI structure, 101, 102
Single-mode fiber (SMF), 10, 11
core diameter, 14
fundamental mode of, 14
polarization maintaining (PMFs), 14
See also Optical fibers
Single-point extrinsic sensors, 40
Single-point intrinsic sensors, 40
Singular value decomposition (SVD), 156
Skin Project, 252
Slab waveguides
cross section of, 8
illustrated, 7
mode profiles for, 10
numerical aperture of, 11
Slope efficiency, 24
Smart medical textiles
biochemical detection, 251–52
defined, 250
embedding FBG sensors, 252
future prospects, 254
OFSETH project, 250, 251
Skin Project, 252
SPR detector, 252
temperature detection, 251
Smartphones, 253–54
Snell's law, 129–31
Spatial division multiplexing (SDM), 93–94, 262
Spectrometers, 35–36
SPR sensors
advantages, 118
application and detection of, 231–37
in bulk optics, 118–20
celiac disease diagnostic, 235–36
DNA detection, 235
genesis and present status, 231–32
lab setup to test, 122
low cost interrogation setup, 122
lung cancer biomarker detection, 232–33
multisensing capabilities, 123
in optical fibers, 121–23
perspectives, 236–37
in photonic crystal fibers, 123–24
schematic representation, 121
for smartphones, 253–54
thrombin detection, 233–35
See also Surface plasmon resonance (SPR)
Strain sensitivity
FBG, 64–65
TFBG, 74
Superluminescent LEDs (SLEDs), 19, 22, 144
Surface functionalization, 116
Surface plasmon resonance (SPR)
in bulk optics, 118–20
as evanescent wave absorption, 118
in fiber structures, 121–24
functionalization, 235
localized (LSPR), 120
occurrence, 118
in semidistributed sensing, 44
spectral response, 120
See also SPR sensors

T

TE modes, 6, 13
Temperature detection, 251

Thermal ablation (TA)
  chirped FBG (CFBG), 222–24
  conclusions and perspectives, 227–29
  cryoablation, 217
  defined, 215
  devices, 215
  distributed sensing, 224–26
  as extreme sensing environment, 218
  FBG arrays, 219–22
  goal in, 215
  high-density temperature sensing, 222–26
  high-intensity focused ultrasound (HIFU), 217
  laser ablation (LA), 217
  LCFBG and, 71
  microwave ablation (MWA), 217
  pressure sensing, 226–27
  procedures and rationale, 215
  radio frequency ablation (RFA), 216
  sensing in, 217–19
  smart concept, 228
Thermal ablation phantom, 178–79
Thermal compensation
  FBG, 66
  FBG arrays, 68
Thermocouples (TCs), 218
Thermo-electric controller (TEC), 134, 135–36
Three-dimensional (3-D) printed phantoms, 179
Thrombin-binding aptamer (TBA), 234
Thrombin detection, 233–35
Tilted FBG (TFBG)
  in biodetection, 75
  broadband response, 76
  in curvature sensing, 74
  defined, 44, 73
  dual configuration, 75
  fiber jacket, 74
  interrogation software, 159–60
  as sensing architecture, 74–75
  spectrum, 73
  SPR sensors, 232–33
  strain sensitivity, 74
  transmission spectrum, 73
  weaknesses, 75–76
  See also Fiber Bragg gratings (FBGs)
Time division multiplexing (TDM), 139–40, 262
Tissue scanner, portable, 258
TM modes, 13
Tools
  anatomical model platforms, 179
  cardiovascular simulators, 178
  laser ablation phantoms, 179
  overview, 177–78
  PhysioBank database, 178
  respiration actuators, 179
  thermal ablation phantom, 178–79
  three-dimensional (3-D) printed phantoms, 179
Tool-shaft force detection, 243–44
Tool-tip FBG force detection, 241–42
Total internal reflection (TIR), 6
Transimpedance amplifiers (TIAs), 28–29
Transverse electric (TE) plane waves, 6
Transverse force protection, 243–44
Transverse magnetic (TM) plane waves, 6
Tunable laser diodes, 27

U

Uniform FBGs, 59–63
Urodynamics, 206–7
Urologic catheter, 185
Urology
  current outlook and future prospects, 210
  differential diagnostic, 208–9
  sensing in, 206
  traditional diagnostic, 207–8
  urodynamics, 206–7

V

Validation protocols and tools
  definitions, 171–75
  investigational device exemption, 176–77
  moral norms, 171
  research and test methodologies, 175–76
  useful tools, 177–79
Vector network analyzer (VNA), 91
Ventricular assist device (VAD), 198
Vertical-cavity surface-emitting laser (VCSEL), 26, 27

Vitreoretinal surgery (VRS)
defined, 241
graphical representation, 243
multi-FPI force detection, 242–43
tool-shaft and transverse force detection, 243–44
tool-tip FBG force detection, 241–42

W

Waveguides
defined, 7
dielectric, 7–10
Wavelength-division multiplexers (WDMs), 31
Wavelength-division multiplexing (WDM), 139–40, 262
Wavelength-insensitive couplers (WINCs), 30–31
Wavelength-sensitive couplers, 33
Waveplates, 34
Wearable sensors, 253–54
White light-based setup
cepstrum-division multiplexing (CDM), 140
illustrated, 138
for interrogation, 138
scanning-source interrogators and, 142–43
setup and instrumentation, 137–38
signal-to-noise ratio (SNR), 137–38
time/wavelength division multiplexing, 139–40

Recent Titles in the Artech House Applied Photonics Series

Brian Culshaw and Marc Wuilpart, Series Editors

The ABCs of Fiber Optic Communication, Sudhir Warier

Advanced Optical Communication Systems and Networks, Milorad Cvijetic and Ivan B. Djordjevic

Chemical and Biochemical Sensing with Optical Fibers and Waveguides, Gilbert Boisdé and Alan Harmer

Coherent and Nonlinear Lightwave Communications, Milorad Cvijetic

Coherent Doppler Wind Lidars in a Turbulent Atmosphere, Viktor Banakh and Igor Smalikho

Coherent Lightwave Communication Systems, Shiro Ryu

Digital Optical Measurement Techniques and Applications, Pramond Rastogi, Editor

DWDM Fundamentals, Components, and Applications, Jean-Pierre Laude

Fiber Bragg Gratings: Fundamentals and Applications in Telecommunications and Sensing, Andrea Othonos and Kyriacos Kalli

The Fiber-Optic Gyroscope, Second Edition, Hervé C. Lefèvre

Fiber-Optic Sensors for Biomedical Applications, Daniele Tosi and Guido Perrone

Finite Element Modeling Methods for Photonics, B. M. Azizur Rahman and Arti Agrawal

Frequency Stabilization of Semiconductor Laser Diodes, Tetsuhiko Ikegami, Shoichi Sudo, and Yoshihisa Sakai

Handbook of Distributed Feedback Laser Diodes, Second Edition, Geert Morthier and Patrick Vankwikelberge

Helmet-Mounted Displays and Sights, Mordekhai Velger

Introduction to Infrared and Electro-Optical Systems, Second Edition, Ronald G. Driggers, Melvin H. Friedman, and Jonathan Nichols

Introduction to Lightwave Communication Systems, Rajappa Papannareddy

Introduction to Radiometry and Photometry, Second Edition, William Ross McCluney

Introduction to Semiconductor Integrated Optics, Hans P. Zappe

Liquid Crystal Devices: Physics and Applications, Vladimir G. Chigrinov

Lithium Niobate Photonics, James E. Toney

Military Laser Technology and Systems, David H. Titterton

New Photonics Technologies for the Information Age: The Dream of Ubiquitous Services, Shoichi Sudo and Katsunari Okamoto, editors

Optical Document Security, Third Edition, Rudolf L. van Renesse

Optical FDM Network Technologies, Kiyoshi Nosu

Optical Fiber Amplifiers: Materials, Devices, and Applications, Shoichi Sudo, editor

Optical Fiber Communication Systems, Leonid Kazovsky, Sergio Benedetto and Alan Willner

Optical Fiber Sensors, Volume Three: Components and Subsystems, John Dakin and Brian Culshaw, editors

Optical Fiber Sensors, Volume Four: Applications, Analysis, and Future Trends, John Dakin and Brian Culshaw, editors

Optical Measurement Techniques and Applications, Pramod Rastogi

Optical Transmission Systems Engineering, Milorad Cvijetic

Optoelectronic Techniques for Microwave and Millimeter-Wave Engineering, William M. Robertson

Reliability and Degradation of III-V Optical Devices, Osamu Ueda

Signal Processing and Performance Analysis for Imaging Systems, S. Susan Young, Ronald G. Driggers, and Eddie L. Jacobs

Smart Structures and Materials, Brian Culshaw

Substrate Surface Preparation Handbook, Max Robertson

Surveillance and Reconnaissance Imaging Systems: Modeling and Performance Prediction, Jon C. Leachtenauer and Ronald G. Driggers

Wavelength Division Multiple Access Optical Networks, Andrea Borella, Giovanni Cancellieri, and Franco Chiaraluce